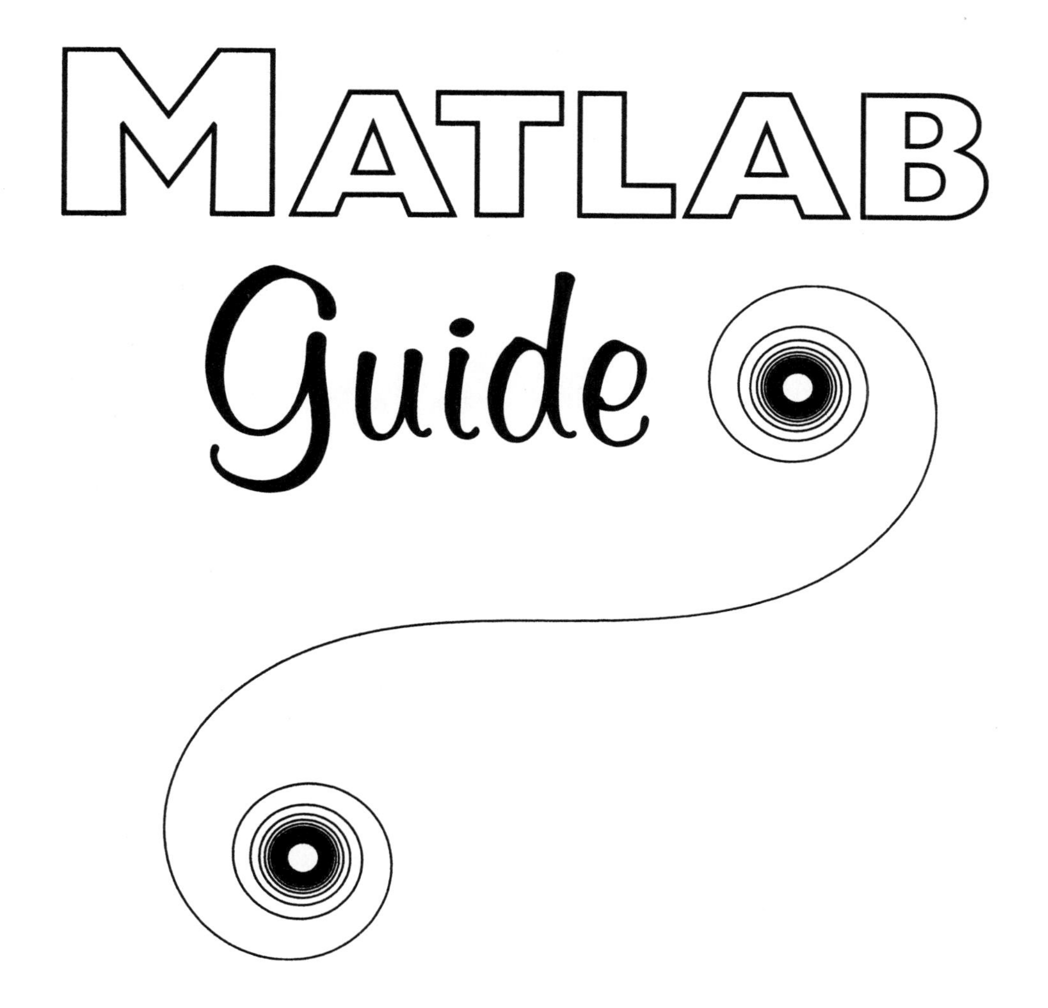
MATLAB
Guide

MATLAB Guide

Second Edition

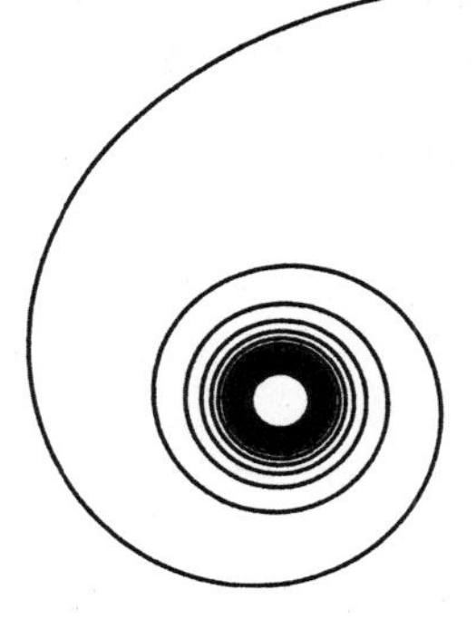

Desmond J. Higham
University of Strathclyde
Glasgow, Scotland

Nicholas J. Higham
University of Manchester
Manchester, England

Society for Industrial and Applied Mathematics   Philadelphia

10 9 8 7 6 5 4 3 2

Figure 8.28 appears courtesy of Oak Ridge National Laboratory.

Library of Congress Cataloging-in-Publication Data

Higham, D. J. (Desmond J.)
MATLAB guide / Desmond J. Higham, Nicholas J. Higham.-- 2nd ed.
p. cm.
Includes bibliographical references and index.
ISBN-10: 0-89871-578-4
ISBN-13: 978-0-89871 5-78-1
1. MATLAB. 2. Numerical analysis--Data processing. I. Higham, Nicholas J., 1961- II. Title.

QA297.H5217 2005
518'.0285--dc22

2004065563

To Lucas, Sophie, Theo, Frederic, and Thomas

Contents

List of Figures

List of Tables

List of M-Files

Preface

MATLAB®[1] is an interactive system for numerical computation. Numerical analyst Cleve Moler wrote the initial Fortran version of MATLAB in the late 1970s as a teaching aid. It became popular for both teaching and research and evolved into a commercial software package written in C. For many years now, MATLAB has been widely used in universities and industry.

MATLAB has several advantages over more traditional means of numerical computing (e.g., writing Fortran or C programs and calling numerical libraries):

- It allows quick and easy coding in a very high-level language.
- Data structures require minimal attention; in particular, arrays need not be declared before first use.
- An interactive interface allows rapid experimentation and easy debugging.
- High-quality graphics and visualization facilities are available.
- MATLAB M-files are completely portable across a wide range of platforms.
- Toolboxes can be added to extend the system, giving, for example, specialized signal processing facilities and a symbolic manipulation capability.
- A wide range of user-contributed M-files is freely available on the Internet.

Furthermore, MATLAB is a modern programming language and problem solving environment: it has sophisticated data structures, contains built-in editing and debugging tools, and supports object-oriented programming. These factors make MATLAB an excellent language for teaching and a powerful tool for research and practical problem solving. Being interpreted, MATLAB inevitably suffers some loss of efficiency compared with compiled languages, but built-in performance acceleration techniques reduce the inefficiencies and users have the possibility of linking to compiled Fortran or C code using MEX files.

This book has two purposes. First, it aims to give a lively introduction to the most popular features of MATLAB, covering all that most users will ever need to know. We assume no prior knowledge of MATLAB, but the reader is expected to be familiar with the basics of programming and with the use of the operating system under which MATLAB is being run. We describe how and why to use MATLAB functions but do not explain the mathematical theory and algorithms underlying them; instead, references are given to the appropriate literature.

The second purpose of the book is to provide a compact reference for all MATLAB users. The scope of MATLAB has grown dramatically as the package has been developed (see Table 0.1), and even experienced MATLAB users may be unaware of

[1] MATLAB is a registered trademark of The MathWorks, Inc.

some of the functionality of the latest versions. Indeed the PDF documentation for MATLAB runs to several thousand pages. Hence we believe that there is a need for a manual that is wide-ranging yet concise. We hope that our approach of focusing on the most important features of MATLAB, combined with the book's logical organization and detailed index, will make *MATLAB Guide* a useful reference.

The book is intended to be used by students, researchers, and practitioners alike. Our philosophy is to teach by giving informative examples rather than to treat every function comprehensively. Full documentation is available in MATLAB's online help and we pinpoint where to look for further details.

Our treatment includes many "hidden" or easily overlooked features of MATLAB and we provide a wealth of useful tips, covering such topics as customizing graphics, M-file style, code optimization, and debugging.

The main subjects omitted are object-oriented programming, Graphical User Interface (GUI) tools, and the publishing features. Every MATLAB user benefits, perhaps unknowingly, from its object-oriented nature, but we think that the typical user does not need to program in an object-oriented fashion. (For excellent examples of the use of object orientation, see [10] and [95].) GUIs can be useful as front-ends to MATLAB computations, but again, most users will not need to write them (for an excellent example of a GUI, see `eigtool` [128]). The facility to publish an M-file to HTML, XML, LaTeX, etc. (new to Release 14) is best learned by using it and by viewing the online documentation. Other areas not covered include MATLAB's Java interface and some of the more advanced visualization features.

We have not included exercises; MATLAB is often taught in conjunction with particular subjects, and exercises are best tailored to the context.

We have been careful to show complete, undoctored MATLAB output and to test every piece of MATLAB code listed. The only editing we have done of output has been to break overlong lines that continued past our right margin—in these cases we have manually inserted the continuation periods "`...`" at the line break.

MATLAB runs on several operating systems and we concentrate on features common to all. We do not describe how to install or run MATLAB, or how to customize it—the manuals, available in both printed and online form, should be consulted for this system-specific information.

A Web page for the book can be found at

`http://www.siam.org/books/ot92`

It includes

- All the M-files used as examples in the book.
- Updates relating to material in the book.
- Links to various MATLAB-related Web resources.

What This Book Describes

This book describes MATLAB 7.01 (Release 14 with Service Pack 1), although most of the examples work with at most minor modification in MATLAB 6 (Release 12) and MATLAB 6.5 (Release 13). If you are not sure which version of MATLAB you are using, type `ver` or `version` at the MATLAB prompt.

All the output shown was generated on a Pentium 4 machine running MATLAB under Windows XP.

Table 0.1. *Versions of MATLAB.*

Year	Version	Notable features
1978	Classic MATLAB	Original Fortran version.
1984	MATLAB 1	Rewritten in C.
1985	MATLAB 2	30% more commands and functions, typeset documentation.
1987	MATLAB 3	Faster interpreter, color graphics, high-resolution graphics printing.
1992	MATLAB 4	Sparse matrices, animation, visualization, user interface controls, debugger, Handle Graphics®,* Microsoft Windows support.
1997	MATLAB 5	Profiler, object-oriented programming, multidimensional arrays, cell arrays, structures, more sparse linear algebra, new ordinary differential equation solvers, browser-based help.
2000	MATLAB 6 (R12)	MATLAB desktop including Help browser, matrix computations based on LAPACK with optimized BLAS, function handles, `eigs` interface to ARPACK, boundary value problem and partial differential equation solvers, graphics object transparency, Java support.
2002	MATLAB 6.5 (R13)	Performance acceleration, improved speed in core linear algebra functions for Pentium 4, more control in warning and error handling.
2004	MATLAB 7.0 (R14)	Mathematics on nondouble operands (single precision, integer), anonymous functions, nested functions, publishing an M-file to HTML, LaTeX, etc., enhanced plot annotation.

* Handle Graphics is a registered trademark of The MathWorks, Inc.

How This Book Is Organized

The book begins with a tutorial that provides a quick tour of MATLAB. The rest of the book is independent of the tutorial, so the tutorial can be skipped—for example, by readers already familiar with MATLAB.

The chapters are ordered so as to introduce topics in a logical fashion, with the minimum of forward references. A principal aim was to cover M-files and graphics as early as possible, subject to being able to provide meaningful examples. Later chapters contain material that is more advanced or less likely to be needed by the beginner.

Using the Book

Readers new to MATLAB should begin by working through the tutorial in Chapter 1. The tutorial gives a fast-paced overview of MATLAB's capabilities, with all its topics being covered in greater detail in subsequent chapters. Although it is designed to be read sequentially, with most chapters building on material from earlier ones, the book can be read in a nonsequential fashion by following cross-references and making use of the index. It is difficult to do serious MATLAB computation without a knowledge of arithmetic, matrices, the colon notation, operators, flow control, and M-files, so Chapters 4–7 contain information essential for all users.

Appendix A lists our choice of the top 111 MATLAB functions—those that we think every MATLAB user should know about. The beginner may like to tick off these functions as they are learned, while the intermediate user can pick out for study those functions with which they are not already familiar.

Experienced MATLAB users who are upgrading from versions earlier than version 7 should refer to Appendix B, which lists some of the main changes in recent releases.

What's New in the Second Edition

The first edition of this book described MATLAB 6 (Release 12). This second edition, which is more than 30 percent longer, differs from the first in several respects.

1. Changes and new features introduced in Releases 13 and 14 of MATLAB are incorporated. Of these, the introduction of anonymous functions and nested functions has produced the most modifications to the existing material from the first edition.

2. Our continuing experience in using MATLAB for teaching and research has led to numerous improvements and additions—in particular, more examples.

3. A new chapter, "Case Studies" (Chapter 22), presents more substantial examples of the use of MATLAB in a variety of problem areas.

4. A new appendix contains a list of the 111 most useful MATLAB functions.

5. Many sections contain new, or reorganized, material.

The main new MATLAB features described here are as follows (a page reference is given for each feature, but see the index for full references):

- `logical` data type (p. 63);

- enhanced control of the `error` and `warning` functions (pp. 221, 223);
- `&&` and `||` operators with short-circuiting (p. 66);
- block commenting (p. 81);
- evaluation of passed functions without the use of `feval` (p. 144);
- `mlint` and related tools (p. 238);
- data types `single`, `int*`, `uint*` (p. 40);
- anonymous functions and nested functions (pp. 144, 151);
- new functions `linsolve` (p. 124), `ode15i` (fully implicit differential equations and differential-algebraic equations) (p. 193), `dde23` (delay differential equation solver) (p. 202), `pchip` (p. 162).

Acknowledgments

We are grateful to a number of people who offered helpful advice and comments during the preparation of the book.

For the first edition:

Penny Anderson, Christian Beardah, Tom Bryan, Brian Duffy, Cleve Moler, Damian Packer, Harikrishna Patel, Larry Shampine, Françoise Tisseur, Nick Trefethen, Jack Williams.

For the second edition:

Penny Anderson, Paolo Bientinesi, David Carlisle, Jacek Kierzenka, Cleve Moler, Jorge Moré, Jim Nagy, Larry Shampine.

It has been a pleasure working with the SIAM staff, namely with Sara Murphy, Linda Thiel, and Kelly Thomas for the second edition, and with Vickie Kearn, Michelle Montgomery, Deborah Poulson, Lois Sellers, Kelly Thomas, and Marianne Will for the first edition. For both editions we were delighted to work with our long-time copy editor Beth Gallagher.

For those of you that have not experienced MATLAB,
we would like to try to show you what everybody is excited about . . .
The best way to appreciate PC-MATLAB is, of course, to try it yourself.
— JOHN LITTLE and CLEVE B. MOLER, *A Preview of PC-MATLAB* (1985)

In teaching, writing and research,
there is no greater clarifier than
a well-chosen example.
— CHARLES F. VAN LOAN, *Using Examples to Build Computational Intuition* (1995)

A new era in scientific computing
has been ushered in by the development of MATLAB.
— LLOYD N. TREFETHEN, *Spectral Methods in MATLAB* (2000)

Chapter 1
A Brief Tutorial

The best way to learn MATLAB is by trying it yourself, and hence we begin with a whirlwind tour. Working through the examples below will give you a feel for the way that MATLAB operates and an appreciation of its power and flexibility.

The tutorial is entirely independent of the rest of the book—all the MATLAB features introduced are discussed in greater detail in the subsequent chapters. Indeed, in order to keep this chapter brief, we have not explained all the functions used here. You can use the index to find out more about particular topics that interest you.

The tutorial contains commands for you to type at the command line. In the last part of the tutorial we give examples of script and function files—MATLAB's versions of programs and functions, subroutines, or procedures in other languages. These files are short, so you can type them in quickly. Alternatively, you can download them from the Web site mentioned in the preface on p. xx. You should experiment as you proceed, keeping the following points in mind.

- Upper and lower case characters are not equivalent (MATLAB is case sensitive).
- Typing the name of a variable will cause MATLAB to display its current value.
- A semicolon at the end of a command suppresses the screen output.
- MATLAB uses parentheses, `()`, square brackets, `[]`, and curly braces, `{}`, and these are not interchangeable.
- The up arrow and down arrow keys can be used to scroll through your previous commands. Also, an old command can be recalled by typing the first few characters followed by up arrow.
- You can type `help topic` to access online help on the command, function, or symbol `topic`. Note that hyperlinks, indicated by underlines, are provided that will take you to related help items and the Help browser.
- If you press the tab key after partially typing a function or variable name, MATLAB will attempt to complete it, offering you a selection of choices if there is more than one possible completion.
- You can quit MATLAB by typing `exit` or `quit`.

Having entered MATLAB, you should work through this tutorial by typing in the text that appears after the MATLAB prompt, `>>`, in the Command Window. After showing you what to type, we display the output that is produced. We begin with

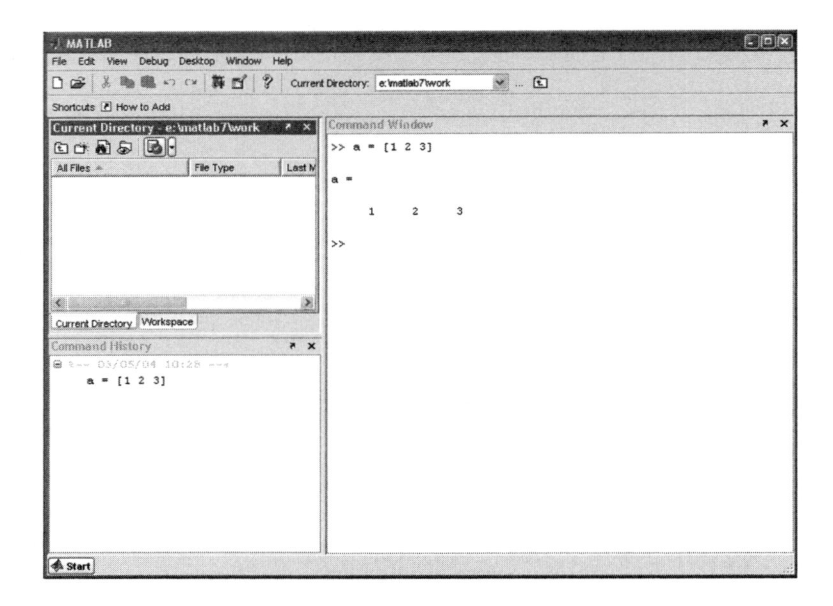

Figure 1.1. *MATLAB desktop at start of tutorial.*

```
>> a = [1 2 3]

a =

     1     2     3
```

This means that you are to type "`a = [1 2 3]`", after which you will see MATLAB's output "`a =`" and "`1     2     3`" on separate lines separated by a blank line. See Figure 1.1. (To save space we will subsequently omit blank lines in MATLAB's output. You can tell MATLAB to suppress blank lines by typing **`format compact`**.) This example sets up a 1-by-3 array `a` (a row vector). In the next example, semicolons separate the entries:

```
>> c = [4; 5; 6]
c =
     4
     5
     6
```

A semicolon tells MATLAB to start a new row, so `c` is 3-by-1 (a column vector). Now you can multiply the arrays `a` and `c`:

```
>> a*c
ans =
    32
```

Here, you performed an inner product: a 1-by-3 array multiplied into a 3-by-1 array. MATLAB automatically assigned the result to the variable `ans`, which is short for answer. An alternative way to compute an inner product is with the `dot` function:

```
>> dot(a,c)
ans =
    32
```

Inputs to MATLAB functions are specified after the function name and within parentheses. You may also form the outer product:

```
>> A = c*a
A =
     4     8    12
     5    10    15
     6    12    18
```

Here, the answer is a 3-by-3 matrix that has been assigned to `A`.

The product `a*a` is not defined, since the dimensions are incompatible for matrix multiplication:

```
>> a*a
??? Error using ==> mtimes
Inner matrix dimensions must agree.
```

Arithmetic operations on matrices and vectors come in two distinct forms. Matrix sense operations are based on the normal rules of linear algebra and are obtained with the usual symbols `+`, `-`, `*`, `/`, and `^`. Array sense operations are defined to act elementwise and are generally obtained by preceding the symbol with a dot. Thus if you want to square each element of `a` you can write

```
>> b = a.^2
b =
     1     4     9
```

Since the new vector `b` is 1-by-3, like `a`, you can form the array product of it with `a`:

```
>> a.*b
ans =
     1     8    27
```

MATLAB has many mathematical functions that operate in the array sense when given a vector or matrix argument. For example,

```
>> exp(a)
ans =
    2.7183    7.3891   20.0855

>> log(ans)
ans =
     1     2     3

>> sqrt(a)
ans =
    1.0000    1.4142    1.7321
```

MATLAB displays floating point numbers to 5 decimal digits, by default, but always stores numbers and computes to the equivalent of 16 decimal digits. The output format can be changed using the `format` command:

```
>> format long

>> sqrt(a)
ans =
   1.00000000000000   1.41421356237310   1.73205080756888

>> format
```

The last command reinstates the default output format of 5 digits. Large or small numbers are displayed in exponential notation, with a power of 10 scale factor preceded by `e`:

```
>> 2^(-24)
ans =
  5.9605e-008
```

Various data analysis functions are also available:

```
>> sum(b), mean(c)
ans =
    14
ans =
     5
```

As this example shows, you may include more than one command on the same line by separating them with commas. If a command is followed by a semicolon then MATLAB suppresses the output:

```
>> pi
ans =
    3.1416

>> y = tan(pi/6);
```

The variable `pi` is a permanent variable with value π. The variable `ans` always contains the most recent unassigned expression, so after the assignment to `y`, `ans` still holds the value π.

You may set up a two-dimensional array by using spaces to separate entries within a row and semicolons to separate rows:

```
>> B = [-3 0 -1; 2 5 -7; -1 4 8]
B =
    -3     0    -1
     2     5    -7
    -1     4     8
```

At the heart of MATLAB is a powerful range of linear algebra functions. For example, recalling that `c` is a 3-by-1 vector, you may wish to solve the linear system `B*x = c`. This can be done with the backslash operator:

```
>> x = B\c
x =
   -1.2995
    1.3779
   -0.1014
```

You can check the result by computing the Euclidean norm of the relative residual:

```
>> norm(B*x-c)/(norm(B)*norm(x))
ans =
  9.6513e-017
```

While nonzero because of rounding errors in the computations, this residual is about as small as we can expect, given that MATLAB computes to the equivalent of about 16 decimal digits.

The eigenvalues of B can be found using eig:

```
>> e = eig(B)
e =
  -3.1361
   6.5680 + 5.1045i
   6.5680 - 5.1045i
```

Here, i is the imaginary unit, $\sqrt{-1}$. You may also specify two output arguments for the function eig:

```
>> [V,D] = eig(B)
V =
  -0.9829              0.0385 + 0.0393i   0.0385 - 0.0393i
   0.1266              0.8005             0.8005
  -0.1337             -0.1683 - 0.5725i  -0.1683 + 0.5725i
D =
  -3.1361                    0                  0
        0              6.5680 + 5.1045i         0
        0                    0            6.5680 - 5.1045i
```

In this case the columns of V are eigenvectors of B and the diagonal elements of D are the corresponding eigenvalues.

The colon notation is useful for constructing vectors of equally spaced values. For example,

```
>> v = 1:6
v =
     1     2     3     4     5     6
```

Generally, m:n generates the vector with entries m, m+1, ..., n. Nonunit increments can be specified with m:s:n, which generates entries that start at m and increase (or decrease) in steps of s as far as n:

```
>> w = 2:3:10, y = 1:-0.25:0
w =
     2     5     8
y =
    1.0000    0.7500    0.5000    0.2500         0
```

You may construct big matrices out of smaller ones by following the conventions that (a) square brackets enclose an array, (b) spaces or commas separate entries in a row, and (c) semicolons separate rows:

```
>> C = [A,[8;9;10]], D = [B;a]
C =
     4     8    12     8
     5    10    15     9
     6    12    18    10

D =
    -3     0    -1
     2     5    -7
    -1     4     8
     1     2     3
```

The element in row `i` and column `j` of the matrix `C` (where `i` and `j` always start at 1) can be accessed as `C(i,j)`:

```
>> C(2,3)
ans =
    15
```

More generally, `C(i1:i2,j1:j2)` picks out the submatrix formed by the intersection of rows `i1` to `i2` and columns `j1` to `j2`:

```
>> C(2:3,1:2)
ans =
     5    10
     6    12
```

You can build certain types of matrix automatically. For example, identities and matrices of 0s and 1s can be constructed with **eye**, **zeros**, and **ones**:

```
>> I3 = eye(3,3), Y = zeros(3,5), Z = ones(2)
I3 =
     1     0     0
     0     1     0
     0     0     1
Y =
     0     0     0     0     0
     0     0     0     0     0
     0     0     0     0     0
Z =
     1     1
     1     1
```

Note that for these functions the first argument specifies the number of rows and the second the number of columns; if both numbers are the same then only one need be given. The functions **rand** and **randn** work in a similar way, generating random entries from the uniform distribution over $[0,1]$ and the normal $(0,1)$ distribution, respectively. The numbers generated depend on the **state** of each generator. By setting the states you can make your experiments repeatable. Here, both states are set to 20:

```
>> rand('state',20), randn('state',20)
>> F = rand(3), G = randn(1,5)
```

```
F =
     0.7062    0.3586    0.8468
     0.5260    0.8488    0.3270
     0.2157    0.0426    0.5541
G =
     1.4051    1.1780   -1.1142    0.2474   -0.8169
```

Single (closing) quotes act as string delimiters, so `'state'` is a string. Many MATLAB functions take string arguments.

By this point several variables have been created in the workspace. You can obtain a list with the `who` command:

```
>> who

Your variables are:

A    C    F    I3   Y    a    b    e    w    y
B    D    G    V    Z    ans  c    v    x
```

Alternatively, type `whos` for a more detailed list showing the size and class of each variable, too.

Like most programming languages, MATLAB has loop constructs. The following example uses a `for` loop to evaluate the continued fraction

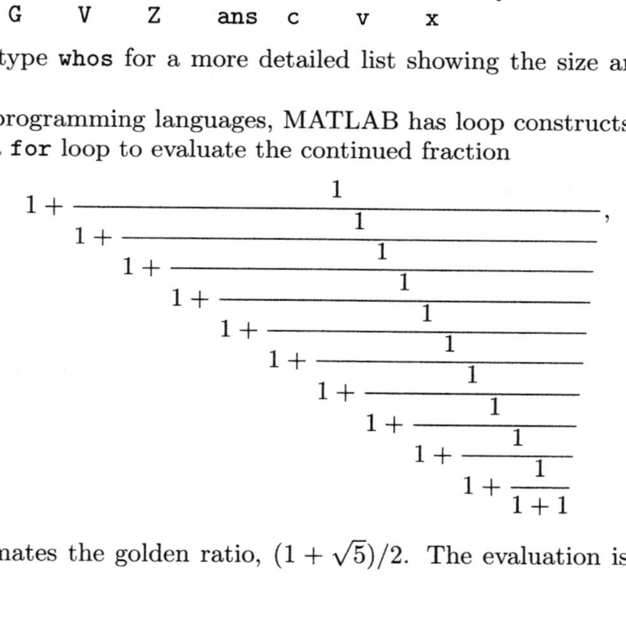

$$1+\cfrac{1}{1+\cfrac{1}{1+\cfrac{1}{1+\cfrac{1}{1+\cfrac{1}{1+\cfrac{1}{1+\cfrac{1}{1+\cfrac{1}{1+\cfrac{1}{1+\cfrac{1}{1+1}}}}}}}}}},$$

which approximates the golden ratio, $(1+\sqrt{5})/2$. The evaluation is done from the bottom up:

```
>> g = 2;
>> for k=1:10, g = 1 + 1/g; end
>> g
g =
    1.6181
```

Loops involving `while` can be found later in this tutorial.

The `plot` function produces two-dimensional (2D) pictures:

```
>> t = 0:0.005:1; z = exp(10*t.*(t-1)).*sin(12*pi*t);
>> plot(t,z)
```

Here, `plot(t,z)` joins the points `t(i),z(i)` using the default solid linetype. MATLAB opens a figure window in which the picture is displayed. Figure 1.2 shows the result. You can close a figure window by typing `close` at the command line.

You can produce a histogram with the function `hist`:

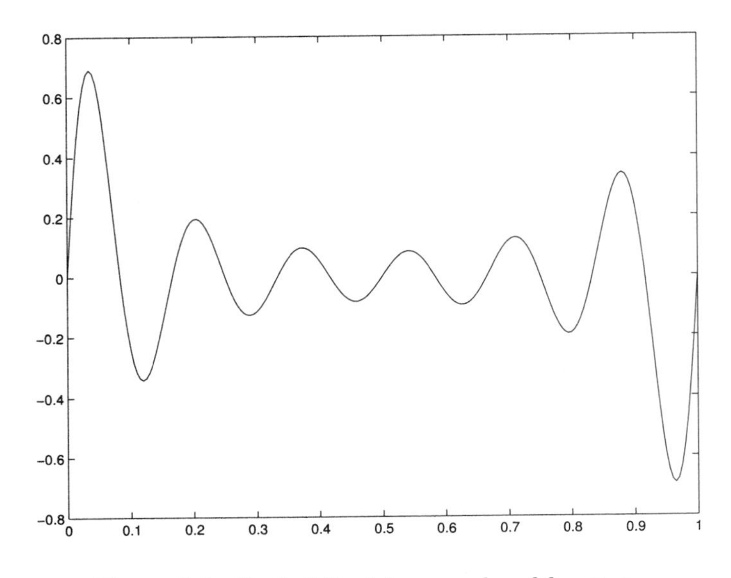

Figure 1.2. *Basic 2D picture produced by* `plot`.

```
>> hist(randn(1000,1))
```

Here, `hist` is given 1000 points from the normal (0,1) random number generator. The result is shown in Figure 1.3.

You are now ready for more challenging computations. A random Fibonacci sequence $\{x_n\}$ is generated by choosing x_1 and x_2 and setting

$$x_{n+1} = x_n \pm x_{n-1}, \quad n \geq 2.$$

Here, the $\pm$ indicates that $+$ and $-$ must have equal probability of being chosen. Viswanath [121] analyzed this recurrence and showed that, with probability 1, for large n the quantity $|x_n|$ increases like a multiple of c^n, where $c = 1.13198824\ldots$ (see also [25]). You can test Viswanath's result as follows:

```
>> clear
>> rand('state',100)
>> x = [1 2];
>> for n = 2:999, x(n+1) = x(n) + sign(rand-0.5)*x(n-1); end
>> semilogy(1:1000,abs(x))
>> c = 1.13198824;
>> hold on
>> semilogy(1:1000,c.^[1:1000])
>> hold off
```

Here, `clear` removes all variables from the workspace. The `for` loop stores a random Fibonacci sequence in the array `x`; MATLAB automatically extends `x` each time a new element `x(n+1)` is assigned. The `semilogy` function then plots `n` on the x-axis against `abs(x)` on the y-axis, with logarithmic scaling for the y-axis. Typing `hold on` tells MATLAB to superimpose the next picture on top of the current one. The second `semilogy` plot produces a line of slope `c`. The overall picture, shown in Figure 1.4, is consistent with Viswanath's theory.

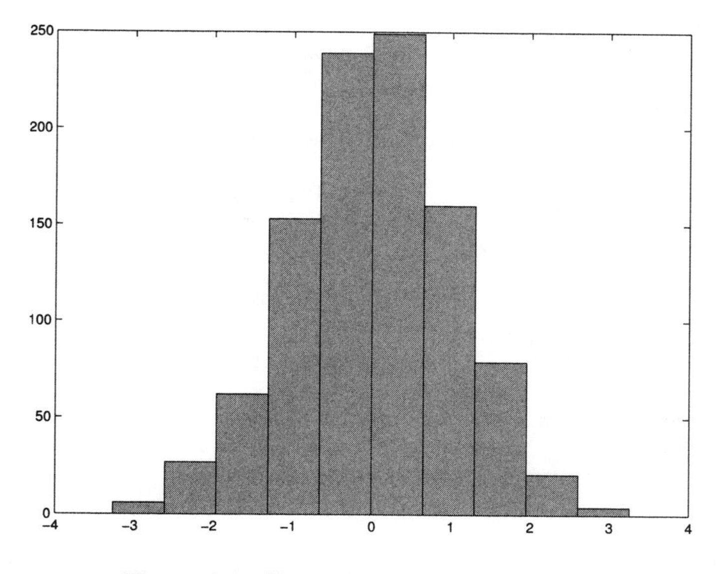

Figure 1.3. *Histogram produced by* `hist`.

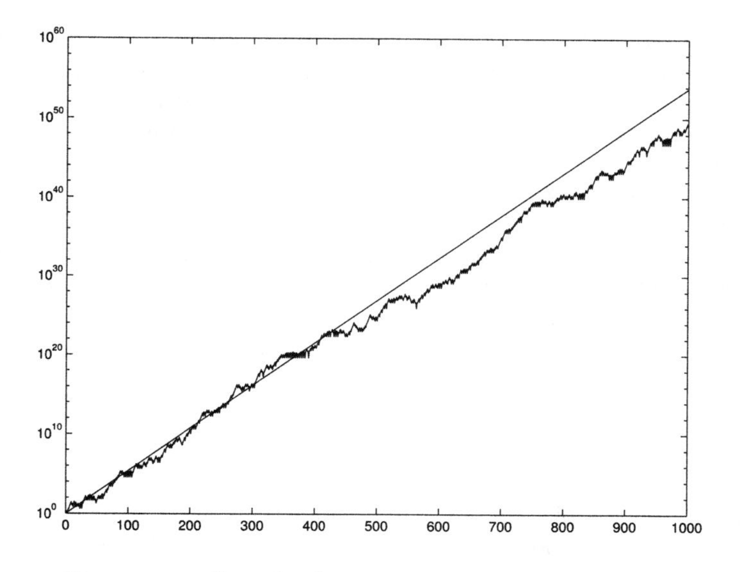

Figure 1.4. *Growth of a random Fibonacci sequence.*

Listing 1.1. *Script M-file* `rfib.m`.

```
%RFIB                          Random Fibonacci sequence.

rand('state',100)              % Set random number state.
m = 1000;                      % Number of iterations.

x = [1 2];                     % Initial conditions.
for n = 2:m-1                  % Main loop.
     x(n+1) = x(n) + sign(rand-0.5)*x(n-1);
end

semilogy(1:m,abs(x))
c = 1.13198824;                % Viswanath's constant.
hold on
semilogy(1:m,c.^(1:m))
hold off
```

The MATLAB commands to generate Figure 1.4 stretched over several lines. This is inconvenient for a number of reasons, not least because if a change is made to the experiment then it is necessary to reenter all the commands. To avoid this difficulty you can employ a script M-file. Create an ASCII file named `rfib.m` identical to Listing 1.1 in your current directory. (Typing `edit` calls up MATLAB's Editor/Debugger; `pwd` displays the current directory and `ls` or `dir` lists its contents.) Now type

```
>> rfib
```

at the command line. This will reproduce the picture in Figure 1.4. Running `rfib` in this way is essentially the same as typing the commands in the file at the command line, in sequence. Note that in Listing 1.1 blank lines and indentation are used to improve readability, and we have made the number of iterations a variable, `m`, so that it can be more easily changed. The script also contains helpful comments—all text on a line after the `%` character is ignored by MATLAB. Having set up these commands in an M-file you are now free to experiment further. For example, changing `rand('state',100)` to `rand('state',101)` generates a different random Fibonacci sequence, and adding the line `title('Random Fibonacci Sequence')` at the end of the file will put a title on the graph.

Our next example involves the Collatz iteration, which, given a positive integer x_1, has the form $x_{k+1} = f(x_k)$, where

$$f(x) = \begin{cases} 3x+1, & \text{if } x \text{ is odd,} \\ x/2, & \text{if } x \text{ is even.} \end{cases}$$

In words: if x is odd, replace it by $3x+1$, and if x is even, halve it. It has been conjectured that this iteration will always lead to a value of 1 (and hence thereafter cycle between 4, 2, and 1) whatever starting value x_1 is chosen. There is ample computational evidence to support this conjecture, which is variously known as the Collatz problem, the $3x+1$ problem, the Syracuse problem, Kakutani's problem, Hasse's algorithm, and Ulam's problem. However, a rigorous proof has so far eluded mathematicians. For further details, see [63] or type "Collatz problem" into your favorite Web search engine. You can investigate the conjecture by creating the script M-file `collatz.m` shown in Listing 1.2. In this file a `while` loop and an `if` statement are used to implement the iteration. The `input` command prompts you for a starting value. The appropriate response is to type an integer and then hit return or enter:

```
>> collatz
Enter an integer bigger than 2:   27
```

Here, the starting value 27 has been entered. The iteration terminates and the resulting picture is shown in Figure 1.5.

To investigate the Collatz problem further, the script `collbar` in Listing 1.3 plots a bar graph of the number of iterations required to reach the value 1, for starting values $1, 2, \ldots, 29$. The result is shown in Figure 1.6. For this picture, the function `grid` adds grid lines that extend from the axis tick marks, and `title`, `xlabel`, and `ylabel` add further information.

The well-known and much studied Mandelbrot set can be approximated graphically in just a few lines of MATLAB. It is defined as the set of points c in the complex plane for which the sequence generated by the map $z \mapsto z^2 + c$, starting with $z = c$, remains bounded [91, Chap. 14]. The script `mandel` in Listing 1.4 produces the plot of the Mandelbrot set shown in Figure 1.7. The script contains calls to `linspace` of the form `linspace(a,b,n)`, which generate an equally spaced vector of `n` values between `a` and `b`. The `meshgrid` and `complex` functions are used to construct a matrix `C` that represents the rectangular region of interest in the complex plane. The `waitbar` function plots a bar showing the progress of the computation and illustrates MATLAB's Handle Graphics (the variable `h` is a "handle" to the bar). The plot itself is produced by `contourf`, which plots a filled contour. The expression `abs(Z)<Z_max` in the call to `contourf` detects points that have not exceeded the threshold `Z_max` and that are therefore assumed to lie in the Mandelbrot set; the `double` function is applied in order to convert the resulting logical array to numeric form. You can experiment with `mandel` by changing the region that is plotted, via the `linspace` calls, the number of iterations `it_max`, and the threshold `Z_max`.

Next we solve the ordinary differential equation (ODE) system

$$\begin{aligned}
\frac{d}{dt}y_1(t) &= 10(y_2(t) - y_1(t)),\\
\frac{d}{dt}y_2(t) &= 28y_1(t) - y_2(t) - y_1(t)y_3(t),\\
\frac{d}{dt}y_3(t) &= y_1(t)y_2(t) - 8y_3(t)/3.
\end{aligned}$$

This is an example from the Lorenz equations family; see [111]. We take initial conditions $y(0) = [0, 1, 0]^T$ and solve over $0 \le t \le 50$. The M-file `lorenzde` in Listing 1.5 is an example of a MATLAB function. Given `t` and `y`, this function returns the right-hand side of the ODE as the vector `yprime`. This is the form required by MATLAB's ODE solving functions. The script `lrun` in Listing 1.6 uses the MATLAB function `ode45` to solve the ODE numerically and then produces the (y_1, y_3) phase plane plot shown in Figure 1.8. You can see an animated plot of the solution by typing `lorenz`, which calls one of MATLAB's demonstrations (type `demos` for the complete list).

Listing 1.2. *Script M-file* `collatz.m`.

```
%COLLATZ              Collatz iteration.

n = input('Enter an integer bigger than 2:    ');
narray = n;

count = 1;
while n ~= 1
  if rem(n,2) == 1   % Remainder modulo 2.
     n = 3*n+1;
  else
     n = n/2;
  end
  count = count + 1;
  narray(count) = n; % Store the current iterate.
end

plot(narray,'*-')    % Plot with * marker and solid line style.
title(['Collatz iteration starting at ' int2str(narray(1))],'FontSize',16)
```

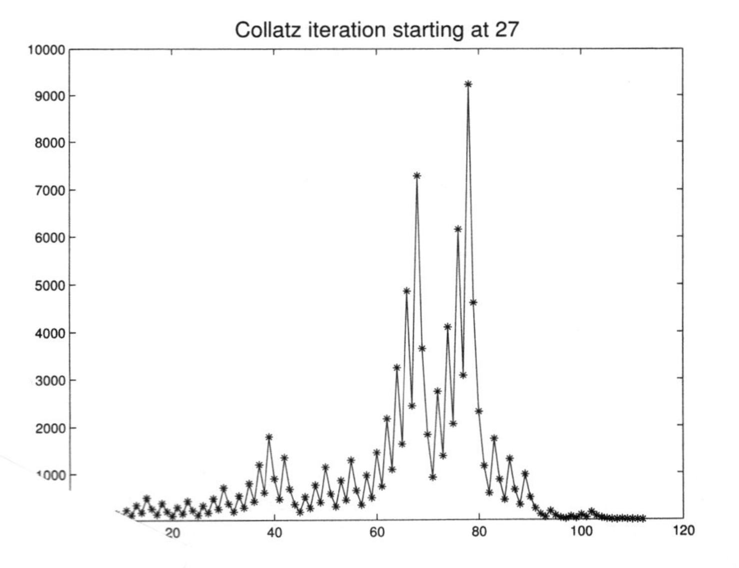

Figure 1.5. *Plot produced by* `collatz.m`.

Listing 1.3. *Script M-file* `collbar.m`.

```
%COLLBAR                  Collatz iteration bar graph.

N = 29;                   % Use starting values 1,2,...,N.
niter = zeros(N,1);       % Preallocate array.
for i = 1:N
    count = 0;
    n = i;
    while n ~= 1
        if rem(n,2) == 1
           n = 3*n+1;
        else
           n = n/2;
        end
        count = count + 1;
    end
    niter(i) = count;
end
bar(niter)      % Bar graph.
grid            % Add horizontal and vertical grid lines.
title('Collatz iteration counts','FontSize',16)
xlabel('Starting value','FontSize',16)          % Label x-axis.
ylabel('Number of iterations','FontSize',16)    % Label y-axis.
```

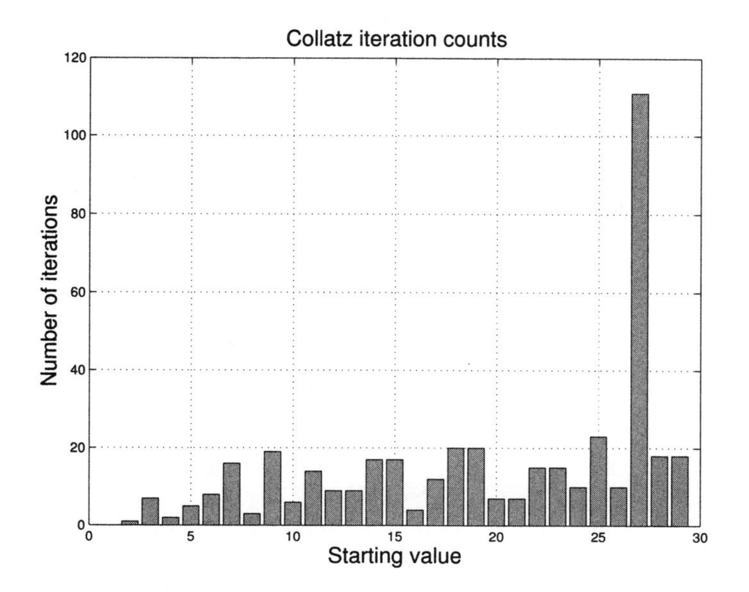

Figure 1.6. *Plot produced by* `collbar.m`.

Listing 1.4. *Script M-file* mandel.m.

```
%MANDEL      Mandelbrot set.

h = waitbar(0,'Computing...');
x = linspace(-2.1,0.6,301);
y = linspace(-1.1,1.1,301);
[X,Y] = meshgrid(x,y);
C = complex(X,Y);

Z_max = 1e6; it_max = 50;
Z = C;
for k = 1:it_max
    Z = Z.^2 + C;
    waitbar(k/it_max)
end
close(h)

contourf(x,y,double(abs(Z)<Z_max))
title('Mandelbrot Set','FontSize',16)
```

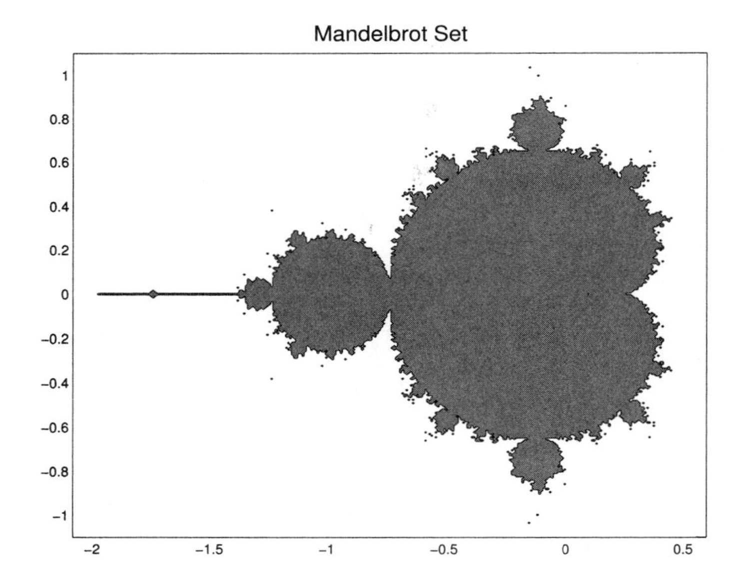

Figure 1.7. *Mandelbrot set approximation produced by* mandel.m.

Listing 1.5. *Function M-file* `lorenzde.m`.

```
function yprime = lorenzde(t,y)
%LORENZDE    Lorenz equations.
%            YPRIME  = LORENZDE(T,Y).

yprime = [10*(y(2)-y(1))
          28*y(1)-y(2)-y(1)*y(3)
          y(1)*y(2)-8*y(3)/3];
```

Listing 1.6. *Script M-file* `lrun.m`.

```
%LRUN     ODE solving example: Lorenz.

tspan = [0 50];                         % Solve for 0 <= t <= 50.
yzero = [0;1;0];                        % Initial conditions.
[t,y] = ode45(@lorenzde,tspan,yzero);
plot(y(:,1),y(:,3))                     % (y_1,y_3) phase plane.
xlabel('y_1','FontSize',14)
ylabel('y_3 ','FontSize',14,'Rotation',0,'HorizontalAlignment','right')
title('Lorenz equations','FontSize',16)
```

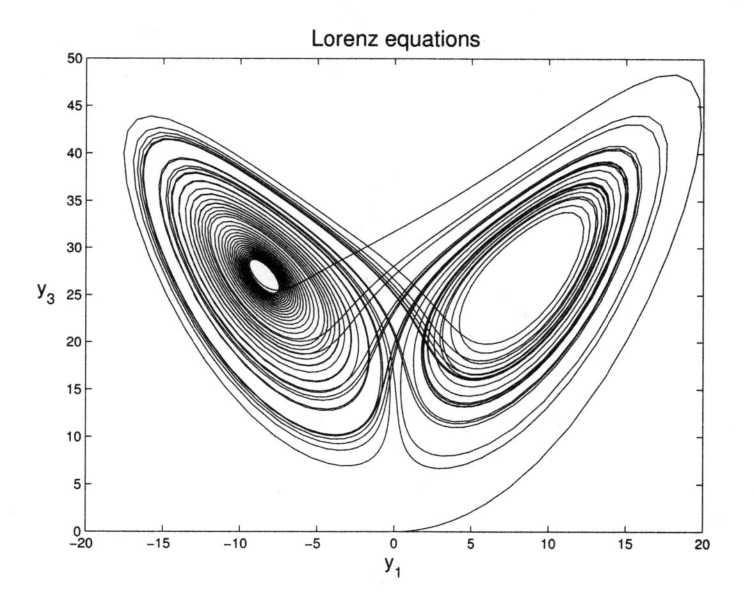

Figure 1.8. *Phase plane plot from* `ode45`.

Now we give an example of a recursive function, that is, a function that calls itself. The Sierpinski gasket [90, Sec. 2.2] is based on the following process. Given a triangle with vertices P_a, P_b, and P_c, we remove the triangle with vertices at the midpoints of the edges, $(P_a + P_b)/2$, $(P_b + P_c)/2$, and $(P_c + P_a)/2$. This removes the "middle quarter" of the triangle, as illustrated in Figure 1.9. Effectively, we have replaced the original triangle with three "subtriangles". We can now apply the middle quarter removal process to each of these subtriangles to generate nine subsubtriangles, and so on. The Sierpinski gasket is the set of all points that are never removed by repeated application of this process. The function `gasket` in Listing 1.7 implements the removal process. The input arguments `Pa`, `Pb`, and `Pc` define the vertices of the triangle and `level` specifies how many times the process is to be applied. If `level` is nonzero then `gasket` calls itself three times with `level` reduced by 1, once for each of the three subtriangles. When `level` finally reaches zero, the appropriate triangle is drawn. The following code generates Figure 1.10.

```
>> level = 5;
>> Pa = [0;0];
>> Pb = [1;0];
>> Pc = [0.5;sqrt(3)/2];
>> gasket(Pa,Pb,Pc,level)
>> hold off
>> title(['Gasket level = ' num2str(level)],'FontSize',16)
>> axis('equal','off')
```

(Figure 1.9 was generated in the same way with `level = 1`.) In the last line, the call to `axis` makes the units of the x- and y-axes equal and turns off the axes and their labels. You should experiment with different initial vertices `Pa`, `Pb`, and `Pc`, and different levels of recursion, but keep in mind that setting `level` bigger than 8 may overstretch either your patience or your computer's resources.

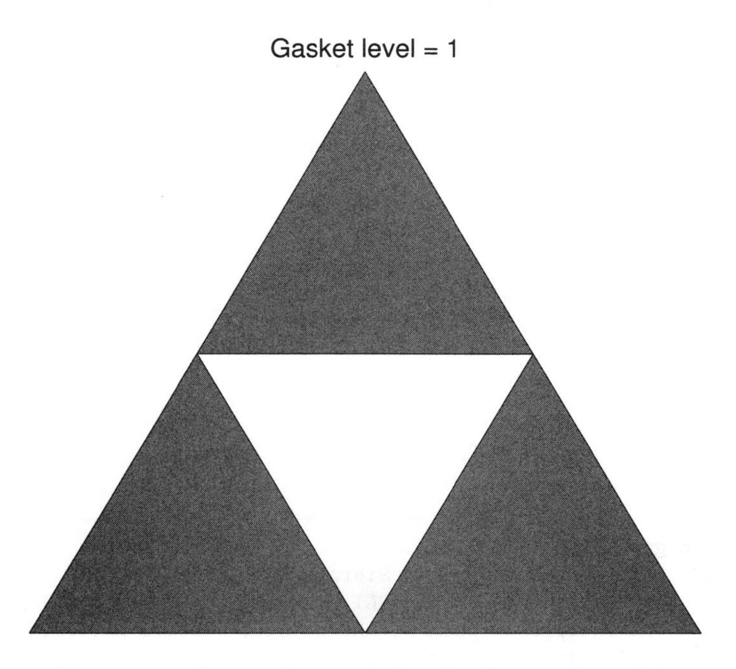

Figure 1.9. *Removal process for the Sierpinski gasket.*

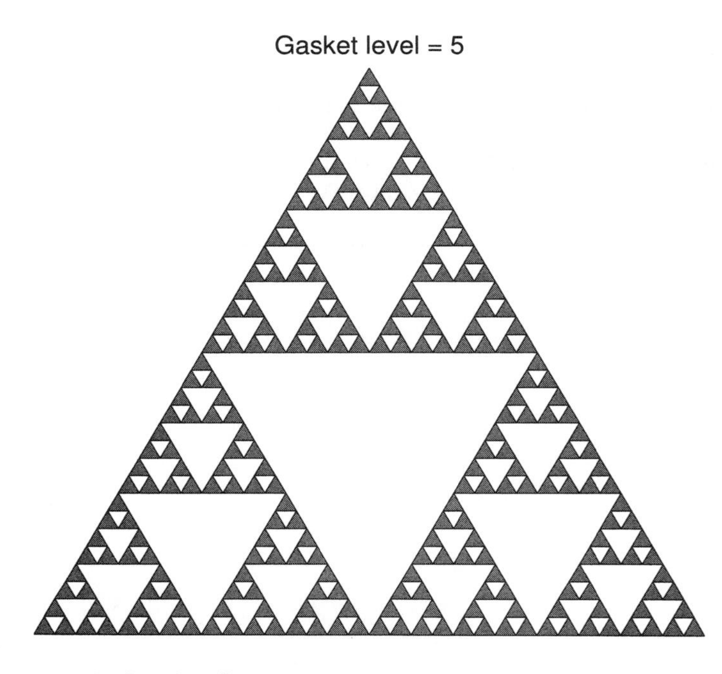

Figure 1.10. *Level* 5 *Sierpinski gasket approximation from* `gasket.m`.

Listing 1.7. *Function M-file* gasket.m.

```
function gasket(Pa,Pb,Pc,level)
%GASKET  Recursively generated Sierpinski gasket.
%        GASKET(PA, PB, PC, LEVEL) generates an approximation to
%        the Sierpinski gasket, where the 2-vectors PA, PB, and PC
%        define the triangle vertices.
%        LEVEL is the level of recursion.

if level == 0
  % Fill the triangle with vertices Pa, Pb, Pc.
  fill([Pa(1),Pb(1),Pc(1)],[Pa(2),Pb(2),Pc(2)],[0.5 0.5 0.5]);
  hold on
else
  % Recursive calls for the three subtriangles.
  gasket(Pa,(Pa+Pb)/2,(Pa+Pc)/2,level-1)
  gasket(Pb,(Pb+Pa)/2,(Pb+Pc)/2,level-1)
  gasket(Pc,(Pc+Pa)/2,(Pc+Pb)/2,level-1)
end
```

The Sierpinski gasket can also be generated by playing Barnsley's "chaos game" [90, Sec. 1.3]. We choose one of the vertices of a triangle as a starting point. Then we pick one of the three vertices at random, take the midpoint of the line joining this vertex with the starting point and plot this new point. Then we take the midpoint of the line joining this point and a randomly chosen vertex as the next point, which is plotted, and the process continues. The script `barnsley` in Listing 1.8 implements the game. Figure 1.11 shows the result of choosing 1000 iterations:

```
>> barnsley
Enter number of points (try 1000) 1000
```

Try experimenting with the number of points, `n`, the type and size of marker in the `plot` command, and the location of the starting point.

We finish with the script `sweep` in Listing 1.9, which generates a volume-swept three-dimensional (3D) object; see Figure 1.12. Here, the command `surf(X,Y,Z)` creates a 3D surface where the height `Z(i,j)` is specified at the point `(X(i,j),Y(i,j))` in the x-y plane. The script is not written in the most obvious fashion, which would use two nested `for` loops. Instead it is vectorized. To understand how it works you will need to be familiar with Chapter 5 and Section 21.4. You can experiment with the script by changing the parameter `N` and the function that determines the variable `radius`: try replacing `sqrt` by other functions, such as `log`, `sin`, or `abs`.

Listing 1.8. *Script M-file* `barnsley.m`.

```
%BARNSLEY          Barnsley's game to compute Sierpinski gasket.

rand('state',1)                     % Set random number state.
V = [0, 1, 0.5; 0, 0, sqrt(3)/2]; % Columns give triangle vertices.

point = V(:,1);                     % Start at a vertex.
n = input('Enter number of points (try 1000) ');

for k = 1:n
    node = ceil(3*rand);            % node is 1, 2, or 3 with equal prob.
    point = (V(:,node) + point)/2;
    plot(point(1),point(2),'.','MarkerSize',15)
    hold on
end

axis('equal','off')
hold off
```

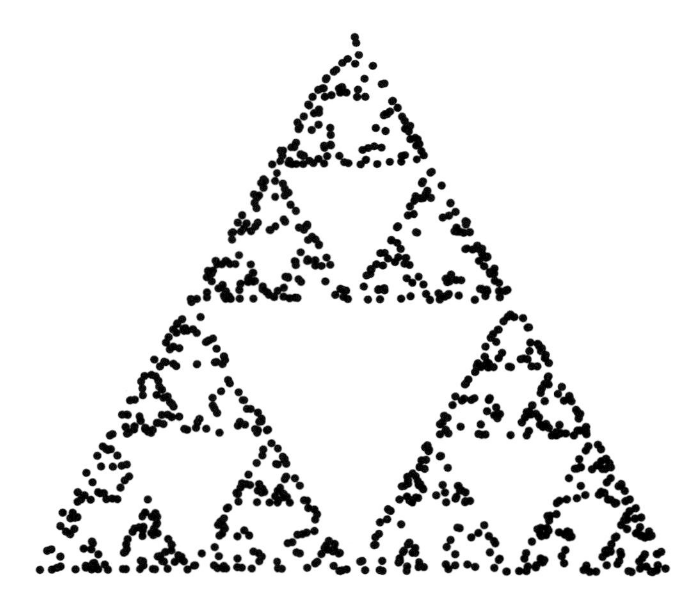

Figure 1.11. *Sierpinski gasket approximation from* `barnsley.m`.

Listing 1.9. *Script M-file* sweep.m.

```
%SWEEP          Generates a volume-swept 3D object.

N = 10;                    % Number of increments - try increasing.

z  = linspace(-5,5,N)';
radius = sqrt(1+z.^2);     % Try changing SQRT to some other function.
theta = 2*pi*linspace(0,1,N);
X = radius*cos(theta);
Y = radius*sin(theta);
Z = z(:,ones(1,N));

surf(X,Y,Z)
axis equal
```

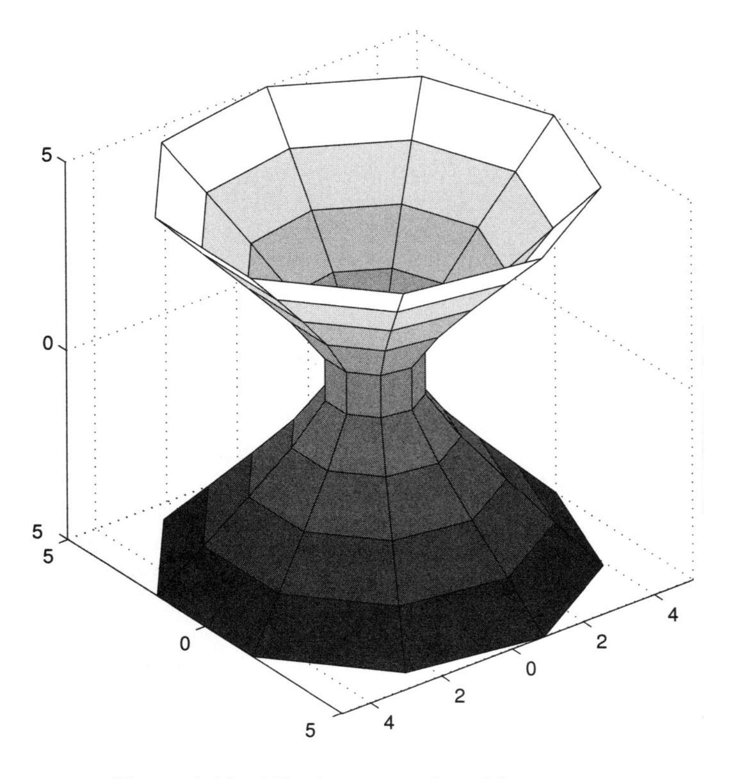

Figure 1.12. *3D picture produced by* sweep.m.

If you are one of those experts who
wants to see something from MATLAB right now
and would rather read the instructions later,
this page is for you.

— 386-MATLAB User's Guide (1989)

Do not be too timid and squeamish about your actions.
All life is an experiment.
The more experiments you make the better.

— RALPH WALDO EMERSON

Here is an interesting number:
0.95012928514718
duced by the MATLAB random number generator
with its default settings.
Start up a fresh MATLAB,
set `format long`, *type* `rand`,
and it's the number you get.

MOLER, *Numerical Computing with MATLAB* (2004)

Chapter 2
Basics

2.1. Interaction and Script Files

MATLAB is an interactive system. You type commands at the prompt (>>) in the Command Window and computations are performed when you press the enter or return key. At its simplest level, MATLAB can be used like a pocket calculator:

```
>> (1+sqrt(5))/2
ans =
     1.6180

>> 2^(-53)
ans =
   1.1102e-016
```

The first example computes $(1+\sqrt{5})/2$ and the second 2^{-53}. Note that the second result is displayed in exponential notation: it represents 1.1102×10^{-16}. The variable ans is created (or overwritten, if it already exists) when an expression is not assigned to a variable. It can be referenced later, just like any other variable. Unlike in most programming languages, variables are not declared prior to use but are created by MATLAB when they are assigned:

```
>> x = sin(22)
x =
   -0.0089
```

Here we have assigned to x the sine of 22 radians. The printing of output can be suppressed by appending a semicolon. The next example assigns a value to y without displaying the result:

```
>> y = 2*x + exp(-3)/(1+cos(.1));
```

Commas or semicolons are used to separate statements that appear on the same line:

```
>> x = 2, y = cos(.3), z = 3*x*y
x =
     2
y =
    0.9553
z =
    5.7320

>> x = 5; y = cos(.5); z = x*y^2
```

```
z =
    3.8508
```

Note again that the semicolon causes output from the preceding command to be suppressed.

MATLAB is case sensitive. This means, for example, that `x` and `X` are distinct variables.

To perform a sequence of related commands, you can write them into a script M-file, which is a text file with a `.m` filename extension. For example, suppose you wish to process a set of exam marks using the MATLAB functions `sort`, `mean`, `median`, and `std`, which, respectively, sort into increasing order and compute the arithmetic mean, the median, and the standard deviation. You can create a file, say `marks.m`, of the form

```
%MARKS
exmark = [12 0 5 28 87 3 56];
exsort = sort(exmark)
exmean = mean(exmark)
exmed  = median(exmark)
exstd  = std(exmark)
```

The `%` denotes a comment line. Typing

```
>> marks
```

at the command line then produces the output

```
exsort =
     0     3     5    12    28    56    87
exmean =
   27.2857
exmed =
    12
exstd =
   32.8010
```

Note that calling `marks` is entirely equivalent to typing each of the individual commands in sequence at the command line. More details on creating and using script files can be found in Chapter 7.

Throughout this book, unless otherwise indicated, the prompt `>>` signals an example that has been typed at the command line and it is immediately followed by MATLAB's output (if any). A sequence of MATLAB commands without the prompt should be interpreted as forming a script file (or part of one).

To quit MATLAB type `exit` or `quit`.

2.2. More Fundamentals

MATLAB has many useful functions in addition to the usual ones found on a pocket calculator. For example, you can set up a random matrix of order 3 by typing

```
>> A = rand(3)
A =
```

```
       0.9501      0.4860      0.4565
       0.2311      0.8913      0.0185
       0.6068      0.7621      0.8214
```

Here each entry of `A` is chosen independently from the uniform distribution on the interval $[0, 1]$. The `inv` command inverts `A`:

```
>> inv(A)
ans =
       1.6740     -0.1196     -0.9276
      -0.4165      1.1738      0.2050
      -0.8504     -1.0006      1.7125
```

The inverse has the property that its product with the matrix is the identity matrix. We can check this property for our example by typing

```
>> ans*A
ans =
       1.0000      0.0000           0
       0.0000      1.0000           0
            0     -0.0000      1.0000
```

The product has 1s on the diagonal, as expected. The off-diagonal elements displayed as plus or minus `0.0000`, are, in fact, not exactly zero. MATLAB stores numbers and computes to a relative precision of about 16 decimal digits. By default it displays numbers in a 5-digit fixed point format. While concise, this is not always the most useful format. The `format` command can be used to set a 5-digit floating point format (also known as scientific or exponential notation):

```
>> format short e
>> ans
ans =
   1.0000e+000  1.1102e-016            0
   2.7756e-017  1.0000e+000            0
             0 -2.2204e-016  1.0000e+000
```

Now we see that some of the off-diagonal elements of the product are nonzero but tiny—the result of rounding errors. The default format can be reinstated by typing `format short`, or simply `format`. The `format` command has many options, which can be seen by typing `help format`. See Table 2.1 for some examples. All the MATLAB output shown in this book was generated with `format compact` in effect, which suppresses blank lines.

Generally, `help foo` displays information on the command or function named `foo`. For example:

```
>> help sqrt

 SQRT   Square root.
    SQRT(X) is the square root of the elements of X. Complex
    results are produced if X is not positive.

    See also sqrtm.
```

Table 2.1. `10*exp(1)` *displayed in several output formats.*

`format short`	27.1828
`format long`	27.18281828459045
`format short e`	2.7183e+001
`format long e`	2.718281828459045e+001
`format short g`	27.183
`format long g`	27.1828182845905
`format hex`	403b2ecd2dd96d44
`format bank`	27.18
`format rat`	2528/93

```
    Reference page in Help browser
       doc sqrt
```

The terms `sqrtm` and `doc sqrt` are underlined in the Command Window to indicate that they are hyperlinks: clicking on them takes you to the relevant entry.

Note that it is a convention that function names are capitalized within help lines, in order to make them easy to identify. The names of all functions that are part of MATLAB or one of its toolboxes should be typed in lower case, however. The names of user-written M-files should be typed to match the case of the name of the `.m` file.

Typing `help` by itself produces the list of directories shown in Table 2.2 (extra directories will be shown for any toolboxes that are available, and if you have added your own directories to the path they will be shown as well). This list provides an overview of how MATLAB functions are organized. Typing `help` followed by a directory name (e.g., `help general`) gives a list of functions in that directory. Type `help help` for further details on the `help` command.

The most comprehensive documentation is available in the Help browser (see Figure 2.1), which provides help for all MATLAB functions, release and upgrade notes, and online versions of the complete MATLAB documentation in html and PDF format (for the latter, links are given to files on The MathWorks' Web site). The Help browser includes a Help Navigator pane containing tabs for a Contents listing, an Index listing, a Search facility, and Demos. The attached display pane displays html documentation containing links to related subjects and allows you to move back or forward a page and to search or print the current page. The Help browser is accessed by clicking the "?" icon on the toolbar of the MATLAB desktop, by selecting MATLAB Help from the Help menu, or by typing `helpbrowser` in the Command Window. From the Command Window, you can type `doc foo` to call up help on function `foo` directly in the Help browser, and `docsearch('phrase')` to execute a full text search of the Help browser documentation for the indicated phrase. Typing `helpwin` calls up the Help browser with the same list of directories produced by `help`; clicking on a directory takes you to a list of M-files in that directory and you can click on an M-file name to obtain help on that M-file.

A useful search facility is provided by the `lookfor` command. Type `lookfor keyword` to search for functions relating to the keyword. Example:

```
>> lookfor elliptic
ELLIPJ Jacobi elliptic functions.
```

Table 2.2. *MATLAB directory structure (under Windows).*

```
>> help

HELP topics

matlab\general        -  General purpose commands.
matlab\ops            -  Operators and special characters.
matlab\lang           -  Programming language constructs.
matlab\elmat          -  Elementary matrices and matrix manipulation.
matlab\elfun          -  Elementary math functions.
matlab\specfun        -  Specialized math functions.
matlab\matfun         -  Matrix functions - numerical linear algebra.
matlab\datafun        -  Data analysis and Fourier transforms.
matlab\polyfun        -  Interpolation and polynomials.
matlab\funfun         -  Function functions and ODE solvers.
matlab\sparfun        -  Sparse matrices.
matlab\scribe         -  Annotation and Plot Editing.
matlab\graph2d        -  Two dimensional graphs.
matlab\graph3d        -  Three dimensional graphs.
matlab\specgraph      -  Specialized graphs.
matlab\graphics       -  Handle Graphics.
matlab\uitools        -  Graphical user interface tools.
matlab\strfun         -  Character strings.
matlab\imagesci       -  Image and scientific data input/output.
matlab\iofun          -  File input and output.
matlab\audiovideo     -  Audio and Video support.
matlab\timefun        -  Time and dates.
matlab\datatypes      -  Data types and structures.
matlab\verctrl        -  Version control.
matlab\codetools      -  Commands for creating and debugging code.
matlab\helptools      -  Help commands.
matlab\winfun         -  Windows Operating System Interface Files
                         (COM/DDE)
matlab\demos          -  Examples and demonstrations.
toolbox\local         -  Preferences.
MATLAB7\work          -  (No table of contents file)
```

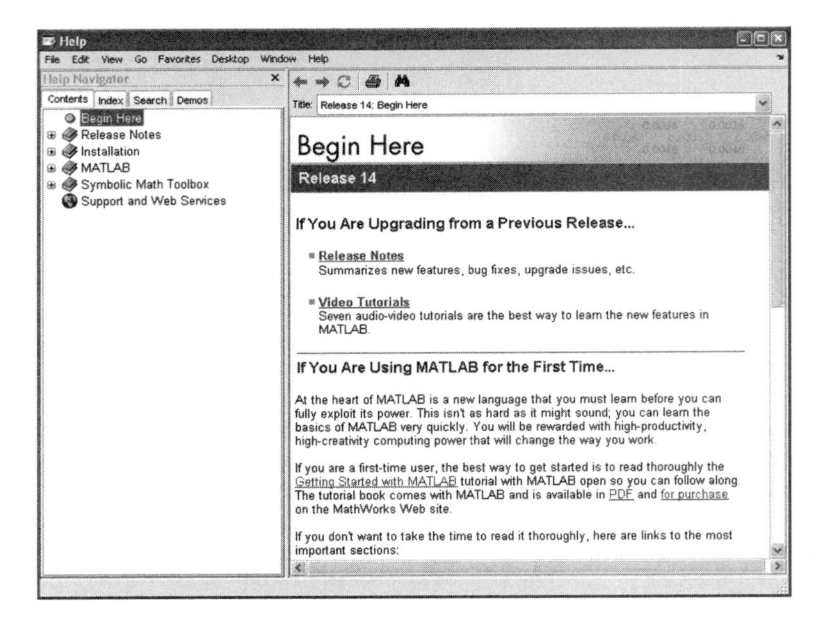

Figure 2.1. *Help browser.*

```
ELLIPKE Complete elliptic integral.
PDEPE   Solve initial-boundary value problems for parabolic-elliptic
        PDEs in 1-D.
```

If you make an error when typing at the prompt you can correct it using the arrow keys and the backspace or delete keys. Previous command lines can be recalled using the up arrow key, and the down arrow key takes you forward through the command list. If you type a few characters before hitting up arrow then the most recent command line beginning with those characters is recalled. A number of the Emacs control key commands for cursor movement are also supported. Table 2.3 summarizes the command line editing keypresses. You can scroll through commands previously typed in the current and past sessions in the Command History window. Double-clicking on a command in this window executes it.

Tab completion is a valuable command line time-saver and is also useful when you don't remember the precise name of a function or variable. If you press the tab key after typing a few characters of a function or variable name then MATLAB will attempt to complete the name, and it will offer you a menu of choices if there is more than one possible completion.

It is possible to enter multiple lines at the command line and run them all at once: press Shift-Enter at the end of each line and then press Enter at the end of the last line to run all of the lines.

Type `clc` to clear the Command Window.

A MATLAB computation can be aborted by pressing ctrl-c. If MATLAB is executing a built-in function it may take some time to respond to this keypress.

A line can be terminated with three periods (`...`), which causes the next line to be a continuation line:

```
>> x = 1 + 1/2 + 1/3 + 1/4 + 1/5 + ...
```

Table 2.3. *Command line editing key*
"hold down the control key and press t

Key	Control equi	
Up arrow	Ctrl-p	
Down arrow	Ctrl-n	
Left arrow	Ctrl-b	
Right arrow	Ctrl-f	
Ctrl left arrow		
Ctrl right arrow		
Home	Ctrl-a	
End	Ctrl-e	
Esc	Ctrl-u	
Del	Ctrl-d	
Backspace	Ctrl-h	
	Ctrl-k	
Insert		ggle insert mode
Shift-home		Select to beginning of line
Shift-end		Select to end of line

```
        1/6 + 1/7 + 1/8 + 1/9 + 1/10
x =
     2.9290
```

The value of x illustrates the fact that, unlike in some other programming languages, arithmetic on integers is done in floating point arithmetic and so can be written in the natural way.

Several functions provide special values:

- pi is $\pi = 3.14159\ldots$;
- i is the imaginary unit, $\sqrt{-1}$, as is j. Complex numbers are entered as, for example, 2-3i, 2-3*i, 2-3*sqrt(-1), or complex(2,-3). Note that the form 2-3*i may not produce the intended results if i is being used as a variable, so the other forms are generally preferred.

Functions generating constants related to floating point arithmetic are described in Chapter 4. It is possible to override existing variables and functions by creating new ones with the same names. This practice should be avoided, as it can lead to confusion. However, the use of i and j as counting variables is widespread.

MATLAB fully supports complex arithmetic, with conj, real, and imag taking the conjugate and the real and imaginary parts, respectively. For example,

```
>> w = (-1)^0.25
w =
   0.7071 + 0.7071i

>> z = conj(w)
z =
   0.7071 -0.7071i
```

```
.7071

1)

0000 + 0.0000i
```

list of variables in the workspace can be obtained by typing `who`, while `whos` ows the size and class of each variable as well. For example, after executing the commands so far in this chapter, `whos` produces

```
  Name          Size                      Bytes  Class

  A             3x3                          72  double array
  ans           1x1                          16  double array (complex)
  exmark        1x7                          56  double array
  exmean        1x1                           8  double array
  exmed         1x1                           8  double array
  exsort        1x7                          56  double array
  exstd         1x1                           8  double array
  w             1x1                          16  double array (complex)
  x             1x1                           8  double array
  y             1x1                           8  double array
  z             1x1                          16  double array (complex)

Grand total is 31 elements using 272 bytes
```

An existing variable `var` can be removed from the workspace by typing `clear var`, while `clear` clears all existing variables.

The workspace can also be examined via the Workspace browser (see Figure 2.2), which is invoked by the Desktop-Workspace menu option or by typing `workspace`. A variable, `A`, say, can be edited interactively in spreadsheet format in the Array Editor by double-clicking on the variable name (see Figure 2.3); alternatively, typing `openvar('A')` calls up the Array Editor on `A`.

Variable names are case sensitive and consist of a letter followed by any combination of letters, digits, and underscores, up to a maximum number of characters given by the value returned by the `namelengthmax` function:

```
>> namelengthmax
ans =
    63
```

To save variables for recall in a future MATLAB session type `save filename`; all variables in the workspace are saved to `filename.mat`. Alternatively, select the File-Save Workspace As menu option. The command

```
save filename A x
```

saves just the variables `A` and `x`. To create a MAT-file that can be read by MATLAB 6 append `-v6` to the `save` command. The command `load filename` loads in the variables from `filename.mat`, and individual variables can be loaded using the same

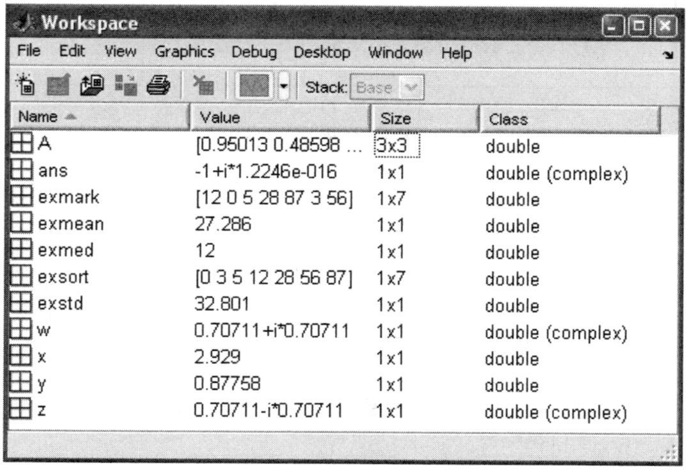
Workspace
File Edit View Graphics Debug Desktop Window Help
Stack: Base
Name Value Size Class
A [0.95013 0.48598 ... 3x3 double
ans -1+i*1.2246e-016 1x1 double (complex)
exmark [12 0 5 28 87 3 56] 1x7 double
exmean 27.286 1x1 double
exmed 12 1x1 double
exsort [0 3 5 12 28 56 87] 1x7 double
exstd 32.801 1x1 double
w 0.70711+i*0.70711 1x1 double (complex)
x 2.929 1x1 double
y 0.87758 1x1 double
z 0.70711-i*0.70711 1x1 double (complex)

Figure 2.2. *Workspace browser.*

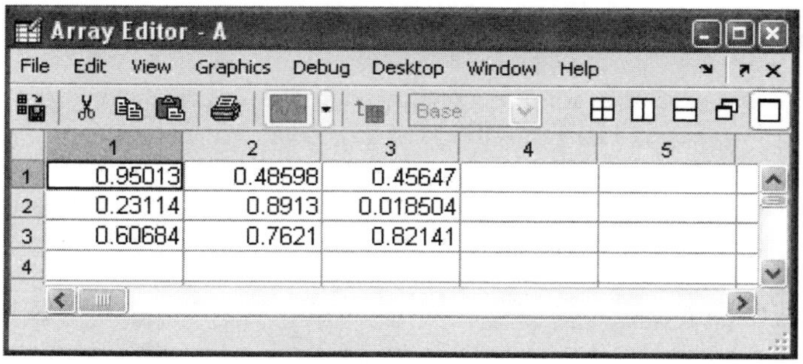
Array Editor - A
File Edit View Graphics Debug Desktop Window Help
Base
1 2 3 4 5
1 0.95013 0.48598 0.45647
2 0.23114 0.8913 0.018504
3 0.60684 0.7621 0.82141
4

Figure 2.3. *Array Editor.*

syntax as for **save**. The default is to save and load variables in binary form, but options allow ASCII form to be specified. MAT-files can be ported between MATLAB implementations running on different computer systems. An Import Wizard, accessible from the File-Import Data menu option, or by typing **uiimport**, provides a graphical interface to MATLAB's import functions.

Often you need to capture MATLAB output for incorporation into a report. This is most conveniently done with the **diary** command. If you type **diary filename** then all subsequent input and (most) text output is copied to the specified file; **diary off** turns off the diary facility. After typing **diary off** you can later type **diary on** to cause subsequent output to be appended to the same diary file.

To print the value of a variable or expression without the name of the variable or **ans** being displayed, you can use **disp**:

```
>> A = eye(2); disp(A)
     1     0
     0     1
```

Table 2.4. *Information and demonstrations.*

`bench`	Benchmarks to test the speed of your computer
`demo`	A collection of demonstrations
`info`	Contact information for The MathWorks
`ver`	Version number and release dates of MATLAB and toolboxes
`version`	Version number and release dates of MATLAB.
`whatsnew`	`whatsnew` brings up the Release Notes in the Help browser. `whatsnew matlab` displays the `readme` file for MATLAB, which explains the new features introduced in the most recent version. `whatsnew toolbox` displays the `readme` file for the specified toolbox

```
>> disp('Result:'), disp(1/7)
Result:
    0.1429
```

Commands are available for interacting with the operating system, including `cd` (change directory), `copyfile` (copy file), `mkdir` (make directory), `pwd` (print working directory), `dir`, or `ls` (list directory), and `delete` (delete file). A command can be issued to the operating system by preceding it with an exclamation mark, `!`. For example, you might type `!emacs myscript.m` to edit `myscript` with the Emacs editor.

Some MATLAB commands giving access to information and demonstrations are listed in Table 2.4.

Help!

— Title of a song by LENNON and MCCARTNEY (1965)

If ifs and ans were pots and pans,
there'd be no trade for tinkers.

— Proverb

Chapter 3
Distinctive Features of MATLAB

MATLAB has three features that distinguish it from most other modern programming languages and problem solving environments. We introduce them in this chapter and elaborate on them later in the book.

3.1. Automatic Storage Allocation

As we saw in Chapter 2, variables are not declared prior to being assigned. This applies to arrays as well as scalars. Moreover, MATLAB automatically expands the dimensions of arrays in order for assignments to make sense. Thus, starting with an empty workspace, we can set up a 1-by-3 vector x of zeros with

```
>> x(3) = 0
x =
     0     0     0
```

and then expand it to length 6 with

```
>> x(6) = 0
x =
     0     0     0     0     0     0
```

MATLAB's automatic allocation of storage is one of its most convenient and distinctive features.

3.2. Variable Arguments Lists

MATLAB contains a large (and user-extendible) collection of functions. They take zero or more input arguments and return zero or more output arguments. MATLAB enforces a clear distinction between input and output: input arguments appear on the right of the function name, within parentheses, and output arguments appear on the left, within square brackets. Functions can support a variable number of input and output arguments, so that on a given call not all arguments need be supplied. Functions can even vary their behavior depending on the precise number and type of arguments supplied. We illustrate with some examples.

The norm function computes the Euclidean norm, or 2-norm, of a vector (the square root of the sum of squares of the absolute values of the elements):

```
>> x = [3 4];

>> norm(x)
```

```
ans =
      5
```

A different norm can be obtained by supplying `norm` with a second input argument. For example, the 1-norm (the sum of the absolute values of the elements) is obtained with

```
>> norm(x,1)
ans =
      7
```

If the second argument is not specified then it defaults to `2`, giving the 2-norm. The `max` function has a variable number of output arguments. With an input vector and one output argument it returns the largest element of the vector:

```
>> m = max(x)
m =
      4
```

If a second output argument is supplied then the index of the largest element is assigned to it:

```
>> [m,k] = max(x)
m =
      4
k =
      2
```

As a final illustration of the versatility of MATLAB functions consider `size`, which returns the dimensions of an array. In the following example we set up a 5-by-3 random matrix and then request its dimensions:

```
>> A = rand(5,3);

>> s = size(A)
s =
      5     3
```

With one output argument, `size` returns a 1-by-2 vector with first element the number of rows of the input argument and second element the number of columns. However, `size` can also be given two output arguments, in which case it sets them to the number of rows and columns individually:

```
>> [m,n] = size(A)
m =
      5
n =
      3
```

3.3. Complex Arrays and Arithmetic

The fundamental data type in MATLAB is a multidimensional array of complex numbers, with real and imaginary parts stored in double precision floating point arithmetic. Important special cases are matrices (two-dimensional arrays), vectors, and scalars. Most computation in MATLAB is performed in floating point arithmetic, and complex arithmetic is used automatically when the data is complex. There is no separate real data type (though for reals the imaginary part is not stored). This can be contrasted with Fortran, in which different data types are used for real and complex numbers, and with C, C++, and Java, which support only real numbers and real arithmetic.

MATLAB also has integer data types, which are intended mainly for memory-efficient storage, rather than computation; see Section 4.4.

The guts of MATLAB are written in C.
Much of MATLAB is also written in MATLAB,
because it's a programming language.
— CLEVE B. MOLER[2] (1999)

In some ways, MATLAB resembles SPEAKEASY and, to a lesser extent, APL.
All are interactive terminal languages that ordinarily
accept single-line commands or statements,
process them immediately,
and print the results.
All have arrays as the principal data type.
— CLEVE B. MOLER, *Demonstration of a Matrix Laboratory* (1982)

[2]In [69].

Chapter 4
Arithmetic

4.1. IEEE Arithmetic

By default MATLAB carries out all its arithmetic computations in double precision floating point arithmetic conforming to the IEEE standard [44]. The function `computer` returns the type of computer on which MATLAB is running. The machine used to produce all the output shown in this book gives

```
>> computer
ans =
PCWIN
```

In MATLAB's `double` data type each number occupies a 64-bit word. Nonzero numbers range in magnitude between approximately 10^{-308} and 10^{+308} and the unit roundoff is $2^{-53} \approx 1.11 \times 10^{-16}$. (See [42, Chap. 2] or [89] for a detailed explanation of floating point arithmetic.) The significance of the unit roundoff is that it is a bound for the relative error in converting a real number to floating point form and also a bound for the relative error in adding, subtracting, multiplying, or dividing two floating point numbers or taking the square root of a floating point number. In simple terms, MATLAB stores floating point numbers and carries out elementary operations to an accuracy of about 16 significant decimal digits.

The function `eps` returns the distance from 1.0 to the next larger floating point number:

```
>> eps
ans =
   2.2204e-016
```

This distance, 2^{-52}, is *twice* the unit roundoff. More generally, `eps(x)` returns the (positive) distance from `x` to the next larger (in magnitude) floating point number:

```
>> eps(1/2)
ans =
   1.1102e-016

>> eps(2)
ans =
   4.4409e-016
```

Because MATLAB implements the IEEE standard, every computation produces a floating point number, albeit possibly one of a special type. If the result of a computation is larger than the value returned by the function `realmax` then overflow occurs and the result is `Inf` (also written `inf`), representing infinity. Similarly, a result more negative than `-realmax` produces `-inf`. Example:

```
>> realmax
ans =
  1.7977e+308

>> -2*realmax
ans =
  -Inf

>> 1.1*realmax
ans =
   Inf
```

A computation whose result is not mathematically defined produces a NaN, standing for Not a Number. The result `NaN` (also written `nan`) is generated by expressions such as `0/0`, `inf/inf`, and `0 * inf`:

```
>> 0/0
Warning: Divide by zero.
ans =
   NaN

>> inf/inf
ans =
   NaN

>> inf-inf
ans =
   NaN
```

Once generated, a NaN propagates through all subsequent computations:

```
>> NaN-NaN
ans =
   NaN

>> 0*NaN
ans =
   NaN
```

The function `realmin` returns the smallest positive normalized floating point number. Any computation whose result is smaller than `realmin` either underflows to zero if it is smaller than `eps*realmin` or produces a subnormal number—one with leading zero bits in its mantissa. To illustrate:

```
>> realmin
ans =
  2.2251e-308

>> realmin*eps
```

```
ans =
  4.9407e-324

>> realmin*eps/2
ans =
     0
```

To obtain further insight, repeat all the above computations after typing **`format hex`**, which displays the binary floating point representation of the numbers in hexadecimal format.

4.2. Precedence

MATLAB's arithmetic operators obey the same precedence rules as those in most calculators and computer languages. The rules are shown in Table 4.1. (For a more complete table, showing the precedence of all MATLAB operators, see Table 6.2.) For operators of equal precedence evaluation is from left to right. Parentheses can always be used to overrule priority, and their use is recommended to avoid ambiguity. Examples:

```
>> 2^10/10
ans =
  102.4000

>> 2 + 3*4
ans =
    14

>> -2 - 3*4
ans =
   -14

>> 1 + 2/3*4
ans =
    3.6667

>> 1 + 2/(3*4)
ans =
    1.1667

>> [2^2^3 2^(2^3)]
ans =
    64   256
```

4.3. Mathematical Functions

MATLAB contains a large set of mathematical functions. Typing **`help elfun`** and **`help specfun`** calls up full lists of elementary and special functions. A selection is listed in Table 4.2. The trigonometric functions take arguments in radians, with "**`*d`**"

Table 4.1. *Arithmetic operator precedence.*

Precedence level	Operator
1 (highest)	Exponentiation (^)
2	Unary plus (+), unary minus (-)
3	Multiplication (*), division (/)
4 (lowest)	Addition (+), subtraction (-)

Table 4.2. *Elementary and special mathematical functions* ("`fun*`" *indicates that more than one function name begins* "`fun`").

Functions	Category
`cos, sin, tan, csc, sec, cot`	Trigonometric
`cosd, sind, tand, cscd, secd, cotd`	
`acos, asin, atan, atan2, asec, acsc, acot`	Inverse trigonometric
`acosd, asind, atand, asecd, acscd, acotd`	
`cosh, sinh, tanh, sech, csch, coth`	Hyperbolic
`acosh, asinh, atanh, asech, acsch, acoth`	Inverse hyperbolic
`log, log2, log10, log1p, exp, expm1, pow2, nextpow2, nthroot`	Exponential
`ceil, fix, floor, round`	Rounding
`abs, angle, conj, imag, real`	Complex
`mod, rem, sign`	Remainder, sign
`airy, bessel*, beta*, ellipj, ellipke, erf*, expint, gamma*, legendre, psi`	Mathematical
`factor, gcd, isprime, lcm, primes, nchoosek, perms, rat, rats`	Number theoretic
`cart2sph, cart2pol, pol2cart, sph2cart`	Coordinate transforms

versions taking arguments in degrees. Of particular note are `expm1` and `log1p`, which accurately compute $e^x - 1$ and $\log(1 + x)$, respectively, for $|x| \ll 1$, avoiding the cancellation that would affect the direct calculation. (Explanations of the formulas underlying these two functions can be found in [42, Sec. 1.14.1, Problem 1.5].)

The Airy and Bessel functions are evaluated using a MEX interface to a library of Amos [2]. (You can view the Fortran source in the file `specfun\besselmx.f`.)

4.4. Other Data Types

MATLAB's `double` is not the only data type on which arithmetic can be performed. A `single` data type, corresponding to IEEE single precision arithmetic, also exists. The two main reasons for working with single precision numbers are

- To save storage. A scalar `single` occupies a 32-bit word rather than the 64-bit word occupied by a `double`. Hence, for example, a `single` vector of length 2000 can be stored in the same space as a `double` vector of length 1000.
- To explore the accuracy of a numerical algorithm. It is a standard technique to apply an algorithm in both single and double precision arithmetic and use

Table 4.3. *Parameters for single and double precision data types.*

Data type	Size	eps	Range
`single`	32 bits	$2^{-23} \approx 1.19 \times 10^{-7}$	$10^{\pm 38}$
`double`	64 bits	$2^{-52} \approx 2.22 \times 10^{-16}$	$10^{\pm 308}$

the difference of the results as an estimate of the error in the single precision solution.

Note that there is no intrinsic speed advantage to using single precision rather than double precision on most current processors. However, the lower storage requirement of single over double may allow better use of cache memory and thereby lead to faster execution.

The drawbacks of single precision are that it has half the precision of double precision (about 8 significant decimal digits versus 16) and it has a much smaller range, so overflow and underflow are much more likely to occur. The parameters of single precision arithmetic can be obtained as follows:

```
>> eps('single')
ans =
  1.1921e-007
>> realmax('single')
ans =
  3.4028e+038
>> realmin('single')
ans =
  1.1755e-038
```

Table 4.3 compares the key parameters for the single and double precisions.

Single precision numbers can be created in MATLAB in two ways: by conversion from another arithmetic data type using the `single` function, or by using the functions `eye`, `ones`, or `zeros` with an extra `'single'` argument (these functions are described in Section 5.1). For example,

```
>> format long, pi_s = single(pi), pi_d = pi
pi_s =
   3.1415927
pi_d =
   3.14159265358979

>> delta = pi_s - pi_d
delta =
   8.7422777e-008

>> A = ones(2,'single')
A =
      1     1
      1     1
```

```
>> whos
  Name        Size                    Bytes  Class

  A           2x2                        16  single array
  delta       1x1                         4  single array
  pi_d        1x1                         8  double array
  pi_s        1x1                         4  single array

Grand total is 7 elements using 32 bytes
```

This example shows several things. First, the single precision version of `pi` differs from the double precision version by about 10^{-7}, which is consistent with the respective precisions. Second, the `whos` output confirms that a `single` occupies half the storage of a `double` of the same dimension. Third, `delta`, which is the difference of a `single` and `double`, has the type `single`. This illustrates the rule that when `single` and `double` arrays interact in arithmetic, the type of the result is `single`.

The support for arithmetic on single precision numbers is new to MATLAB 7. Most, but not all, built-in functions support inputs of type `single`, while most M-file functions require `double` inputs.

Single and double precision numbers can be distinguished with the `class` function:

```
>> class(pi)
ans =
double

>> class(single(pi))
ans =
single
```

MATLAB is consistent in its use of these two precisions. For example, if a single precision computation overflows, then the resulting `inf` or `-inf` has the class `single`.

MATLAB also has eight integer data types: `int8`, `int16`, `int32`, and `int64` store signed integers of 8, 16, 32, and 64 bits, respectively, and their analogues `uint8`, `uint16`, `uint32`, and `uint64` store unsigned integers. One of the main uses of these data types is for efficient storage of image data. Limited arithmetic operations are supported. Variables can be created using `eye`, `ones`, and `zeros` with an extra argument:

```
>> E = zeros(1,3,'int8')
E =
     0     0     0

>> A = 5*ones(1,3,'uint16')
A =
     5     5     5

>> whos
  Name       Size                    Bytes  Class

  A          1x3                         6  uint16 array
  E          1x3                         3  int8 array

Grand total is 6 elements using 9 bytes
```

Functions of the same name as the data type convert into these storage formats, and the range of numbers supported can be obtained with `intmin` and `intmax`:

```
>> int8(30.8)
ans =
    31

>> [intmin('int8') intmax('int8')]
ans =
  -128   127

>> int8(128)
ans =
   127

>> [intmin('uint8') intmax('uint8')]
ans =
     0   255

>> [uint8(-1) uint8(256)]
ans =
     0   255
```

As these examples show, numbers outside the range are mapped to one of the ends of the range.

For a full list of MATLAB's data types and their interrelations, see Figure 18.1.

Round numbers are always false.
— SAMUEL JOHNSON, Boswell's *Life of Johnson* (1791)

Minus times minus is plus.
The reason for this we need not discuss.
— W. H. AUDEN, *A Certain World* (1971)

The only feature of classic MATLAB that is
not present in modern MATLAB is the "chop" function
which allows the simulation of shorter precision arithmetic.
It is an interesting curiosity,
but it is no substitute for roundoff error analysis
and it makes execution very slow, even when it isn't used.
— CLEVE B. MOLER, *MATLAB Digest Volume 3, Issue 1* (1991)

Chapter 5
Matrices

An m-by-n matrix is a two-dimensional array of numbers consisting of m rows and n columns. Special cases are a column vector ($n = 1$) and a row vector ($m = 1$).

Matrices are fundamental to MATLAB, and even if you are not intending to use MATLAB for linear algebra computations you need to become familiar with matrix generation and manipulation. In versions 3 and earlier of MATLAB there was only one data type: the complex matrix.[3] Nowadays MATLAB has several data types (see Chapter 18) and matrices are special cases of a multidimensional numeric array.

5.1. Matrix Generation

Matrices can be generated in several ways. Many elementary matrices can be constructed directly with a MATLAB function; see Table 5.1. The matrix of zeros, the matrix of ones, and the identity matrix (which has ones on the diagonal and zeros elsewhere) are returned by the functions **zeros**, **ones**, and **eye**, respectively. All have the same syntax. For example, **zeros**(m,n) or **zeros**([m,n]) produces an m-by-n matrix of zeros, while **zeros**(n) produces an n-by-n matrix. Examples:

```
>> zeros(2)
ans =
     0     0
     0     0

>> ones(2,3)
ans =
     1     1     1
     1     1     1

>> eye(3,2)
ans =
     1     0
     0     1
     0     0
```

A common requirement is to set up an identity matrix whose dimensions matc[illegible] those of a given matrix A. This can be done with eye(size(A)), where **size** [illegible] the function introduced in Section 3.2. Related to **size** is the **length** fun[illegible] **length**(A) is the larger of the two dimensions of A. Thus for an n-by-1 o[illegible] vector x, **length**(x) returns n.

[3]Cleve Moler used to joke that "MATLAB is a strongly typed language: it only h[illegible]

Table 5.1. *Elementary matrices.*

`zeros`	Zeros array
`ones`	Ones array
`eye`	Identity matrix
`repmat`	Replicate and tile array
`rand`	Uniformly distributed random numbers
`randn`	Normally distributed random numbers
`linspace`	Linearly spaced vector
`logspace`	Logarithmically spaced vector
`meshgrid`	X and Y arrays for 3D plots
`:`	Regularly spaced vector and index into matrix

Two other very important matrix generation functions are **rand** and **randn**, which generate matrices of (pseudo-)random numbers using the same syntax as **eye**. The function **rand** produces a matrix of numbers from the uniform distribution over the interval $[0, 1]$. For this distribution the proportion of numbers in an interval $[a, b]$ with $0 < a < b < 1$ is $b - a$. The function **randn** produces a matrix of numbers from the standard normal (0,1) distribution. Called without any arguments, both functions produce a single random number.

```
>> rand
ans =
    0.9528

>> rand(3)
ans =
    0.7041    0.8407    0.5187
    0.9539    0.4428    0.0222
    0.5982    0.8368    0.3759
```

In carrying out experiments with random numbers it is often important to be able to regenerate the same numbers on a subsequent occasion. The numbers produced by a call to **rand** depend on the state of the generator. The state can be set using the command **rand('state',j)**. For j=0 the **rand** generator is set to its initial state (the state it has when MATLAB starts). For a nonzero integer j, the generator is set to its jth state. The state of **randn** is set in the same way. The periods of **rand** and
... the number of terms generated before the sequences start to repeat,

... citly using the square bracket notation. For example,
... first 9 primes can be set up with the command

The end of a row can be specified by a semicolon instead of a carriage return, so a more compact command with the same effect is

```
>> A = [2 3 5; 7 11 13; 17 19 23]
```

Within a row, elements can be separated by spaces or by commas. In the former case, if numbers are specified with a plus or minus sign take care not to leave a space after the sign, else MATLAB will interpret the sign as an addition or subtraction operator. To illustrate with vectors:

```
>> v = [-1 2 -3 4]
v =
    -1     2    -3     4

>> w = [-1, 2, -3, 4]
w =
    -1     2    -3     4

>> x = [-1 2 - 3 4]
x =
    -1    -1     4
```

Matrices can be constructed in block form. With `B` defined by `B = [1 2; 3 4]`, we may create

```
>> C = [B        zeros(2)
        ones(2)  eye(2)]
C =
     1     2     0     0
     3     4     0     0
     1     1     1     0
     1     1     0     1
```

Block diagonal matrices can be defined using the function `blkdiag`, which is easier than using the square bracket notation. Example:

```
>> A = blkdiag(2*eye(2),ones(2))
A =
     2     0     0     0
     0     2     0     0
     0     0     1     1
     0     0     1     1
```

Useful for constructing "tiled" block matrices is `repmat`: `repmat(A,m,n)` creates a block `m`-by-`n` matrix in which each block is a copy of `A`. If `m` is omitted, it defaults to `n`. Example:

```
>> A = repmat(eye(2),2)
A =
     1     0     1     0
     0     1     0     1
     1     0     1     0
     0     1     0     1
```

Table 5.2. *Special matrices.*

`compan`	Companion matrix
`gallery`	Large collection of test matrices
`hadamard`	Hadamard matrix
`hankel`	Hankel matrix
`hilb`	Hilbert matrix
`invhilb`	Inverse Hilbert matrix
`magic`	Magic square
`pascal`	Pascal matrix
`rosser`	Classic symmetric eigenvalue test problem
`toeplitz`	Toeplitz matrix
`vander`	Vandermonde matrix
`wilkinson`	Wilkinson's eigenvalue test matrix

MATLAB provides a number of special matrices; see Table 5.2. These matrices have interesting properties that make them useful for constructing examples and for testing algorithms. One of the most famous is the Hilbert matrix, whose (i, j) element is $1/(i + j - 1)$. The matrix is generated by `hilb` and its inverse (which has integer entries) by `invhilb`. The function `magic` generates magic squares, which are fun to investigate using MATLAB [80]. The `toeplitz` function constructs a Toeplitz matrix: one for which the elements down each diagonal are constant. Similarly, `hankel` constructs a Hankel matrix: one for which the elements down each *anti*diagonal are constant. To construct a Toeplitz matrix, specify the first column and first row:

```
>> toeplitz([1 0 -1 -2],[1 2 4 8])
ans =
     1     2     4     8
     0     1     2     4
    -1     0     1     2
    -2    -1     0     1
```

For a Hankel matrix it is the first column and last row that are specified:

```
>> hankel([3 1 2 0],[0 -1 -2 -3])
ans =
     3     1     2     0
     1     2     0    -1
     2     0    -1    -2
     0    -1    -2    -3
```

The function `gallery` provides access to a large collection of test matrices created by N. J. Higham [41] (an earlier version of the collection was published in [40]). Table 5.3 lists the matrices; more information is obtained by typing `help private\matrix_name`. As indicated in the table, some of the matrices in `gallery` are returned in the `sparse` data type—see Chapter 15. Example:

```
>> help private\moler

 MOLER  Moler matrix (symmetric positive definite).
```

```
   A = GALLERY('MOLER',N,ALPHA) is the symmetric positive definite
   N-by-N matrix U'*U, where U = GALLERY('TRIW',N,ALPHA).

   For the default ALPHA = -1, A(i,j) = MIN(i,j)-2, and A(i,i) = i.
   One of the eigenvalues of A is small.

>> A = gallery('moler',5)
A =
     1    -1    -1    -1    -1
    -1     2     0     0     0
    -1     0     3     1     1
    -1     0     1     4     2
    -1     0     1     2     5
```

Table 5.4 lists matrices from Tables 5.2 and 5.3 having certain properties; in most cases the matrix has the property for the default arguments, but in some cases, such as for `gallery`'s `randsvd`, the arguments must be suitably chosen. For definitions of these properties see Chapter 9 and the textbooks listed at the start of that chapter.

Another way to generate a matrix is to load it from a file using the `load` command (see p. 30).

Table 5.3. *Matrices available through* `gallery`.

`cauchy`	Cauchy matrix
`chebspec`	Chebyshev spectral differentiation matrix
`chebvand`	Vandermonde-like matrix for the Chebyshev polynomials
`chow`	Chow matrix—a singular Toeplitz lower Hessenberg matrix
`circul`	Circulant matrix
`clement`	Clement matrix—tridiagonal with zero diagonal entries
`compar`	Comparison matrices
`condex`	Counterexamples to matrix condition number estimators
`cycol`	Matrix whose columns repeat cyclically
`dorr`	Dorr matrix—diagonally dominant, ill-conditioned, tridiagonal (one or three output arguments, `sparse`)
`dramadah`	Matrix of 1s and 0s whose inverse has large integer entries
`fiedler`	Fiedler matrix—symmetric
`forsythe`	Forsythe matrix—a perturbed Jordan block
`frank`	Frank matrix—ill-conditioned eigenvalues
`gearmat`	Gear matrix
`grcar`	Grcar matrix—a Toeplitz matrix with sensitive eigenvalues
`hanowa`	Matrix whose eigenvalues lie on a vertical line in the complex plane
`house`	Householder matrix (three output arguments)
`invhess`	Inverse of an upper Hessenberg matrix
`invol`	Involutory matrix
`ipjfact`	Hankel matrix with factorial elements (two output arguments)
`jordbloc`	Jordan block matrix
`kahan`	Kahan matrix—upper trapezoidal
`kms`	Kac–Murdock–Szego Toeplitz matrix
`krylov`	Krylov matrix

Table 5.3. (*continued*)

`lauchli`	Läuchli matrix—rectangular
`lehmer`	Lehmer matrix—symmetric positive definite
`leslie`	Leslie matrix
`lesp`	Tridiagonal matrix with real, sensitive eigenvalues
`lotkin`	Lotkin matrix
`minij`	Symmetric positive definite matrix $\min(i, j)$
`moler`	Moler matrix—symmetric positive definite
`neumann`	Singular matrix from the discrete Neumann problem (`sparse`)
`orthog`	Orthogonal and nearly orthogonal matrices
`parter`	Parter matrix—a Toeplitz matrix with singular values near π
`pei`	Pei matrix
`poisson`	Block tridiagonal matrix from Poisson's equation (`sparse`)
`prolate`	Prolate matrix—symmetric, ill-conditioned Toeplitz matrix
`randcolu`	Random matrix with normalized columns and specified singular values
`randcorr`	Random correlation matrix with specified eigenvalues
`randhess`	Random, orthogonal upper Hessenberg matrix
`randjorth`	Random J-orthogonal matrix
`rando`	Random matrix with elements -1, 0, or 1
`randsvd`	Random matrix with preassigned singular values and specified bandwidth
`redheff`	Matrix of 0s and 1s of Redheffer
`riemann`	Matrix associated with the Riemann hypothesis
`ris`	Ris matrix—a symmetric Hankel matrix
`smoke`	Smoke matrix—complex, with a "smoke ring" pseudospectrum
`toeppd`	Symmetric positive definite Toeplitz matrix
`toeppen`	Pentadiagonal Toeplitz matrix (`sparse`)
`tridiag`	Tridiagonal matrix (`sparse`)
`triw`	Upper triangular matrix discussed by Wilkinson and others
`wathen`	Wathen matrix—a finite element matrix (`sparse`, random entries)
`wilk`	Various specific matrices devised/discussed by Wilkinson (two output arguments)
`gallery(3)`	Badly conditioned 3-by-3 matrix
`gallery(5)`	Interesting eigenvalue problem

5.2. Subscripting and the Colon Notation

To enable access and assignment to submatrices MATLAB has a powerful notation based on the colon character. The colon is used to define vectors that can act as subscripts. For integers `i` and `j`, `i:j` denotes the row vector of integers from `i` to `j` (in steps of 1). A nonunit step (or stride) `s` is specified as `i:s:j`. This notation is valid even for noninteger `i`, `j`, and `s`. Examples:

```
>> 1:5
ans =
     1     2     3     4     5
```

Table 5.4. *Matrices classified by property. Most of the matrices listed here are accessed through* `gallery`.

Defective	`chebspec`, `gallery(5)`, `gearmat`, `jordbloc`, `triw`
Hankel	`hilb`, `ipjfact`, `ris`
Hessenberg	`chow`, `frank`, `grcar`, `randhess`, `randsvd`
Idempotent	`invol`
Inverse of tridiagonal matrix	`kms`, `lehmer`, `minij`
Involutary	`invol`, `orthog`, `pascal`
Nilpotent	`chebspec`, `gallery(5)`
Normal*	`circul`
Orthogonal	`hadamard`, `orthog`, `randhess`, `randsvd`
Rectangular	`chebvand`, `cycol`, `kahan`, `krylov`, `lauchli`, `rando`, `randsvd`, `triw`
Symmetric indefinite	`clement`, `fiedler`
Symmetric positive definite	`hilb`, `invhilb`, `ipjfact`, `kms`, `lehmer`, `minij`, `moler`, `pascal`, `pei`, `poisson`, `prolate`, `randcorr`, `randsvd`, `toeppd`, `tridiag`, `wathen`
Toeplitz	`chow`, `dramadah`, `grcar`, `kms`, `parter`, `prolate`, `toeppd`, `toeppen`
Totally positive/nonnegative	`cauchy`,[†] `hilb`, `lehmer`, `pascal`
Tridiagonal	`clement`, `dorr`, `lesp`, `randsvd`, `tridiag`, `wilk`, `wilkinson`
Triangular	`dramadah`, `jordbloc`, `kahan`, `pascal`, `triw`

* But not symmetric or orthogonal.
† `cauchy(x,y)` is totally positive if $0 < x_1 < \cdots < x_n$ and $0 < y_1 < \cdots < y_n$ [42].

```
>> 4:-1:-2
ans =
     4     3     2     1     0    -1    -2

>> 0:.75:3
ans =
         0    0.7500    1.5000    2.2500    3.0000
```

Single elements of a matrix are accessed as `A(i,j)`, where $\mathtt{i} \geq 1$ and $\mathtt{j} \geq 1$ (zero or negative subscripts are not supported in MATLAB). The submatrix comprising the intersection of rows `p` to `q` and columns `r` to `s` is denoted by `A(p:q,r:s)`. As a special case, a lone colon as the row or column specifier covers all entries in that row or column; thus `A(:,j)` is the `j`th column of `A` and `A(i,:)` the `i`th row. The keyword **`end`** used in this context denotes the last index in the specified dimension; thus `A(`**`end`**`,:)` picks out the last row of `A`. In effect, a lone colon is short for `1:`**`end`**. Finally, an arbitrary submatrix can be selected by specifying the individual row and column indices. For example, `A([i j k],[p q])` produces the submatrix given by the intersection of rows `i`, `j`, and `k` and columns `p` and `q`. Here are some examples, using the matrix of primes set up above:

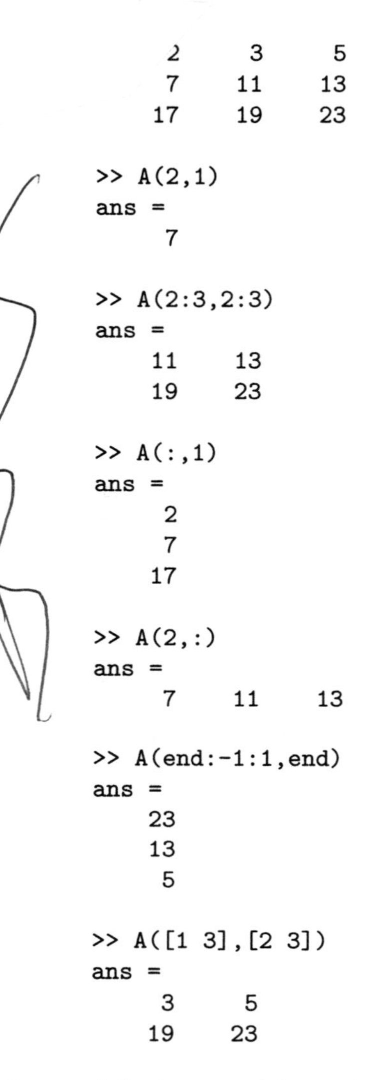

```
      2      3      5
      7     11     13
     17     19     23

>> A(2,1)
ans =
      7

>> A(2:3,2:3)
ans =
     11     13
     19     23

>> A(:,1)
ans =
      2
      7
     17

>> A(2,:)
ans =
      7     11     13

>> A(end:-1:1,end)
ans =
     23
     13
      5

>> A([1 3],[2 3])
ans =
      3      5
     19     23
```

A further special case is `A(:)`, which denotes a vector comprising all the elements of `A` taken down the columns from first to last:

```
>> B = A(:)
B =
      2
      7
     17
      3
     11
     19
      5
     13
     23
```

When placed on the left side of an assignment statement `A(:)` fills `A`, preserving its shape. Using this notation, another way to define our 3-by-3 matrix of primes is

```
>> A = zeros(3); A(:) = primes(23); A = A'
A =
     2     3     5
     7    11    13
    17    19    23
```

The function `primes` returns a vector of the prime numbers less than or equal to its argument. The transposition `A = A'` (see the next section) is necessary to reorder the primes across the rows rather than down the columns.

In one circumstance—when the right-hand side is a single element—the number of elements in a subscripted assignment can be different on the two sides of the assignment. In this case the scalar is "expanded" to match the number of elements on the left:

```
>> A = ones(3);
>> A(2:3,2:3) = 0
A =
     1     1     1
     1     0     0
     1     0     0
```

Related to the colon notation for generating vectors of equally spaced numbers is the function `linspace`, which accepts the number of points rather than the increment: `linspace(a,b,n)` generates `n` equally spaced points between `a` and `b`. If `n` is omitted it defaults to 100. Example:

```
>> linspace(-1,1,9)
ans =
  Columns 1 through 7
   -1.0000   -0.7500   -0.5000   -0.2500         0    0.2500    0.5000
  Columns 8 through 9
    0.7500    1.0000
```

The notation `[]` denotes an empty, 0-by-0 matrix. Assigning `[]` to a row or column is one way to delete that row or column from a matrix:

```
>> A(2,:) = []
A =
     2     3     5
    17    19    23
```

In this example the same effect is achieved by `A = A([1 3],:)`. The empty matrix is also useful as a placeholder in argument lists, as we will see in Section 5.5.

5.3. Matrix and Array Operations

For scalars a and b, the operators +, -, *, /, and ^ produce the obvious results. As well as the usual right division operator, /, MATLAB has a left division operator, \:

MATLAB notation	Mathematical equivalent
Right division: a/b	$\frac{a}{b}$
Left division: a\b	$\frac{b}{a}$

For matrices, all these operations can be carried out in a matrix sense (according to the rules of matrix algebra) or an array sense (elementwise). Table 5.5 summarizes the syntax.

Addition and subtraction, which are identical operations in the matrix and array senses, are defined for matrices of the same dimension. The product A*B is the result of matrix multiplication, defined only when the number of columns of A and the number of rows of B are the same. The backslash and the forward slash define solutions of linear systems: A\B is a solution X of A*X = B, while A/B is a solution X of X*B = A; see Section 9.2 for more details. Examples:

```
>> A = [1 2; 3 4], B = ones(2)
A =
     1     2
     3     4
B =
     1     1
     1     1

>> A+B
ans =
     2     3
     4     5

>> A*B
ans =
     3     3
     7     7

>> A\B
ans =
    -1    -1
     1     1
```

Multiplication and division in the array, or elementwise, sense are specified by preceding the operator with a period. If A and B are matrices of the same dimensions then C = A.*B sets C(i,j) = A(i,j)*B(i,j) and C = A./B sets C(i,j) = A(i,j)/B(i,j). The assignment C = A.\B is equivalent to C = B./A. With the same A and B as in the previous example:

```
>> A.*B
```

Table 5.5. *Elementary matrix and array operations.*

Operation	Matrix sense	Array sense
Addition	+	+
Subtraction	-	-
Multiplication	*	.*
Left division	\	.\
Right division	/	./
Exponentiation	^	.^

```
ans =
     1     2
     3     4

>> B./A
ans =
    1.0000    0.5000
    0.3333    0.2500
```

Exponentiation with ^ is defined as matrix powering, but the dot form exponentiates elementwise. Thus if `A` is a square matrix then `A^2` is the matrix product `A*A`, but `A.^2` is `A` with each element squared:

```
>> A^2, A.^2
ans =
     7    10
    15    22
ans =
     1     4
     9    16
```

The dot form of exponentiation allows the power to be an array when the dimensions of the base and the power agree, or when the base is a scalar:

```
>> x = [1 2 3]; y = [2 3 4]; Z = [1 2; 3 4];

>> x.^y
ans =
     1     8    81

>> 2.^x
ans =
     2     4     8

>> 2.^Z
ans =
     2     4
     8    16
```

Matrix exponentiation is defined for all powers, not just for positive integers. If `n<0` is an integer then `A^n` is defined as `inv(A)^n`. For noninteger `p`, `A^p` is evaluated using the eigensystem of `A`; results can be incorrect or inaccurate when `A` is not diagonalizable or when `A` has an ill-conditioned eigensystem.

The conjugate transpose of the matrix `A` is obtained with `A'`. If `A` is real, this is simply the transpose. The transpose without conjugation is obtained with `A.'`. The functional alternatives `ctranspose(A)` and `transpose(A)` are sometimes more convenient.

For the special case of column vectors `x` and `y`, `x'*y` is the inner, scalar, or dot product, which can also be obtained using the `dot` function as `dot(x,y)`. The vector or cross product of two 3-by-1 or 1-by-3 vectors (as used in mechanics) is produced by `cross`. Example:

```
>> x = [-1 0 1]'; y = [3 4 5]';

>> x'*y
ans =
     2

>> dot(x,y)
ans =
     2

>> cross(x,y)
ans =
    -4
     8
    -4
```

The `kron` function evaluates the Kronecker product of two matrices. The Kronecker product of an m-by-n A and p-by-q B has dimensions mp-by-nq and can be expressed as a block m-by-n matrix with (i,j) block $a_{ij}B$. Example:

```
>> A = [1 10; -10 100]; B = [1 2 3; 4 5 6; 7 8 9];

>> kron(A,B)
ans =
     1     2     3    10    20    30
     4     5     6    40    50    60
     7           9    70    80    90
                 0   100   200   300
                     400   500   600
                     700   800   900
```

matrix MATLAB will expand the scalar into a matrix
at scalar. For example:

```
>> A = [1 -1] - 6
A =
    -5    -7
```

However, if an assignment makes sense without expansion then it will be interpreted in that way. Thus if the previous command is followed by `A = 1` then `A` becomes the scalar 1, not `ones(1,2)`.

If a matrix is multiplied or divided by a scalar, the operation is performed elementwise. For example:

```
>> [3 4 5; 4 5 6]/12
ans =
    0.2500    0.3333    0.4167
    0.3333    0.4167    0.5000
```

Most of the functions described in Section 4.3 can be given a matrix argument, in which case the functions are computed elementwise. Functions of a matrix in the linear algebra sense are signified by names ending in `m` (see Section 9.9): `expm`, `funm`, `logm`, `sqrtm`. For example, for `A = [2 2; 0 2]`,

```
>> sqrt(A)
ans =
    1.4142    1.4142
         0    1.4142

>> sqrtm(A)
ans =
    1.4142    0.7071
         0    1.4142

>> ans*ans
ans =
    2.0000    2.0000
         0    2.0000
```

5.4. Matrix Manipulation

Several commands are available for manipulating matrices (commands more specifically associated with linear algebra are discussed in Chapter 9); see Table 5.6.

The `reshape` function changes the dimensions of a matrix: `reshape(A,m,n)` produces an `m`-by-`n` matrix whose elements are taken columnwise from `A`. For example:

```
>> A = [1 4 9; 16 25 36], B = reshape(A,3,2)
A =
     1     4     9
    16    25    36
B =
     1    25
    16     9
     4    36
```

Table 5.6. *Matrix manipulation functions.*

`reshape`	Change size
`diag`	Diagonal matrices and diagonals of matrix
`blkdiag`	Block diagonal matrix
`tril`	Extract lower triangular part
`triu`	Extract upper triangular part
`fliplr`	Flip matrix in left/right direction
`flipud`	Flip matrix in up/down direction
`rot90`	Rotate matrix 90 degrees

The function `diag` deals with the diagonals of a matrix and can take a vector or a matrix as argument. For a vector `x`, `diag(x)` is the diagonal matrix with main diagonal `x`:

```
>> diag([1 2 3])
ans =
     1     0     0
     0     2     0
     0     0     3
```

More generally, `diag(x,k)` puts `x` on the `k`th diagonal, where `k` > 0 specifies diagonals above the main diagonal and `k` < 0 diagonals below the main diagonal (`k` $= 0$ gives the main diagonal):

```
>> diag([1 2], 1)
ans =
     0     1     0
     0     0     2
     0     0     0

>> diag([3 4], -2)
ans =
     0     0     0     0
     0     0     0     0
     3     0     0     0
     0     4     0     0
```

For a matrix `A`, `diag(A)` is the column vector comprising the main diagonal of `A`. To produce a diagonal matrix with diagonal the same as that of `A` you must therefore write `diag(diag(A))`. Analogously to the vector case, `diag(A,k)` produces a column vector made up from the `k`th diagonal of `A`. Thus if

```
A =
     2     3     5
     7    11    13
    17    19    23
```

then

```
>> diag(A)
ans =
     2
    11
    23

>> diag(A,-1)
ans =
     7
    19
```

Triangular parts of a matrix can be extracted using `tril` and `triu`. The lower triangular part of `A` (the elements on and below the main diagonal) is specified by `tril(A)` and the upper triangular part of `A` (the elements on and above the main diagonal) is specified by `triu(A)`. More generally, `tril(A,k)` gives the elements on and below the `k`th diagonal of `A`, while `triu(A,k)` gives the elements on and above the `k`th diagonal of `A`. With `A` as above:

```
>> tril(A)
ans =
     2     0     0
     7    11     0
    17    19    23

>> triu(A,1)
ans =
     0     3     5
     0     0    13
     0     0     0

>> triu(A,-1)
ans =
     2     3     5
     7    11    13
     0    19    23
```

5.5. Data Analysis

Table 5.7 lists functions for basic data analysis computations. The simplest usage is to apply these functions to vectors. For example:

```
>> x = [4 -8 -2 1 0]
x =
     4    -8    -2     1     0

>> [min(x) max(x)]
ans =
    -8     4

>> sort(x)
```

Table 5.7. *Basic data analysis functions.*

`max`	Largest component
`min`	Smallest component
`mean`	Average or mean value
`median`	Median value
`std`	Standard deviation
`var`	Variance
`sort`	Sort in ascending order
`sum`	Sum of elements
`prod`	Product of elements
`cumsum`	Cumulative sum of elements
`cumprod`	Cumulative product of elements
`diff`	Difference of elements

```
ans =
    -8    -2     0     1     4

>> sum(x)
ans =
    -5
```

The `sort` function sorts into ascending order by default. Descending order is obtained by appending an extra argument `'descend'`. For complex vectors, `sort` sorts by absolute value:

```
>> x = [1+i -3-4i 2i 1];
>> sort(x,'descend')
ans =
  -3.0000 -4.0000i        0 + 2.0000i   1.0000 + 1.0000i   1.0000
```

Any NaN elements are placed by `sort` at the high end, while `max` and `min` ignore NaNs.

For matrices the functions are defined columnwise. Thus `max` and `min` return a vector containing the maximum and minimum element, respectively, in each column, `sum` returns a vector containing the column sums, and `sort` sorts the elements in each column of the matrix into ascending order. The functions `min` and `max` can return a second argument that specifies in which components the minimum and maximum elements are located. For example, if

```
A =
     0    -1     2
     1     2    -4
     5    -3    -4
```

then

```
>> max(A)
ans =
     5     2     2
```

```
>> [m,i] = min(A)
m =
     0    -3    -4
i =
     1     3     2
```

As this example shows, if there are two or more minimal elements in a column then the index of the first is returned. The smallest element in the matrix can be found by applying `min` twice in succession:

```
>> min(min(A))
ans =
    -4
```

An alternative, which has the advantage that it also works for arrays of dimension greater than 2, is

```
>> min(A(:))
ans =
    -4
```

Functions `max` and `min` can be made to act row-wise via a third argument:

```
>> max(A,[],2)
ans =
     2
     2
     5
```

The 2 in `max(A,[],2)` specifies the maximum over the second dimension, that is, over the column index. The empty second argument, `[]`, is needed because with just two arguments `max` and `min` return the elementwise maxima and minima of the two arguments:

```
>> max(A,0)
ans =
     0     0     2
     1     2     0
     5     0     0
```

Functions `sort` and `sum` can also be made to act row-wise, via a second argument. For more on `sort` see Section 21.3.

For complex data, `max` and `min` measure size using the absolute value, like `sort`.

The `diff` function forms differences. Applied to a vector `x` of length `n` it produces the vector `[x(2)-x(1) x(3)-x(2) ... x(n)-x(n-1)]` of length `n-1`. Example:

```
>> x = (1:8).^2
x =
     1     4     9    16    25    36    49    64

>> y = diff(x)
y =
```

```
     3     5     7     9    11    13    15

>> z = diff(y)
z =
     2     2     2     2     2     2
```

In data analysis NaNs are often used to represent "missing values": data that is not available, which could be the result of the failure of a process being measured (see Section 22.4 for an example). Before carrying out any floating point computation with the data it is necessary to remove the NaNs, because any computation involving a NaN produces a NaN. This can be done in several ways, all using the function `isnan` (see Section 6.1 for more on `isnan`). For example:

```
>> x = [2 1 NaN -1 6], y = x;
x =
     2     1   NaN    -1     6

>> mean(x)
ans =
   NaN

>> x = x(~isnan(x)), mean(x)
x =
     2     1    -1     6
ans =
     2

>> y(isnan(y)) = [], mean(y)
y =
     2     1    -1     6
ans =
     2
```

Handled properly,
empty arrays relieve programmers of the
nuisance of special cases at beginnings and ends of
algorithms that construct matrices recursively from submatrices.
— WILLIAM M. KAHAN (1994)

Kirk: "You did all this in a day?"
Carol: "The matrix formed in a day.
The lifeforms grew later at a substantially accelerated rate."
— *Star Trek III: The Search For Spock* (Stardate 8130.4)

I start by looking at a 2 by 2 matrix.
Sometimes I look at a 4 by 4 matrix.
That's when things get out of control and too hard.
Usually 2 by 2 or 3 by 3 is enough, and I look at them,
and I compute with them, and I try to guess the facts.
— PAUL R. HALMOS, in *Paul Halmos: Celebrating 50 Years of Mathematics* (1991)

Chapter 6
Operators and Flow Control

6.1. Relational and Logical Operators

MATLAB has a logical data type, with the possible values 1, representing true, and 0, representing false. Logicals are produced by relational and logical operators/functions and by the functions `true` and `false`:

```
>> a = true
a =
     1

>> b = false
b =
     0

>> c = 1
c =
     1

>> whos
  Name      Size                    Bytes  Class

  a         1x1                         1  logical array
  b         1x1                         1  logical array
  c         1x1                         8  double array

Grand total is 3 elements using 10 bytes
```

As this example shows, logicals occupy one byte, rather than the eight bytes needed by a `double`.

MATLAB's relational operators are

```
==   equal
~=   not equal
 <   less than
 >   greater than
<=   less than or equal
>=   greater than or equal
```

Note that a single = denotes assignment and never a test for equality in MATLAB.

Comparisons between scalars produce logical 1 if the relation is true and logical 0 if it is false. Comparisons are also defined between matrices of the same dimension and between a matrix and a scalar, the result being a matrix of logicals in both cases. For matrix–matrix comparisons corresponding pairs of elements are compared, while for matrix–scalar comparisons the scalar is compared with each matrix element. For example:

```
>> A = [1 2; 3 4]; B = 2*ones(2);

>> A == B
ans =
     0     1
     0     0

>> A > 2
ans =
     0     0
     1     1
```

To test whether arrays `A` and `B` are equal, that is, of the same size with identical elements, the expression `isequal(A,B)` can be used:

```
>> isequal(A,B)
ans =
     0
```

The function `isequal` is one of many useful logical functions whose names begin with `is`, a selection of which is listed in Table 6.1; for a full list type `doc is`. For example, `isinf(A)` returns a logical array of the same size as `A` containing true where the elements of `A` are plus or minus `inf` and false where they are not:

```
>> A = [1 inf; -inf NaN];
>> isinf(A)
ans =
       0           1
       1           0
```

The function `isnan` is particularly important because the test `x == NaN` always produces the result `0` (false), even if `x` is a NaN! (A NaN is defined to compare as unequal and unordered with everything.)

Note that an array can be real in the mathematical sense, but not real as reported by `isreal`. For `isreal(A)` is true if `A` has *no* imaginary part. Mathematically, `A` is real if every component has *zero* imaginary part. How a mathematically real `A` is formed can determine whether it has an imaginary part or not in MATLAB. The distinction can be seen as follows:

```
>> a = 1;
>> b = complex(1,0);
>> c = 1 + 0i;

>> [a b c]
ans =
     1     1     1
```

Table 6.1. *Selected logical* `is*` *functions.*

`ischar`	Test for char array (string)
`isempty`	Test for empty array
`isequal`	Test if arrays are equal
`isequalwithequalnans`	Test if arrays are equal, treating NaNs as equal
`isfinite`	Detect finite array elements
`isfloat`	Test for floating point array (single or double)
`isinf`	Detect infinite array elements
`isinteger`	Test for integer array
`islogical`	Test for logical array
`isnan`	Detect NaN array elements
`isnumeric`	Test for numeric array (integer or floating point)
`isreal`	Test for real array
`isscalar`	Test for scalar array
`issorted`	Test for sorted vector
`isvector`	Test for vector array

```
>> whos a b c
  Name      Size                    Bytes  Class

  a         1x1                         8  double array
  b         1x1                        16  double array (complex)
  c         1x1                         8  double array

Grand total is 3 elements using 32 bytes

>> [isreal(a), isreal(b), isreal(c)]
ans =
     1     0     1
```

MATLAB's logical operators are

`&`	logical and
`&&`	logical and (for scalars) with short-circuiting
`|`	logical or
`||`	logical or (for scalars) with short-circuiting
`~`	logical not
`xor`	logical exclusive or
`all`	true if all elements of vector are nonzero
`any`	true if any element of vector is nonzero

Like the relational operators, the `&`, `|`, and `~` operators produce matrices of logical 0s and 1s when one of the arguments is a matrix. When applied to a vector, the `all` function returns 1 if all the elements of the vector are nonzero and 0 otherwise. The `any` function is defined in the same way, with "any" replacing "all". Examples:

```
>> x = [-1 1 1]; y = [1 2 -3];
```

```
>> x>0 & y>0
ans =
     0     1     0

>> x>0 | y>0
ans =
     1     1     1

>> xor(x>0,y>0)
ans =
     1     0     1

>> any(x>0)
ans =
     1

>> all(x>0)
ans =
     0
```

Note that `xor` must be called as a function: `xor(a,b)`. The **and**, `or`, and **not** operators and the relational operators can also be called in functional form as **and(a,b)**, ..., `eq(a,b)`, ... (see **help ops**).

The operators `&&` and `||` are special in two ways. First, they work with scalar expressions only, and should be used in preference to `&` and `|` for scalar expressions. Continuing the previous example, compare

```
>> any(x>0) && any(y>0)
ans =
     1
>> x>0 && y>0
??? Operands to the || and && operators must be convertible to
    logical scalar values.
```

The second feature of these "double barreled" operators is that they short-circuit the evaluation of the logical expressions, where possible. In the compound expression *expr1* `&&` *expr2*, if *expr1* evaluates to false then *expr2* is not evaluated. Similarly, in *expr1* `||` *expr2*, if *expr1* evaluates to true then *expr2* is not evaluated. Short-circuiting saves computation, but it also enables warnings and errors to be avoided. For example, a statement beginning

```
if x > 0 && sin(1/x) < 0.5
```

avoids a division by zero.

The precedence of arithmetic, relational, and logical operators is summarized in Table 6.2 (which is based on the information provided by **help precedence**). For operators of equal precedence MATLAB evaluates from left to right. Precedence can be overridden by using parentheses. Note, in particular, that **and** has higher precedence than `or`, so a logical expression of the form

```
x | y & z
```

is equivalent to

Table 6.2. *Operator precedence.*

Precedence level	Operator
1 (highest)	Parentheses ()
2	Transpose (.'), power (.^), complex conjugate transpose ('), matrix power (^)
3	Unary plus (+), unary minus (-), logical negation (~)
4	Multiplication (.*), right division (./), left division (.\), matrix multiplication (*), matrix right division (/), matrix left division (\)
5	Addition (+), subtraction (-)
6	Colon operator (:)
7	Less than (<), less than or equal to (<=), greater than (>), greater than or equal to (>=), equal to (==), not equal to (~=)
8	Logical and (&)
9	Logical or (|)
10	Logical short-circuit and (&&)
11 (lowest)	Logical short-circuit or (||)

```
x | (y & z)
```

It is good practice to insert parentheses to make the intention completely clear.

For matrices, `all` returns a row vector containing the result of `all` applied to each column. Therefore `all(all(A==B))` is another way of testing equality of the matrices `A` and `B`. The `any` function works in the corresponding way. Thus, for example, `any(any(A==B))` has the value 1 if `A` and `B` have any equal elements and 0 otherwise. Alternatives are `all(A(:)==B(:))` and `any(A(:)==B(:))`.

The `find` command returns the indices corresponding to the nonzero elements of a vector. For example,

```
>> x = [-3 1 0 -inf 0];
>> f = find(x)
f =
     1     2     4
```

The result of `find` can then be used to extract just those elements of the vector:

```
>> x(f)
ans =
    -3     1  -Inf
```

With `x` as above, we can use `find` to obtain the finite elements of `x`,

```
>> x(find(isfinite(x)))
ans =
    -3     1     0     0
```

and to replace negative components of `x` by zero:

```
>> x(find(x < 0)) = 0
x =
     0     1     0     0     0
```

When `find` is applied to a matrix `A`, the index vector corresponds to `A` regarded as a vector of the columns stacked one on top of the other (that is, `A(:)`), and this vector can be used to index into `A`. In the following example we use `find` to set to zero those elements of `A` that are less than the corresponding elements of `B`:

```
>> A = [4 2 16; 12 4 3], B = [12 3 1; 10 -1 7]
A =
     4     2    16
    12     4     3
B =
    12     3     1
    10    -1     7

>> f = find(A<B)
f =
     1
     3
     6

>> A(f) = 0
A =
     0     0    16
    12     4     0
```

An alternative usage of `find` for matrices is `[i,j] = find(A)`, which returns vectors `i` and `j` containing the row and column indices of the nonzero elements.

The results of MATLAB's logical operators and logical functions are logical arrays of 0s and 1s. Logical arrays can also be created by applying the function `logical` to a numeric array; nonzero values other than 1 that are converted to 1 result in a warning message. Logical arrays and numeric arrays can both be used for subscripting, but with an important difference: logical arrays pick out elements where the subscript is true, whereas numeric arrays pick out elements indexed by the subscript. An example should make this distinction clear.

```
>> clear
          [1 2 0 -3 0]

                          0

                          0

                     1    0
```

```
>> y(i1)
ans =
     1     2    -3

>> y(i2)
??? Subscript indices must either be real positive integers or
    logicals.

>> whos i1 i2
  Name      Size                    Bytes  Class

  i1        1x5                         5  logical array
  i2        1x5                        40  double array

>> isequal(i1,i2)
ans =
     1

>> i3 = [1 2 4]; y(i3)
ans =
     1     2    -3
```

Although the numeric array `i2` has the same elements as the logical array `i1` (and compares as equal with it), only `i1` can be used for subscripting. To achieve the required subscripting effect with a numerical array, `i3` must be used.

A call to `find` can sometimes be avoided when its argument is a logical array. In our example on p. 67, `x(find(isfinite(x)))` can be replaced by `x(isfinite(x))`.

Addition and multiplication can be done on logicals, and they can be used in arithmetic expressions containing doubles. The result is always a double:

```
>> a = true; b = false; c = 2*a+b, class(c)
c =
     2
ans =
double
```

However, many other arithmetic operations fail:

```
>> b/a
??? Function 'mrdivide' is not defined for values of class 'logical'.

Error in ==> mrdivide at 16
  builtin('mrdivide', varargin{:});
```

6.2. Flow Control

MATLAB has four flow control structures: the `if` statement, the `for` loop, the `while` loop, and the `switch` statement. The simplest form of the `if` statement is

```
if expression
   statements
end
```

where the statements are executed if the elements of *expression* are all nonzero. For example, the following code swaps `x` and `y` if `x` is greater than `y`:

```
if x > y
   temp = y;
   y = x;
   x = temp;
end
```

When an `if` statement is followed on its line by further statements, a comma is needed to separate the `if` from the next statement:

```
if x > 0, x = sqrt(x); end
```

Statements to be executed only if *expression* is false can be placed after `else`, as in the example

```
e = exp(1);
if 2^e > e^2
   disp('2^e is bigger')
else
   disp('e^2 is bigger')
end
```

Finally, one or more further tests can be added with `elseif` (note that there must be no space between `else` and `if`):

```
if isnan(x)
   disp('Not a Number')
elseif isinf(x)
   disp('Plus or minus infinity')
else
   disp('A ''regular'' floating point number')
end
```

In the third `disp`, `''` prints as a single quote '.

The `for` loop is one of the most useful MATLAB constructs although, as discussed in Section 20.1, experienced programmers who are concerned with producing compact and fast code try to avoid `for` loops wherever possible. The syntax is

```
for variable = expression
    statements
end
```

Usually, *expression* is a vector of the form `i:s:j` (see Section 5.2). The statements are executed with *variable* equal to each element of *expression* in turn. For example, the sum of the first 25 terms of the harmonic series $1/i$ is computed by

```
>> s = 0;
>> for i = 1:25, s = s + 1/i; end, s
s =
    3.8160
```

Another way to define *expression* is using the square bracket notation:

```
>> for x = [pi/6 pi/4 pi/3], disp([x, sin(x)]), end
    0.5236    0.5000
    0.7854    0.7071
    1.0472    0.8660
```

Multiple `for` loops can of course be nested, in which case indentation helps to improve the readability. The following code forms the 5-by-5 symmetric matrix `A` with (i, j) element i/j for $j \geq i$:

```
n = 5; A = eye(n);
for j = 2:n
    for i = 1:j-1
        A(i,j) = i/j;
        A(j,i) = i/j;
    end
end
```

The *expression* in the `for` loop can be a matrix, in which case *variable* is assigned the columns of *expression* from first to last. For example, to set `x` to each of the unit vectors in turn, we can write `for x=eye(n), ..., end`.

The `while` loop has the form

```
while expression
      statements
end
```

The *statements* are executed as long as *expression* is true. The following example approximates the smallest nonzero floating point number:

```
>> x = 1; while x>0, xmin = x; x = x/2; end, xmin
xmin =
  4.9407e-324
```

A `while` loop can be terminated with the `break` statement, which passes control to the first statement after the corresponding `end`. An infinite loop can be constructed using `while 1, ..., end`, which is useful when it is not convenient to put the exit test at the top of the loop. (Note that, unlike some other languages, MATLAB does not have a "repeat-until" loop.) We can rewrite the previous example less concisely as

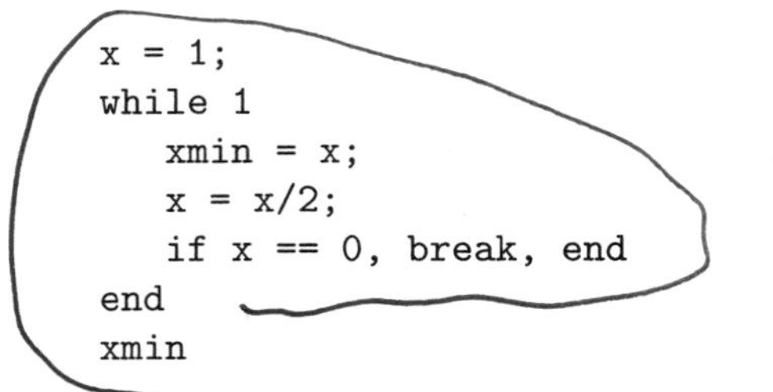

```
x = 1;
while 1
   xmin = x;
   x = x/2;
   if x == 0, break, end
end
xmin
```

The `break` statement can also be used to exit a `for` loop. In a nested loop a `break` exits to the loop at the next higher level.

The `continue` statement causes execution of a `for` or `while` loop to pass immediately to the next iteration of the loop, skipping the remaining statements in the loop. As a trivial example,

```
for i=1:10
    if i < 5, continue, end
    disp(i)
end
```

displays the integers 5 to 10. In more complicated loops the `continue` statement can be useful to avoid long-bodied `if` statements.

The final control structure is the `switch` statement. It consists of "`switch` *expression*" followed by a list of "`case` *expression statements*", optionally ending with "`otherwise` *statements*" and followed by `end`. The switch expression is evaluated and the statements following the first matching `case` expression are executed. If none of the cases produces a match then the statements following `otherwise` are executed. The next example evaluates the p-norm of a vector `x` (i.e., `norm(x,p)`) for just three values of p:

```
switch p
    case 1
         y = sum(abs(x));
    case 2
         y = sqrt(x'*x);
    case inf
         y = max(abs(x));
    otherwise
         error('p must be 1, 2 or inf.')
end
```

(The `error` function is described in Section 14.1.) The expression following `case` can be a list of values enclosed in parentheses (a cell array—see Section 18.3). The switch expression then matches any value in the list:

```
x = input('Enter a real number: ');
switch x
     case {inf,-inf}
          disp('Plus or minus infinity')
     case 0
          disp('Zero')
     otherwise
          disp('Nonzero and finite')
end
```

C programmers should note that MATLAB's `switch` construct behaves differently from that in C: once a MATLAB `case` group expression has been matched and its statements executed, control is passed to the first statement after the `switch`, with no need for `break` statements.

Kirk: "Well, Spock, here we are.
Thanks to your restored memory, a little bit of good luck,
we're walking the streets of San Francicso,
looking for a couple of humpback whales.
How do you propose to solve this minor problem?"
Spock: "Simple logic will suffice."

— *Star Trek IV: The Voyage Home* (Stardate 8390)

Things equally high on the pecking order get evaluated from left to right.
When in doubt, throw in some parentheses and be sure.
Only use good quality parentheses with nice round sides.

— ROGER EMANUEL KAUFMAN, *A FORTRAN Coloring Book* (1978)

Chapter 7
M-Files

7.1. Scripts and Functions

Although you can do many useful computations working entirely at the MATLAB command line, sooner or later you will need to write M-files. These are the equivalents of programs, functions, subroutines, and procedures in other programming languages. Collecting together a sequence of commands into an M-file opens up many possibilities, including

- experimenting with an algorithm by editing a file, rather than retyping a long list of commands,
- making a permanent record of a numerical experiment,
- building up utilities that can be reused at a later date,
- exchanging M-files with others.

Many useful M-files that have been written by enthusiasts can be obtained over the internet; see Appendix C.

An M-file is a text file that has a `.m` filename extension and contains MATLAB commands. There are two types:

Script M-files (or command files) have no input or output arguments and operate on variables in the workspace.

Function M-files contain a `function` definition line and can accept input arguments and return output arguments, and their internal variables are local to the function (unless declared `global`).

A script enables you to store a sequence of commands that are to be used repeatedly or will be needed at some future time. A simple example of a script M-file, `marks.m`, was given in Section 2.1. As another example we describe a script for playing "eigenvalue roulette" [23], which is based on counting how many eigenvalues of a random real matrix are real. If the matrix `A` is real and of dimension 8 then the number of real eigenvalues is 0, 2, 4, 6, or 8 (the number must be even, since nonreal eigenvalues appear in complex conjugate pairs). The short script

```
%SPIN
% Counts number of real eigenvalues of random matrix.
A = randn(8); sum(abs(imag(eig(A)))<.0001)
```

Listing 7.1. *Script* rouldist.

```
%ROULDIST      Empirical distribution of number of real eigenvalues.

k = 1000;
wheel = zeros(k,1);
for i = 1:k
    A = randn(8);
    % Count number of eigenvalues with imag. part < tolerance.
    wheel(i) = sum(abs(imag(eig(A)))<.0001);
end
hist(wheel,[0 2 4 6 8]);
```

creates a random normally distributed 8×8 matrix and counts how many eigenvalues have imaginary parts with absolute value less than the (somewhat arbitrary) threshold 10^{-4}. The first two lines of this script begin with the % symbol and hence are comment lines. Whenever MATLAB encounters a % it ignores the remainder of the line. This allows you to insert text that makes the script easier for humans to understand. Assuming this script exists as a file `spin.m`, typing `spin` is equivalent to typing the two commands `A = randn(8);` and `sum(abs(imag(eig(A)))<.0001)`. This "spins the roulette wheel," producing one of the five answers 0, 2, 4, 6, and 8. Each call to `spin` produces a different random matrix and hence may give a different answer:

```
>> spin
ans =
     2

>> spin
ans =
     4
```

To get an idea of the probability of each of the five outcomes you can run the script `rouldist` in Listing 7.1. It generates 1000 random matrices and plots a histogram of the distribution of the number of real eigenvalues. Figure 7.1 shows a possible result. (The exact probabilities are known and are given in [23], [24].) Note that to make `rouldist` more readable we have used spaces to indent the `for` loop and inserted a blank line before the first command.

Function M-files enable you to extend the MATLAB language by writing your own functions that accept and return arguments. They can be used in exactly the same way as existing MATLAB functions such as `sin`, `eye`, `size`, etc.

Listing 7.2 shows a simple function that evaluates the largest element in absolute value of a matrix. This example illustrates a number of features. The first line begins with the keyword `function` followed by the output argument, `y`, and the = symbol. On the right of = comes the function name, `maxentry`, followed by the input argument, `A`, within parentheses. (In general there can be any number of input and output arguments.) The function name must be the same as the name of the `.m` file in which the function is stored—in this case the file must be named `maxentry.m`.

The second line of a function file is called the H1 (help 1) line. It should be a comment line of a special form: a line beginning with a % character, followed without

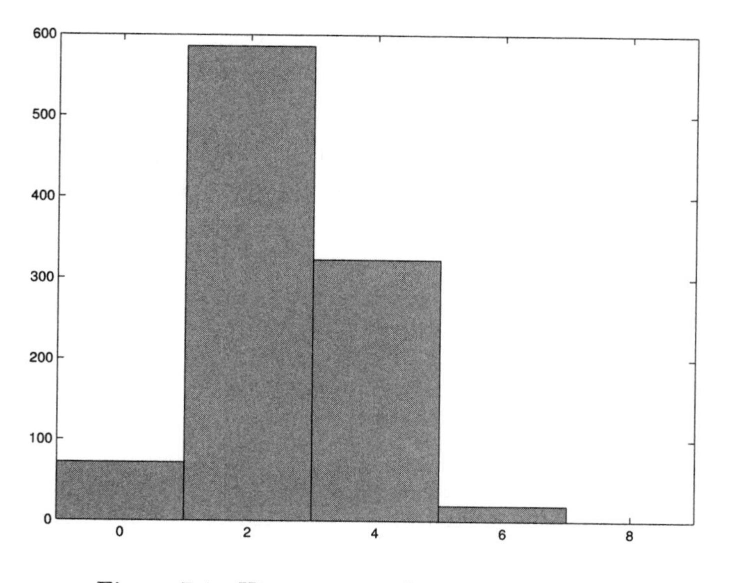

Figure 7.1. *Histogram produced by* `rouldist`.

any space by the function name in capital letters, followed by one or more spaces, and then a brief description. The description should begin with a capital letter, end with a period, and omit the words "the" and "a". All the comment lines from the first comment line up to the first noncomment line (usually a blank line, for readability of the source code) are displayed when `help function_name` is typed. Therefore these lines should describe the function and its arguments. It is conventional to capitalize function and argument names in these comment lines. For the `maxentry.m` example, we have

```
>> help maxentry
 MAXENTRY   Largest absolute value of matrix entries.
            MAXENTRY(A) is the maximum of the absolute values
            of the entries of A.
```

We strongly recommend documenting *all* your function files in this way, however short they may be. It is often useful to record in comment lines the date when the function was first written and to note any subsequent changes that have been made. The `help` command works in a similar manner on script files, displaying the initial sequence of comment lines.

The function `maxentry` is called just like any other MATLAB function:

```
>> maxentry(1:10)
ans =
    10

>> mx = maxentry(magic(4))
mx =
    16
```

The function `flogist` shown in Listing 7.3 illustrates the use of multiple input and output arguments. This function evaluates the scalar logistic function $x(1-ax)$

Listing 7.2. *Function* `maxentry`.

```
function y = maxentry(A)
%MAXENTRY   Largest absolute value of matrix entries.
%           MAXENTRY(A) is the maximum of the absolute values
%           of the entries of A.

y = max(max(abs(A)));
```

Listing 7.3. *Function* `flogist`.

```
function [f,fprime] = flogist(x,a)
%FLOGIST   Logistic function and its derivative.
%          [F,FPRIME] = FLOGIST(X,A) evaluates the logistic
%          function F(X) = X.*(1-A*X) and its derivative FPRIME
%          at the matrix argument X, where A is a scalar parameter.

f = x.*(1-a*x);
fprime = 1-2*a*x;
```

and its derivative with respect to x. The two output arguments `f` and `fprime` are enclosed in square brackets. When calling a function with multiple input or output arguments it is not necessary to request all the output arguments, but arguments must be dropped starting at the end of the list. If more than one output argument is requested the arguments must be listed within square brackets. Examples of usage are

```
>> f = flogist(2,.1)
f =
    1.6000

>> [f,fprime] = flogist(2,.1)
f =
    1.6000
fprime =
    0.6000
```

A technical point of note in function `flogist` is that array multiplication (`.*`) is used in the statement `f = x.*(1-a*x)`. So, if a vector or matrix is supplied for `x`, the function is evaluated at each element simultaneously:

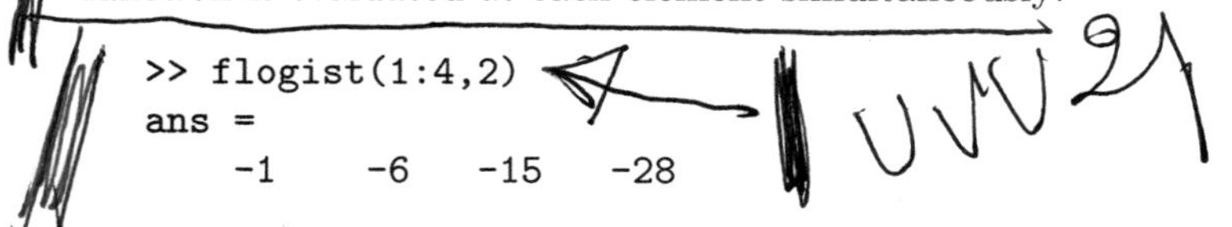

```
>> flogist(1:4,2)
ans =
    -1    -6   -15   -28
```

Another function using array multiplication is `cheby` in Listing 7.4, which is used with MATLAB's `fplot` to produce Figure 8.11. The kth Chebyshev polynomial,

Listing 7.4. *Function* cheby.

```
function Y = cheby(x,p)
%CHEBY    Chebyshev polynomials.
%         Y = CHEBY(X,P) evaluates the first P Chebyshev polynomials
%         at the vector X.  The K'th column of Y contains the
%         Chebyshev polynomial of degree K-1 evaluated at X.

Y = ones(length(x),p);
x = x(:);  % Ensure x is a column vector.
if p == 1, return, end

Y(:,2) = x;
for k = 3:p
  Y(:,k) = 2*x.*Y(:,k-1) - Y(:,k-2);
end
```

$T_k(x)$, can be defined by the recurrence

$$T_k(x) = 2xT_{k-1}(x) - T_{k-2}(x), \quad \text{for } k \geq 2,$$

with $T_0(x) = 1$ and $T_1(x) = x$. The function **cheby** accepts a vector **x** and an integer p and returns a matrix Y whose ith row gives the values of $T_0(x), T_1(x), \ldots, T_{p-1}(x)$ at $x = $ x(i). (This is the form of output argument required by **fplot**.)

Note that **cheby** uses the **return** command, which causes an immediate return from the M-file. It is not necessary (or usual) to put a **return** statement at the end of a function or script, unlike in some other programming languages.

A more complicated function is **sqrtn**, shown in Listing 7.5. Given $a > 0$, it implements the Newton iteration for $\sqrt{a}$,

$$x_{k+1} = \frac{1}{2}\left(x_k + \frac{a}{x_k}\right), \qquad x_1 = a,$$

printing the progress of the iteration. Output is controlled by the **fprintf** command, which is described in Section 13.2. Examples of usage are

```
>> [x,iter] = sqrtn(2)
 k            x_k             rel. change
 1:  1.5000000000000000e+000  3.33e-001
 2:  1.4166666666666665e+000  5.88e-002
 3:  1.4142156862745097e+000  1.73e-003
 4:  1.4142135623746899e+000  1.50e-006
 5:  1.4142135623730949e+000  1.13e-012
 6:  1.4142135623730949e+000  0.00e+000
x =
    1.4142
iter =
     6

>> x = sqrtn(2,1e-4);
```

Listing 7.5. *Function* sqrtn.

```
function [x,iter] = sqrtn(a,tol)
%SQRTN    Square root of a scalar by Newton's method.
%         X = SQRTN(A,TOL) computes the square root of the scalar
%         A by Newton's method (also known as Heron's method).
%         A is assumed to be >= 0.
%         TOL is a convergence tolerance (default EPS).
%         [X,ITER] = SQRTN(A,TOL) returns also the number of
%         iterations ITER for convergence.

if nargin < 2, tol = eps; end

x = a;
iter = 0;
xdiff = inf;
fprintf(' k                x_k               rel. change\n')

while xdiff > tol
    iter = iter + 1;
    xold = x;
    x = (x + a/x)/2;
    xdiff = abs(x-xold)/abs(x);
    fprintf('%2.0f:  %20.16e  %9.2e\n', iter, x, xdiff)
    if iter > 50
       error('Not converged after 50 iterations.')
    end
end
```

```
 k              x_k              rel. change
 1:  1.5000000000000000e+000  3.33e-001
 2:  1.4166666666666665e+000  5.88e-002
 3:  1.4142156862745097e+000  1.73e-003
 4:  1.4142135623746899e+000  1.50e-006
```

This M-file illustrates the use of optional input arguments. The function **nargin** returns the number of input arguments supplied when the function was called and enables default values to be assigned to arguments that have not been specified. In this case, if the call to **sqrtn** does not specify a value for **tol**, then **eps** is assigned to **tol**.

An analogous function **nargout** returns the number of output arguments requested. In this example there is no need to check **nargout**, because **iter** is computed by the function whether or not it is requested as an output argument. Some functions gain efficiency by inspecting **nargout** and computing only those output arguments that are requested (for example, **eig** in Chapter 9). To illustrate, Listing 7.6 shows how the **marks** M-file on p. 24 can be rewritten as a function. Its usage is illustrated by

```
>> exmark = [12 0 5 28 87 3 56];
```

Listing 7.6. *Function* marks2.

```
function [x_sort,x_mean,x_med,x_std] = marks2(x)
%MARKS2  Statistical analysis of marks vector.
%        Given a vector of marks X,
%        [X_SORT,X_MEAN,X_MED,X_STD] = MARKS2(X) computes a
%        sorted marks list and the mean, median, and standard deviation
%        of the marks.

x_sort = sort(x);
if nargout > 1, x_mean = mean(x); end
if nargout > 2, x_med  = median(x); end
if nargout > 3, x_std  = std(x); end
```

```
>> x_sort = marks2(exmark)
x_sort =
     0     3     5    12    28    56    87

>> [x_sort,x_mean,x_med] = marks2(exmark)
x_sort =
     0     3     5    12    28    56    87
x_mean =
   27.2857
x_med =
    12
```

7.2. Naming and Editing M-Files

M-files share with variables the naming restrictions described on p. 30. In particular, M-file names are case sensitive. Case sensitivity of M-file names has always been present in Unix versions of MATLAB, but was introduced for Windows systems only in version 7 (see Section B.5). For how to check whether a tentative name already exists, see the next section.

To create and edit M-files you have two choices. You can use whatever editor you normally use for ASCII files (if it is a word processor you need to ensure that you save the files in standard ASCII form, not in the word processor's own format). Or you can use the built-in MATLAB Editor/Debugger, shown in Figure 7.2. This is invoked by typing `edit` at the command prompt or from the File-New or File-Open menu options. The MATLAB editor has various features to aid in editing M-files, including automatic indentation of loops and `if` structures, commenting out blocks of code, color syntax highlighting, and bracket and quote matching. These and other features can be turned off or customized via the File-Preferences menu of the editor.

A very useful feature introduced in MATLAB 7 is block commenting: a block of code can be commented out (no matter what editor you are using) by surrounding it by two special comment lines:

```
%{
```

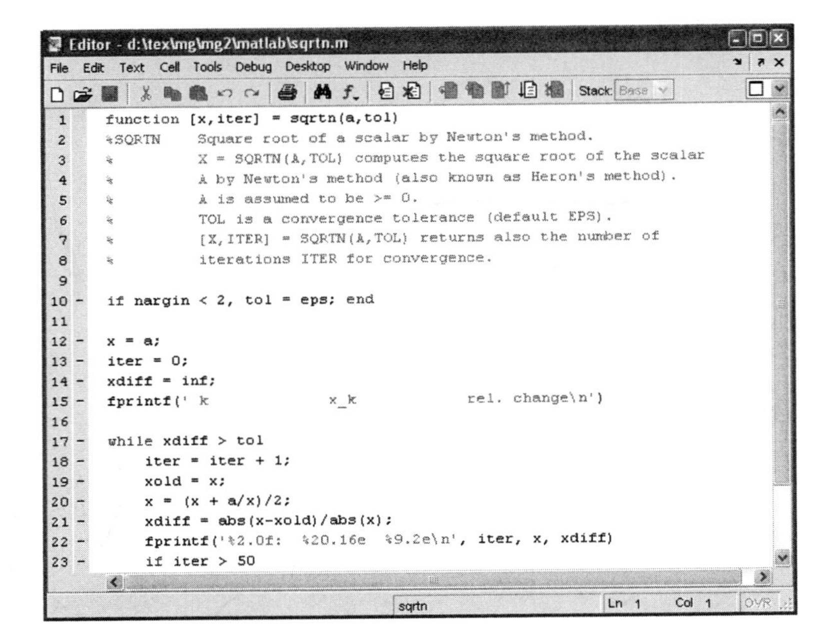

Figure 7.2. *MATLAB Editor/Debugger.*

```
<block of code>
%}
```

Here, `<block of code>` denotes an arbitrary number of lines of code. MATLAB considers all lines between `%{` and `%}` to be comments, even those that are not individually commented out with a leading `%` sign. Block comments can be nested, so that a block comment can be extended without losing the original block comment.

7.3. Working with M-Files and the MATLAB Path

Many MATLAB functions are M-files residing on the disk, while others are built into the MATLAB interpreter. The MATLAB search path is a list of directories that specifies where MATLAB looks for M-files. An M-file is available only if it is on the search path. Type `path` to see the current search path. The path can be set and added to with the `path` and `addpath` commands, or from the Path browser that is invoked by the File-Set Path menu option or by typing `pathtool`.

Several commands can be used to search the path. The `what` command lists the MATLAB files in the current directory, grouped by type; `what dirname` lists the MATLAB files in the directory `dirname` on the path.

The command `lookfor keyword` (illustrated on p. 26) searches the path for M-files containing `keyword` in their H1 line (the first line of help text). All the comment lines displayed by the `help` command can be searched using `lookfor keyword -all`.

Some MATLAB functions use comment lines after the initial block of comment lines to provide further information, such as bibliographic references (an example is `fminsearch`). This information can be accessed using `type` but is not displayed by `help`.

Typing `which foo` displays the pathname of the function `foo` or declares it to be `not found`. This is useful if you want to know in which directory on the path an M-file is located. If you suspect there may be more than one M-file with a given name on the path you can use `which foo -all` to display all of them.

A script (but not a function) not on the search path can be invoked by typing `run` followed by a statement in which the *full pathname* to the M-file is given.

You may list the M-file `foo.m` to the screen with `type foo` or `type foo.m`. (If there is an ASCII file called `foo` then the former command will list `foo` rather than `foo.m`.) Preceding a `type` command with `more on` will cause the listing to be displayed a page at a time; `more off` turns off paging.

Before writing an M-file it is important to check whether the name you are planning to give it is the name of an existing M-file or built-in function. This can be done in several ways: using `which` as just described, using `type` (e.g., `type lu` produces the response that `lu is a built-in function`), using `help`, or using the function `exist`. The command `exist('myname')` tests whether `myname` is a variable in the workspace, a file (with various possible extensions, including `.m`) on the path, or a directory. A result of 0 means no matches were found, while the numbers 1–8 indicate a match; see `help exist` for the precise meaning of these numbers. You should also avoid using MATLAB keywords for M-file or variable names. A list of keywords can be obtained with the `iskeyword` function: these are `break`, `case`, `catch`, `continue`, `else`, `elseif`, `end`, `for`, `function`, `global`, `if`, `otherwise`, `persistent`, `return`, `switch`, `try`, `while`.

When a function residing on the path is invoked for the first time it is compiled into memory (see the chapter "M-File Programming" in [75] for more details). MATLAB can usually detect when a function M-file has changed and then automatically recompiles it when it is invoked.

To clear function `fun` from memory, type `clear fun`. To clear all functions type `clear functions`.

7.4. Startup

When MATLAB starts it executes the M-file `matlabrc.m` (located in the directory `toolbox\local` off the MATLAB root). This M-file sets various defaults and then calls the M-file `startup.m`, if it exists on the MATLAB search path. The `startup` file is the place to make your own default settings and to add directories to the MATLAB path. This file is best placed in your MATLAB startup directory (the directory that is initially the current directory in MATLAB). To find how to change the startup directory, search for the string "startup directory for MATLAB" in the Help browser. A slightly shortened version of our `startup.m` is as follows:

```
%STARTUP       Startup file.

cd d:\matlab

% Save original path, in case want to restore standard setup.
path_org = path;

mypaths = {%
'd:/matlab/matrixcomp'
'd:/matlab/gen'
```

```
'd:/matlab/misc'
'd:/matlab/book'
'd:/matlab/tools'
'd:/matlab/mats'
'd:/matlab/eigtool'};

for i=1:length(mypaths)
    addpath(mypaths{i},'-end')
end
clear i mypaths
format compact
```

7.5. Command/Function Duality

User-written functions are usually called by giving the function name followed by a list of arguments in parentheses. Yet some built-in MATLAB functions, such as `type` and `what` described in the previous section, are normally called with arguments separated from the function name by spaces. This is not an inconsistency but an illustration of command/function duality. Consider the function

```
function comfun(x,y,z)
%COMFUN   Illustrative function with three string arguments.
disp(x), disp(y), disp(z)
```

We can call it with string arguments in parentheses (functional form), or with the string arguments separated by spaces after the function name (command form):

```
>> comfun('ab','cd','ef')
ab
cd
ef

>> comfun ab cd ef
ab
cd
ef
```

The two invocations are equivalent. Other examples of command/function duality are (with the first in each pair being the most commonly used)

```
format long, format('long')
disp('Hello'), disp Hello
diary mydiary, diary('mydiary')
warning off, warning('off')
```

Note, however, that the command form should be used only for functions that take string arguments. In the example

```
>> mean 2
ans =
    50
```

MATLAB interprets 2 as a string and `mean` is applied to the ASCII value of 2, namely 50. Note also that the command form can be used only when no output argument is requested. Thus `x = mean 2` gives an error.

```
>> why
Cleve insisted on it.
>> why
Jack knew it was a good idea.
```

— MATLAB

Replace repetitive expressions by calls to a common function.

— BRIAN W. KERNIGHAN and P. J. PLAUGER, *The Elements of Programming Style* (1978)

Much of MATLAB's power is derived from its extensive set of functions... Some of the functions are intrinsic, or "built-in" to the MATLAB processor itself. Others are available in the library of external M-files distributed with MATLAB... It is transparent to the user whether a function is intrinsic or contained in an M-file.

— 386-MATLAB User's Guide (1989)

Chapter 8
Graphics

MATLAB has powerful and versatile graphics capabilities. Figures of many types can be generated with relative ease and their "look and feel" is highly customizable. In this chapter we cover the basic use of MATLAB's most popular tools for graphing two- and three-dimensional data; Chapter 17 on Handle Graphics delves more deeply into the innards of MATLAB's graphics. Our philosophy of teaching a useful subset of MATLAB's language, without attempting to be exhaustive, is particularly relevant to this chapter. The final section hints at what we have left unsaid.

Our emphasis in this chapter is on generating graphics at the command line or in M-files, but existing figures can also be modified and annotated interactively using the Plot Editor. To use the Plot Editor see **`help plotedit`** and the Tools menu and toolbar of the figure window.

The figures in this chapter—and throughout the book—are the results of saving the figure window generated by the commands shown. We have not postprocessed the MATLAB figures to make them more readable on the printed page. For more on this issue, see Section 8.4.

Note that the graphics output shown in this book is printed in black and white. Most of the output appears as color on the screen and can be printed as color on a color printer.

8.1. Two-Dimensional Graphics

8.1.1. Basic Plots

MATLAB's `plot` function can be used for simple "join-the-dots" x-y plots. Typing

```
>> x = [1.5  2.2  3.1  4.6  5.7  6.3  9.4];
>> y = [2.3  3.9  4.3  7.2  4.5  6.1  1.1];
>> plot(x,y)
```

produces the left-hand picture in Figure 8.1, where the points `x(i)`, `y(i)` are joined in sequence. MATLAB opens a figure window (unless one has already been opened as a result of a previous command) in which to draw the picture. In this example, default values are used for a number of features, including the ranges for the x- and y-axes, the spacing of the axis tick marks, and the color and type of the line used for the plot.

More generally, we could replace `plot(x,y)` with `plot(x,y,`*string*`)`, where *string* combines up to three elements that control the color, marker, and line style. For example, `plot(x,y,'r*--')` specifies that a red asterisk is to be placed at each point `x(i)`, `y(i)` and that the points are to be joined by a red dashed line, whereas

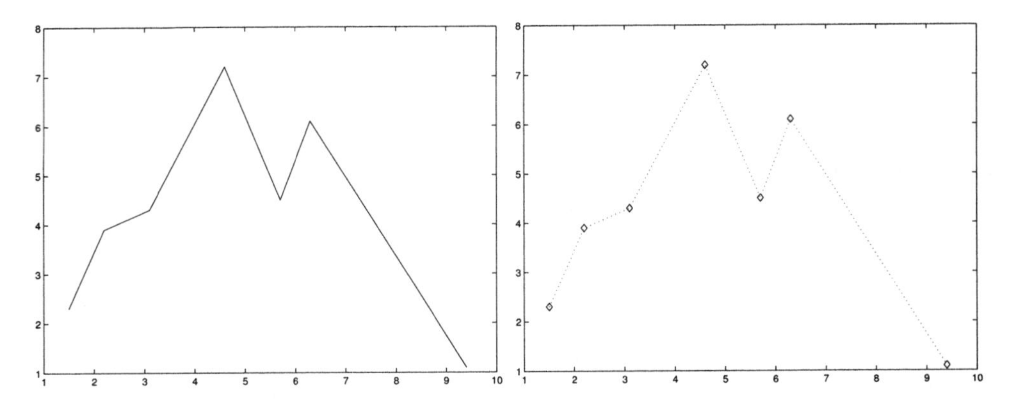

Figure 8.1. *Simple x-y plots. Left: default. Right: nondefault.*

Table 8.1. *Options for the* `plot` *command.*

Color	
`r`	Red
`g`	Green
`b`	Blue
`c`	Cyan
`m`	Magenta
`y`	Yellow
`k`	Black
`w`	White

Marker	
`o`	Circle
`*`	Asterisk
`.`	Point
`+`	Plus
`x`	Cross
`s`	Square
`d`	Diamond
`^`	Upward triangle
`v`	Downward triangle
`>`	Right triangle
`<`	Left triangle
`p`	Five-point star
`h`	Six-point star

Line style	
`-`	Solid line (default)
`--`	Dashed line
`:`	Dotted line
`-.`	Dash-dot line

`plot(x,y,'y+')` specifies a yellow cross marker with no line joining the points. Table 8.1 lists the options available. The right-hand picture in Figure 8.1 was produced with `plot(x,y,'kd:')`, which gives a black dotted line with diamond marker. The three elements in *string* may appear in any order, so, for example, `plot(x,y,'ms--')` and `plot(x,y,'s--m')` are equivalent. Note that more than one set of data can be passed to `plot`. For example,

```
plot(x,y,'g-',b,c,'r--')
```

superimposes plots of `x(i)`, `y(i)` and `b(i)`, `c(i)` with solid green and dashed red line styles, respectively.

The `plot` command also accepts matrix arguments. If `x` is an m-vector and `Y` is an m-by-n matrix, `plot(x,Y)` superimposes the plots created by `x` and each column of `Y`. Similarly, if `X` and `Y` are both m-by-n, `plot(X,Y)` superimposes the plots created by corresponding columns of `X` and `Y`. If nonreal numbers are supplied to `plot` then imaginary parts are generally ignored. The only exception to this rule

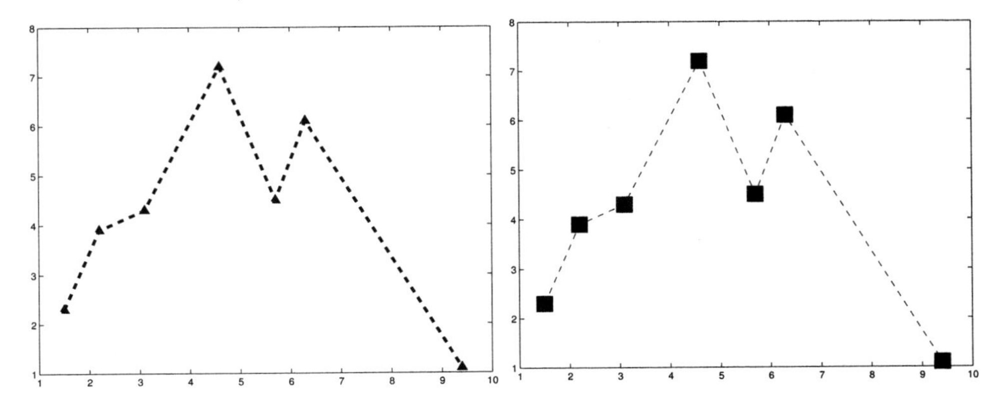

Figure 8.2. *Two nondefault x-y plots.*

arises when `plot` is given a single argument. If `Y` is nonreal, `plot(Y)` is equivalent to `plot(real(Y),imag(Y))`. In the case where `Y` is real, `plot(Y)` plots the columns of `Y` against their index.

You can exert further control by supplying more arguments to `plot`. The properties `LineWidth` (default 0.5 points) and `MarkerSize` (default 6 points) can be specified in points, where a point is 1/72 inch. For example, the commands

```
plot(x,y,'LineWidth',2)
plot(x,y,'p','MarkerSize',10)
```

produce a plot with a 2-point line width and 10-point marker size, respectively. For markers that have a well-defined interior, the `MarkerEdgeColor` and `MarkerFaceColor` can be set to one of the colors in Table 8.1. So, for example,

```
plot(x,y,'o','MarkerEdgeColor','m')
```

gives magenta edges to the circles. The left-hand plot in Figure 8.2 was produced with

```
plot(x,y,'m--^','LineWidth',3,'MarkerSize',5)
```

and the right-hand plot with

```
plot(x,y,'--rs','MarkerSize',20,'MarkerFaceColor','g')
```

Default values for these properties, and for some others to be discussed later in the chapter, are summarized in Table 8.2.

Using `loglog` instead of `plot` causes the axes to be scaled logarithmically. This feature is useful for revealing power-law relationships as straight lines. In the example below we plot $|1+h+h^2/2-\exp(h)|$ against h for $h = 1, 10^{-1}, 10^{-2}, 10^{-3}, 10^{-4}$. This quantity behaves like a multiple of h^3 when h is small, and hence on a log-log scale the values should lie close to a straight line of slope 3. To confirm this, we also plot a dashed reference line with the predicted slope, exploiting the fact that more than one set of data can be passed to the plot commands. The output is shown in Figure 8.3.

```
h = 10.^[0:-1:-4];
```

Table 8.2. *Default values for some properties.*

LineWidth	0.5
MarkerSize	6
MarkerEdgeColor	auto
MarkerFaceColor	none
FontSize	10
FontAngle	normal

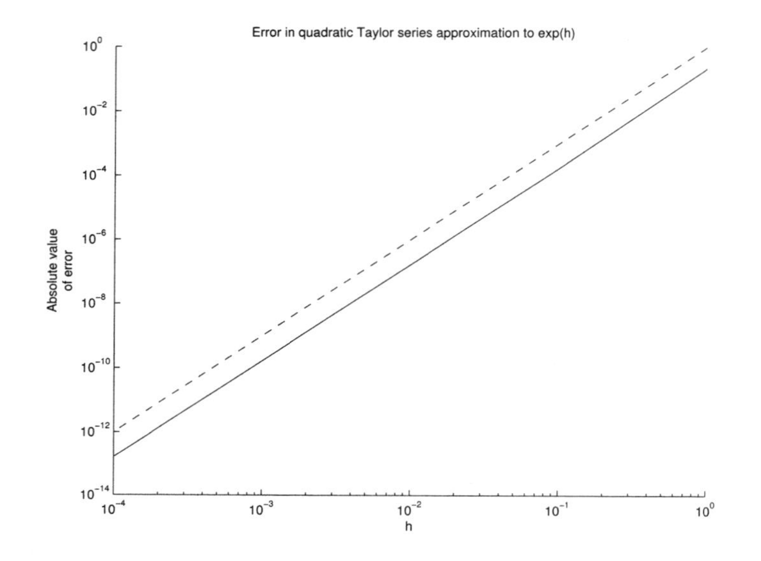

Figure 8.3. `loglog` *example.*

```
taylerr = abs((1+h+h.^2/2) - exp(h));
loglog(h,taylerr,'-',h,h.^3,'--')
xlabel('h')
ylabel({'Absolute value','of error'})
title('Error in quadratic Taylor series approximation to exp(h)')
box off
```

In this example, we used `title`, `xlabel`, and `ylabel`. These functions reproduce their input string above the plot and on the x- and y-axes, respectively. The multiline y-axis label is created by a cell array of strings, one for each line (see Section 18.3 for details of cell arrays). We also used the command `box off`, which removes the box from the current plot, leaving just the x- and y-axes. MATLAB will, of course, complain if nonpositive data is sent to `loglog` (it displays a warning and plots only the positive data). Related functions are `semilogx` and `semilogy`, for which only the x- or y-axis, respectively, is logarithmically scaled.

If one plotting command is later followed by another then the new picture will either replace or be superimposed on the old picture, depending on the current `hold` state. Typing `hold on` causes subsequent plots to be superimposed on the current

Table 8.3. *Some commands fo*

`axis([xmin xmax ymin ymax])`	Set specifie
`axis auto`	Return to de
`axis equal`	Equalize data
`axis off`	Remove axes
`axis square`	Make axis box squa
`axis tight`	Set axis limits to ran
`xlim([xmin xmax])`	Set specified x-axis lim
`ylim([ymin ymax])`	Set specified y-axis limits

one, whereas `hold off` specifies that each new plot should start afresh.
status corresponds to `hold off`.

The command `clf` clears the current figure window, while `close` closes
possible to have several figure windows on the screen. The simplest way to cr
new figure window is to type `figure`. The `n`th figure window (where `n` is displaye
the title bar) can be made current by typing `figure(n)`. The command `close al`
causes all the figure windows to be closed.

Note that many aspects of a figure can be changed interactively, after the figure has been displayed, by using the items on the toolbar of the figure window or on the Tools pull-down menu. In particular, it is possible to zoom in on a particular region of the plot using mouse clicks (see `help zoom`).

8.1.2. Axes and Annotation

Various aspects of the axes of a plot can be controlled with the `axis` command. Some of the options are summarized in Table 8.3. The axes are removed from a plot with `axis off`. The aspect ratio can be set to unity, so that, for example, a circle appears circular rather than elliptical, by typing `axis equal`. The axis box can be made square with `axis square`.

To illustrate, the left-hand plot in Figure 8.4 was produced by

```
plot(fft(eye(17))), axis equal, axis square
```

Since the plot obviously lies inside the unit circle the axes are hardly necessary. The right-hand plot in Figure 8.4 was produced with

```
plot(fft(eye(17))), axis equal, axis off
```

(The meaning of this interesting picture is described in [79].)

Setting `axis([xmin xmax ymin ymax])` causes the x-axis to run from `xmin` to `xmax` and the y-axis from `ymin` to `ymax`. To return to the default axis scaling, which MATLAB chooses automatically based on the data being plotted, type `axis auto`. If you want one of the limits to be chosen automatically by MATLAB, set it to `-inf` or `inf`; for example, `axis([-1 1 -inf 0])`. The x-axis and y-axis limits can be set individually with `xlim([xmin xmax])` and `ylim([ymin ymax])`.

Our next example plots the function $1/(x-1)^2+3/(x-2)^2$ over the interval $[0,3]$:

```
x = linspace(0,3,500);
plot(x,1./(x-1).^2 + 3./(x-2).^2)
grid on
```

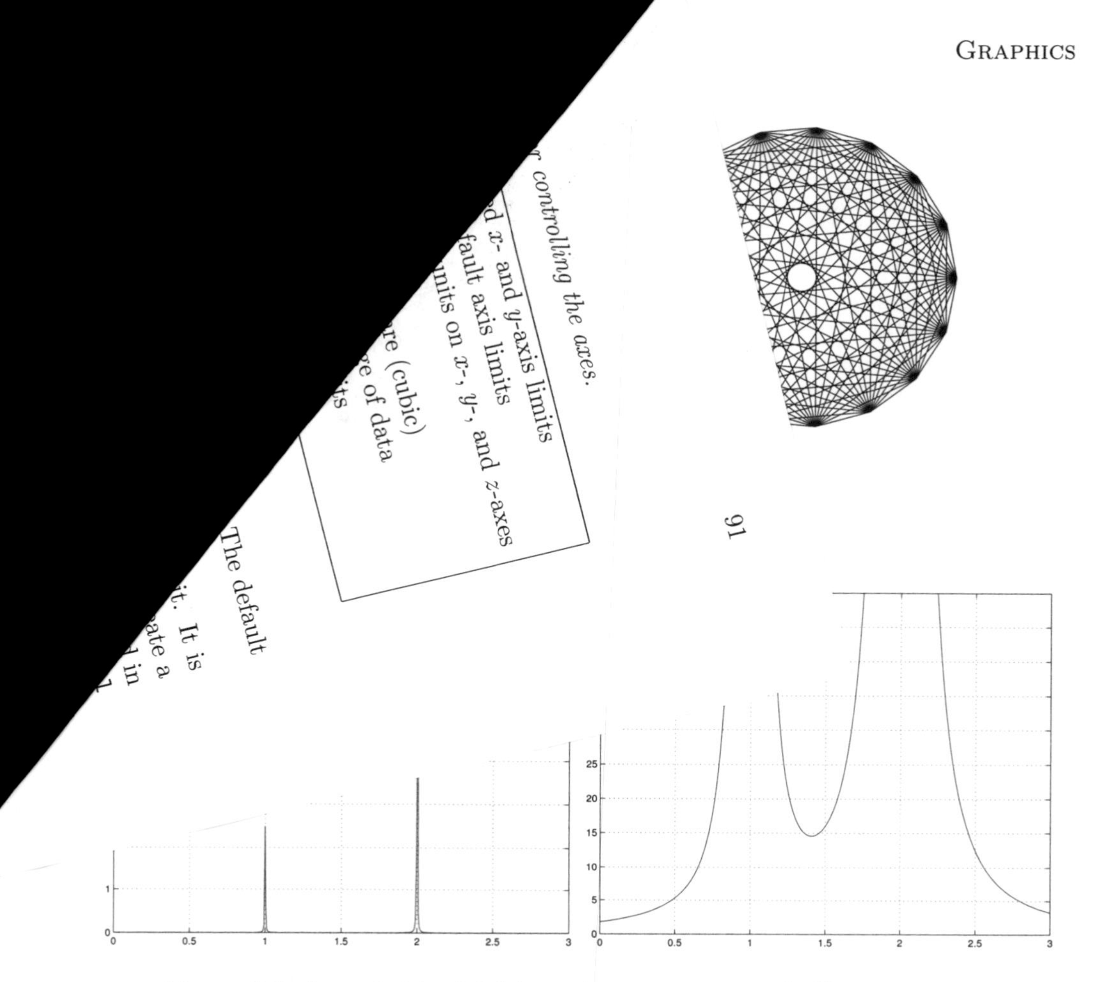

Figure 8.5. *Use of* `ylim` *(right) to change automatic (left) y-axis limits.*

We specified **`grid on`**, which introduces a light horizontal and vertical hashing that extends from the axis ticks. The result is shown in the left-hand plot of Figure 8.5. Because of the singularities at $x = 1, 2$ the plot is uninformative. However, by executing the additional command

```
ylim([0 50])
```

the right-hand plot of Figure 8.5 is produced, which focuses on the interesting part of the first plot.

In the following example we plot the epicycloid

$$\left.\begin{aligned} x(t) &= (a+b)\cos(t) - b\cos((a/b+1)t) \\ y(t) &= (a+b)\sin(t) - b\sin((a/b+1)t) \end{aligned}\right\} \quad 0 \le t \le 10\pi,$$

for $a = 12$ and $b = 5$.

```
a = 12; b = 5;
t = 0:0.05:10*pi;
x = (a+b)*cos(t) - b*cos((a/b+1)*t);
```

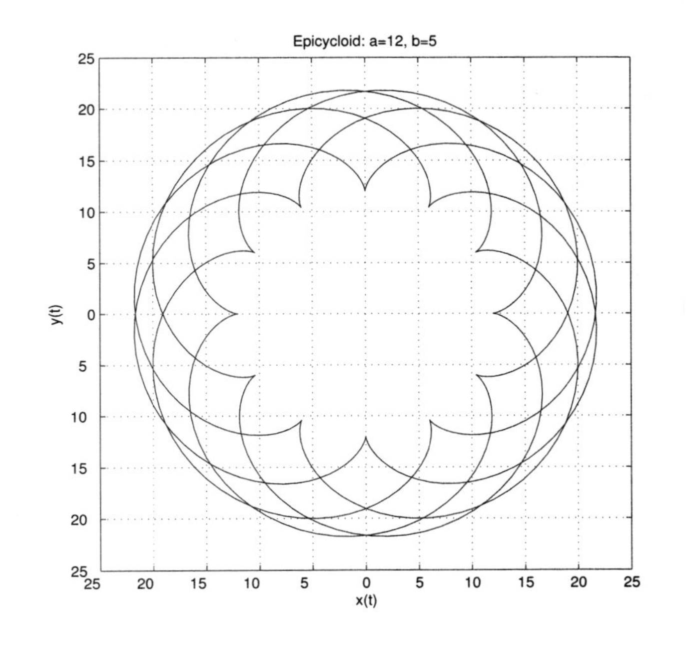

Figure 8.6. *Epicycloid example.*

```
y = (a+b)*sin(t) - b*sin((a/b+1)*t);

plot(x,y)
axis equal
axis([-25 25 -25 25])
grid on

title('Epicycloid: a=12, b=5')
xlabel('x(t)'), ylabel('y(t)')
```

The resulting picture appears in Figure 8.6. The **axis** limits were chosen to put some space around the epicycloid.

Next we plot the Legendre polynomials of degrees 1 to 4 (for the properties of these polynomials, see, for example, [14]) and use the **legend** function to add a box that explains the line styles. The result is shown in Figure 8.7.

```
x = -1:.01:1;
p1 = x;
p2 = (3/2)*x.^2 - 1/2;
p3 = (5/2)*x.^3 - (3/2)*x;
p4 = (35/8)*x.^4 - (15/4)*x.^2 + 3/8;

plot(x,p1,'r:',x,p2,'g--',x,p3,'b-.',x,p4,'m-')
box off

legend('\it n=1','\it n=2','\it n=3','\it n=4','Location','SouthEast')
xlabel('x','FontSize',12,'FontAngle','italic')
```

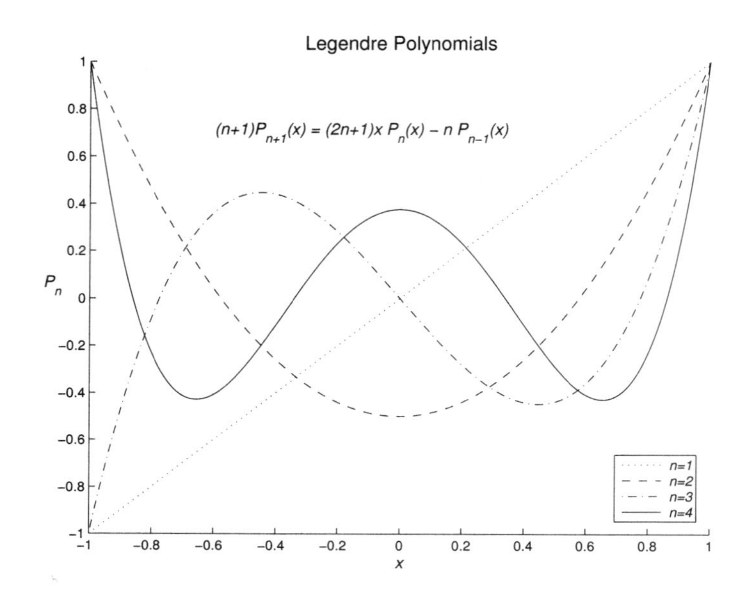

Figure 8.7. *Legendre polynomial example, using* `legend`.

```
ylabel('P_n','FontSize',12,'FontAngle','italic','Rotation',0)
title('Legendre Polynomials','FontSize',14)
text(-.6,.7,'(n+1)P_{n+1}(x) = (2n+1)x P_n(x) - n P_{n-1}(x)',...
      'FontSize',12,'FontAngle','italic')
```

Generally, typing `legend('string1','string2',...,'stringn')` will create a legend box that puts `'stringi'` next to the color/marker/line style information for the corresponding plot. By default, the box appears in the top right-hand (northeast) corner of the axis area. The location of the box can be specified with the syntax `legend(',...,'Location',location)`, where `location` is a string with possible values that include:

`'North'`	inside plot box near top
`'NorthWest'`	inside top left
`'NorthOutside'`	outside plot box near top
`'Best'`	automatically chosen to give least conflict with data
`'BestOutside'`	automatically chosen to leave least unused space outside plot

These values can be abbreviated as `'N'`, `'NW'`, etc. In our example we chose the bottom right-hand corner. Once the plot has been drawn, the legend box can be repositioned by putting the cursor over it and dragging it using the left mouse button. The `legend` function has many other options, which can be seen by typing `doc legend`.

This example uses the `text` command: generally, `text(x,y,'string')` places `'string'` at the position whose coordinates are given by `x` and `y`. (A related function `gtext` allows the text location to be determined interactively via the mouse.) Note that the strings in the `ylabel` and `text` commands use the notation of the typesetting system TeX to specify Greek letters, mathematical symbols, fonts, and superscripts and subscripts [33], [58], [61], [64]. Table 8.4 lists some of the TeX notation supported,

Table 8.4. *Some of the TEX commands supported in text strings.*

Greek letters	
Lower case	
α	\alpha
β	\beta
γ	\gamma
⋮	⋮
ω	\omega
Upper case	
$\varGamma$	\Gamma
$\varDelta$	\Delta
$\varTheta$	\Theta
⋮	⋮
$\varOmega$	\Omega

Selected symbols	
$\approx$	\approx
$\circ$	\circ
$\geq$	\geq
$\Im$	\Im
$\in$	\in
∞	\infty
$\int$	\int
$\leq$	\leq
$\neq$	\neq
$\otimes$	\otimes
∂	\partial
$\pm$	\pm
$\Re$	\Re
$\sim$	\sim
$\surd$	\surd

Fonts	
Normal	\rm
Bold	\bf
Italic	\it

and a full list can be found in the `string` entry under `doc text_props`. Note that curly braces can be used to delimit the range of application of the font commands and of subscripts and superscripts. Thus

```
title('{\itItalic} Normal {\bfBold} \int_{-\infty}^\infty')
```

produces a title of the form "*Italic* Normal **Bold** $\int_{-\infty}^{\infty}$". (Note that, unlike in TEX, if you leave a space after a font command then that space is printed.) If you are unfamiliar with TEX or LATEX you may prefer to use `texlabel('string')`, which allows `'string'` to be given in the style of a MATLAB expression. Thus the following two commands have identical effect:

```
text(5,5,'\alpha^{3/2}+\beta^{12}-\sigma_i')
text(5,5,texlabel('alpha^(3/2)+beta^12-sigma_i'))
```

A final note about the Legendre polynomial example is that we have used the `FontSize` and `FontAngle` properties to adjust the point size and angle of the text produced by the `xlabel`, `ylabel`, `title`, and `text` commands (as Table 8.2 indicates, the default value of `FontSize` is 10 and the default `FontAngle` is `normal`). However, `legend` does not accept these arguments, so we used TEX notation to make the legend italic. For plots that are to be incorporated into a printed document or presentation, increasing the font size can improve readability. We also used the `Rotation` property to rotate the y-axis label, for better readability.

New to MATLAB 7 is a LATEX interpreter, which supports the mathematical typesetting features of LATEX [33], [61], [64]. This facility, invoked by setting the `interpreter` property to `latex`, allows considerable control over the formatting of the axis labels, title, and text placed with the `text` command. The LATEX interpret largely removes the need for the technique, employed by some LATEX users, of u the `psfrag` package to overwrite text in PostScript figures generated with MA by LATEX text. The following script produces Figure 8.8:

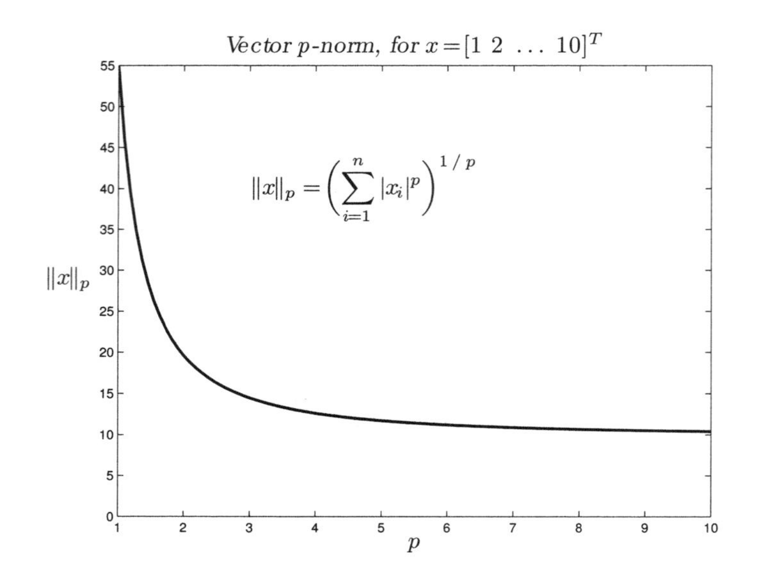

Figure 8.8. *Plot with text produced using MATLAB's LaTeX interpreter.*

```
n = 10; m = 100;
x = 1:n;
y = zeros(m,1);

pvals = linspace(1,10,m);
for i = 1:m
    y(i) = norm(x,pvals(i));
end

plot(pvals,y,'LineWidth',2)
ylim([0 inf])
options = {'Interpreter','latex','FontSize',18};
ylabel('$\|x\|_p$',options{:},'HorizontalAlignment','right')
xlabel('$p$',options{:})
title(['\slshape Vector $p$-norm, for $x =' ...
       '[1~2~\dots~' int2str(n) ']^T$'], options{:})

s = '$$\|x\|_p = \biggl(\sum_{i=1}^n|x_i|^p\biggr)^{1/p}$$';
text(options{:},'String',s,'Position',[3 40])
```

...cell array **options** to avoid repeatedly having to type the argu-
...ex','FontSize',18.
...s in a similar manner to **plot**. Typing **fill(x,y,[r g**
... vertices are specified by the points **x(i)**, **y(i)**. The points
...e last vertex is joined to the first. The color of the shading
...d argument **[r g b]**. The elements **r**, **g**, and **b**, which must
..., 1], determine the level of red, green, and blue, respectively,
...ll(x,y,[0 1 0]) uses pure green and **fill(x,y,[1 0 1])**

uses magenta. Specifying equal amounts of red, green, and blue gives a grey shading that can be varied between black (`[0 0 0]`) and white (`[1 1 1]`). The next example plots a cubic Bezier curve, which is defined by

$$p(u) = (1-u)^3\mathbf{P}_1 + 3u(1-u)^2\mathbf{P}_2 + 3u^2(1-u)\mathbf{P}_3 + u^3\mathbf{P}_4, \quad 0 \le u \le 1,$$

where the four control points, $\mathbf{P}_1$, $\mathbf{P}_2$, $\mathbf{P}_3$, and $\mathbf{P}_4$, have given x and y components. We use `fill` to shade the control polygon, that is, the polygon formed by the control points. The matrix `P` stores the control point $\mathbf{P}_j$ in its `j`th column, and `fill(P(1,:),P(2,:),[.8 .8 .8])` shades the control polygon with light grey. The columns of the matrix `Curve` are closely spaced points on the Bezier curve, and `plot(Curve(1,:),Curve(2,:),'--')` joins these with a dashed line. Figure 8.9 gives the resulting picture.

```
P = [0.1 0.3 0.7 0.8;
     0.3 0.8 0.6 0.1];
plot(P(1,:),P(2,:),'*')
axis([0 1 0 1])
hold on

u = 0:.01:1;
umat = [(1-u).^3; 3.*u.*(1-u).^2; 3.*u.^2.*(1-u); u.^3];
Curve = P*umat;
fill(P(1,:),P(2,:),[.8 .8 .8])
plot(Curve(1,:),Curve(2,:),'--')

text(0.35,0.35,'control polygon')
text(0.05,0.3,'P_1')
text(0.25,0.8,'P_2')
text(0.72,0.6,'P_3')
text(0.82,0.1,'P_4')
hold off
```

The function `annotation` allows the creation of annotation objects, including lines, various types of arrow, rectangles, and ellipses. While powerful, this function is not particularly easy to use because of the need to specify the location of these objects in normalized coordinates, which represent the bottom left corner of the figure by $(0, 0)$ and the top right corner by $(1, 1)$. Annotations are most easily created using the interactive tools in the figure window. See [76] for details.

8.1.3. Multiple Plots in a Figure

MATLAB's `subplot` allows you to place a number of plots in a grid pattern together on the same figure. Typing `subplot(mnp)` or, equivalently, `subplot(m,n,p)`, splits the figure window into an `m`-by-`n` array of regions, each having its own axes. The current plotting commands will then apply to the `p`th of these regions, where the count moves along the first row, and then along the second row, and so on. So, fo example, `subplot(425)` splits the figure window into a 4-by-2 matrix of regions a specifies that plotting commands apply to the fifth region, that is, the first regi the third row. If `subplot(427)` appears later, then the region in the (4,1) p becomes active. Several examples in which `subplot` is used appear below.

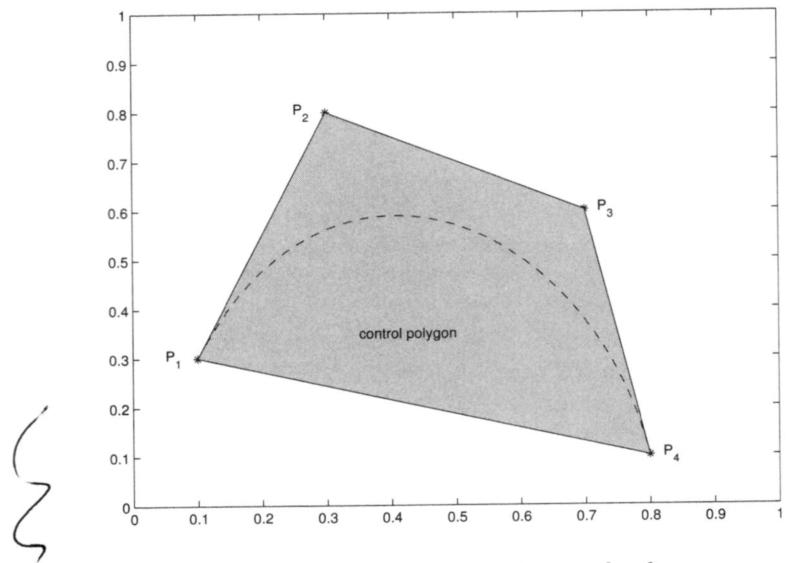

Figure 8.9. *Bezier curve and control polygon.*

For plotting mathematical functions the `fplot` command is useful. It adaptively samples a function at enough points to produce a representative graph. The following example generates the graphs in Figure 8.10.

```
subplot(221), fplot('exp(sqrt(x)*sin(12*x))',[0 2*pi])
subplot(222), fplot('sin(round(x))',[0 10],'--')
subplot(223), fplot('cos(30*x)/x',[0.01 1 -15 20],'-.')
subplot(224), fplot('[sin(x),cos(2*x),1/(1+x)]',[0 5*pi -1.5 1.5])
```

In this example, the first call to `fplot` produces a graph of the function $\exp(\sqrt{x}\sin 12x)$ over the interval $0 \le x \le 2\pi$. In the second call, we override the default solid line style and specify a dashed line with `'--'`. The argument `[0.01 1 -15 20]` in the third call forces limits in both the x and y directions, $0.01 \le x \le 1$ and $-15 \le y \le 20$, and `'-.'` asks for a dash-dot line style. The final `fplot` example illustrates how more than one function can be plotted in the same call.

It is possible to supply further arguments to `fplot`. The general pattern is `fplot(fun,lims,tol,N,'LineSpec',p1,p2,...)`. The argument list works as follows.

`fun` specifies the function to be plotted, e.g., as a string or a function handle ...ction 10.1).

.../or y limits are given by `lims`.

...e error tolerance, the default value of 2×10^{-3} corresponding to

... will be used to produce the plot.

...es the line type.

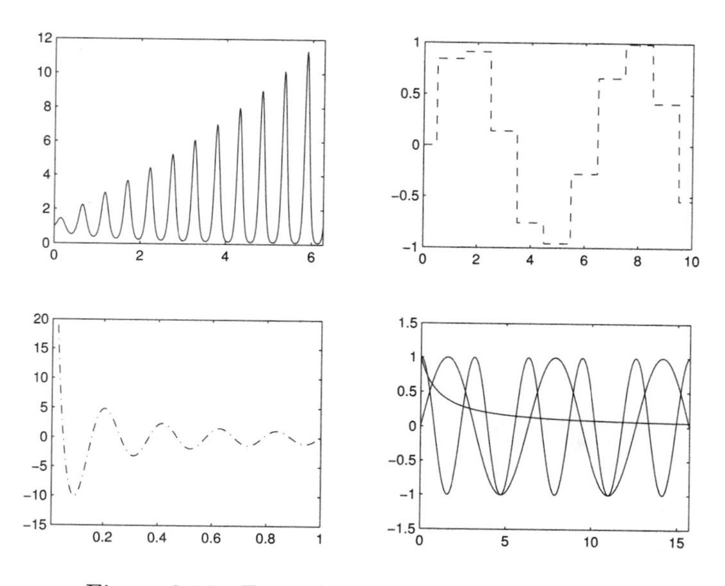

Figure 8.10. *Example with* `subplot` *and* `fplot`.

- `p1, p2, ...` are parameters that are passed to `fun`, which must have input arguments `x,p1,p2,...`.

The arguments `tol`, `N`, and `'LineSpec'` can be specified in any order, and an empty matrix (`[]`) can be passed to obtain the default for any of these arguments.

In Listing 7.4 on p. 79 is a function `cheby(x,p)` that returns the first `p` Chebyshev polynomials evaluated at `x`. Using this function the code

```
subplot(211), fplot(@cheby,[-1 1],[],[],[],5)
subplot(212), fplot(@cheby,[-1 1],[],[],[],35)
```

produces the pictures in Figure 8.11. Here, the first 5 and first 35 Chebyshev polynomials are plotted in the upper and lower regions, respectively. (See Section 10.1 for an explanation of the function handle `@cheby`.)

It is possible to produce irregular grids of plots by invoking `subplot` with different grid patterns. For example, Figure 8.12 was produced as follows:

```
x = linspace(0,15,100);
subplot(2,2,1), plot(x,sin(x))
subplot(2,2,2), plot(x,round(x))
subplot(2,1,2), plot(x,sin(round(x)))
```

The third argument to `subplot` can be a vector specifying several regions, so we could replace the last line by

```
subplot(2,2,3:4), plot(x,sin(round(x)))
```

To complete this section, we list in Table 8.5 the most popular 2D plotting functions in MATLAB. Some of these functions are discussed in Section 8.3.

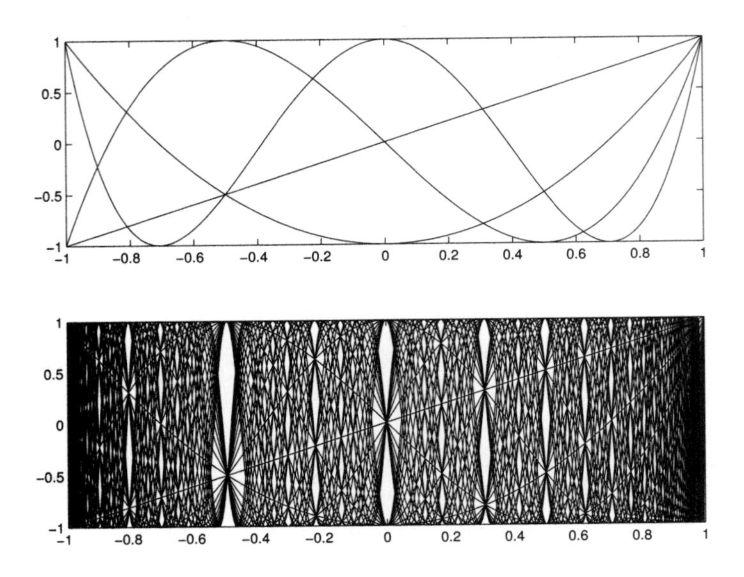

Figure 8.11. *First* 5 *(upper) and* 35 *(lower) Chebyshev polynomials, plotted using* `fplot` *and* `cheby` *in Listing* 7.4.

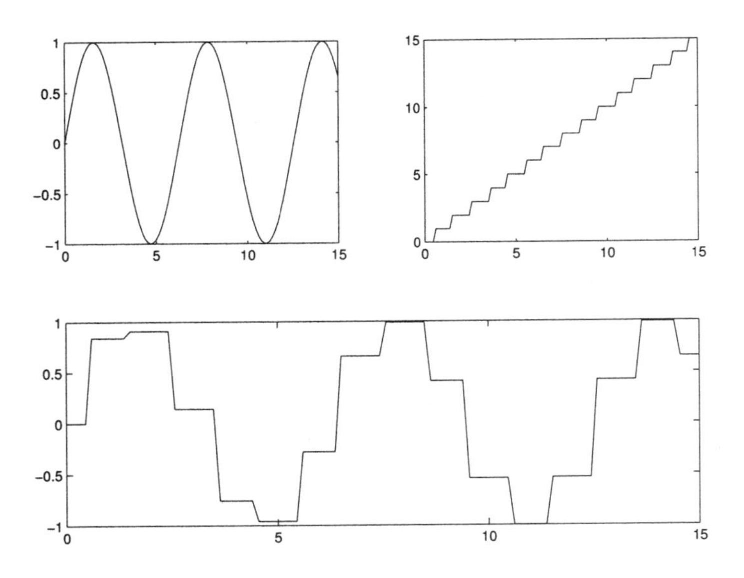

Figure 8.12. *Irregular grid of plots produced with* `subplot`.

Table 8.5. *2D plotting functions.*

`plot`	Simple x-y plot
`loglog`	Plot with logarithmically scaled axes
`semilogx`	Plot with logarithmically scaled x-axis
`semilogy`	Plot with logarithmically scaled y-axis
`plotyy`	x-y plot with y-axes on left and right
`polar`	Plot in polar coordinates
`fplot`	Function plotter
`ezplot`	Easy-to-use function plotter
`ezpolar`	Easy-to-use version of `polar`
`fill`	Polygon fill
`area`	Filled area graph
`bar`	Bar graph
`barh`	Horizontal bar graph
`hist`	Histogram
`pie`	Pie chart
`comet`	Animated, comet-like, x-y plot
`errorbar`	Error bar plot
`quiver`	Quiver (velocity vector) plot
`scatter`	Scatter plot
`stairs`	Stairstep plot

8.2. Three-Dimensional Graphics

The function `plot3` is the three-dimensional analogue of `plot`. The following example illustrates the simplest usage: `plot3(x,y,z)` draws a "join-the-dots" curve by taking the points `x(i)`, `y(i)`, `z(i)` in order. The result is shown in Figure 8.13.

```
t = -5:.005:5;
x = (1+t.^2).*sin(20*t);
y = (1+t.^2).*cos(20*t);
z = t;

plot3(x,y,z)
grid on
FS = 'FontSize';
xlabel('x(t)',FS,14), ylabel('y(t)',FS,14)
zlabel('z(t)',FS,14,'Rotation',0)
title('\it{plot3 example}',FS,14)
```

This example also uses the functions `xlabel`, `ylabel`, and `title`, which were discussed in the previous section, and the analogous `zlabel`. Note that we have used the TEX notation `\it` in the `title` command to produce italic text. The color, marker, and line styles for `plot3` can be controlled in the same way as for `plot`. So, for example, `plot3(x,y,z,'rx--')` would use a red dashed line and place a cross at each point. Note that for 3D plots the default is `box off`; specifying `box on` adds a box that bounds the plot.

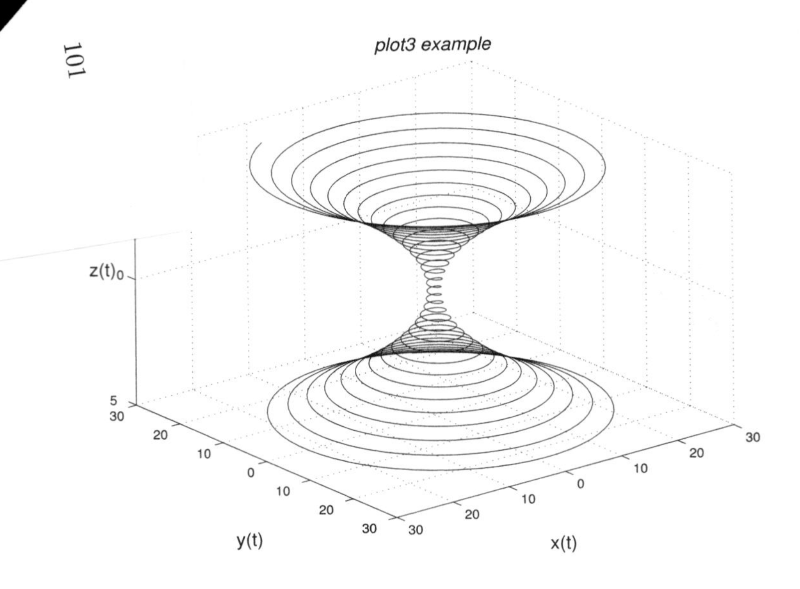

Figure 8.13. *3D plot created with* `plot3`.

A simple contour plotting facility is provided by `ezcontour`. The call to `ezcontour` in the following example produces contours for the function $\sin(3y-x^2+1)+\cos(2y^2-2x)$ over the range $-2 \leq x \leq 2$ and $-1 \leq y \leq 1$; the result can be seen in the upper half of Figure 8.14.

```
subplot(211)
ezcontour('sin(3*y-x^2+1)+cos(2*y^2-2*x)',[-2 2 -1 1]);

x = -2:.01:2; y = -1:.01:1;
[X,Y] = meshgrid(x,y);
Z = sin(3*Y-X.^2+1)+cos(2*Y.^2-2*X);

subplot(212)
contour(x,y,Z,20)
```

Note that the contour levels have been chosen automatically. For the lower half of Figure 8.14 we use the more general function `contour`. We first assign `x = -2:.01:2` and `y = -1:.01:1` to obtain closely spaced points in the appropriate range. We then set `[X,Y] = meshgrid(x,y)`, which produces matrices `X` and `Y` such that each row of `X` is a copy of the vector `x` and each column of `Y` is a copy of the vector `y`. (The function `meshgrid` is extremely useful for setting up data for many of MATLAB's 3D plotting tools.) The matrix `Z` is then generated from array operations on `X` and `Y`, with the result that `Z(i,j)` stores the function value corresponding to `x(j)`, `y(i)`. This is precisely the form required by `contour`. Typing `contour(x,y,Z,20)` tells MATLAB to regard `Z` as defining heights above the x-y plane with spacing given by `x` and `y`. The final input argument specifies that 20 contour levels are to be used; if this argument is omitted MATLAB automatically chooses the number of contour levels.

The next example illustrates the use of `clabel` to label contours, with the result shown in Figure 8.15.

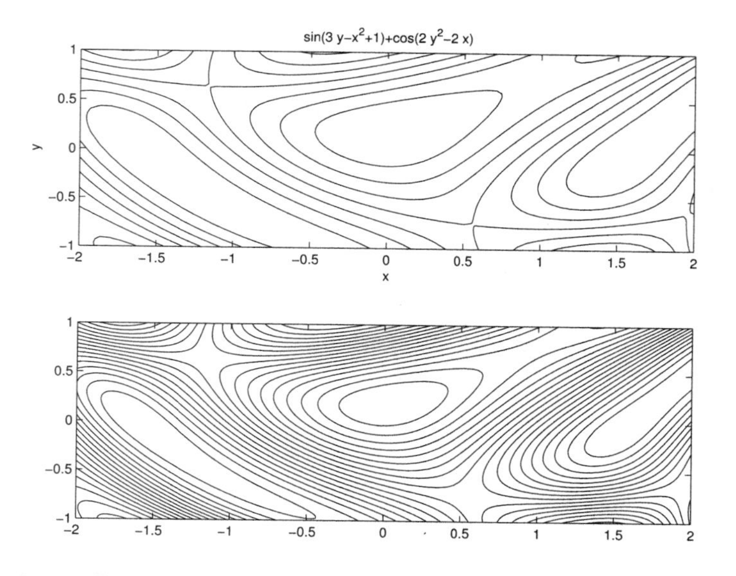

Figure 8.14. *Contour plots with* `ezcontour` *(upper) and* `contour` *(lower).*

```
[X,Y] = meshgrid(-3:.05:3, -1.5:.025:1.5);
Z = 4*X.^2 - 2.1*X.^4 + X.^6/3 + X.*Y - 4*Y.^2 + 4*Y.^4;
cvals = [-2:.5:2 2.3 3:5 6:2:10];
[C,h] = contour(X,Y,Z,cvals);
clabel(C,h,cvals([1:2:9 10 11 14 16]))
xlabel('x'), ylabel('y')
title('Six hump camel back function','Fontsize',16)
```

Here, we are using an interesting function having a number of maxima, minima, and saddle points. MATLAB's default choice of contour levels does not produce an attractive picture, so we specify the levels (chosen by trial and error) in the vector `cvals`. The `clabel` command takes as input the output from `contour` (`C` contains the contour data and `h` is a graphics object handle) and adds labels to the contour levels specified in its third input argument. Again the contour levels need not be specified, but the default of labelling all contours produces a cluttered plot in this example. An alternative form of `clabel` is `clabel(C,h,'manual')`, which allows you to specify with the mouse the contours to be labelled: click to label a contour and press return to finish. The `h` argument of `clabel` can be omitted, in which case the labels are placed unrotated and close to each contour, with a plus sign marking the contour.

The function `mesh` accepts data in a similar form to `contour` and produces wire-frame surface plots. If `meshc` is used in place of `mesh`, a contour plot is appended below the surface. The example below, which produces Figure 8.16, involves the surface defined by $\sin(y^2+x)-\cos(y-x^2)$ for $0 \le x, y \le \pi$. The first subplot is produced by `mesh(Z)`. Since no x, y information is supplied to `mesh`, row and column indices are used for the axis ranges. The second subplot shows the effect of `meshc(Z)`. For the third subplot, we use `mesh(x,y,Z)`, so the tick labels on the x- and y-axes correspond to the values of `x` and `y`. We also specify the axis limits with `axis([0 pi 0 pi -5 5])`, which gives $0 \le x, y \le \pi$ and $-5 \le z \le 5$. For the final subplot, we use `mesh(Z)` again, followed by `hidden off`, which causes hidden lines to be shown.

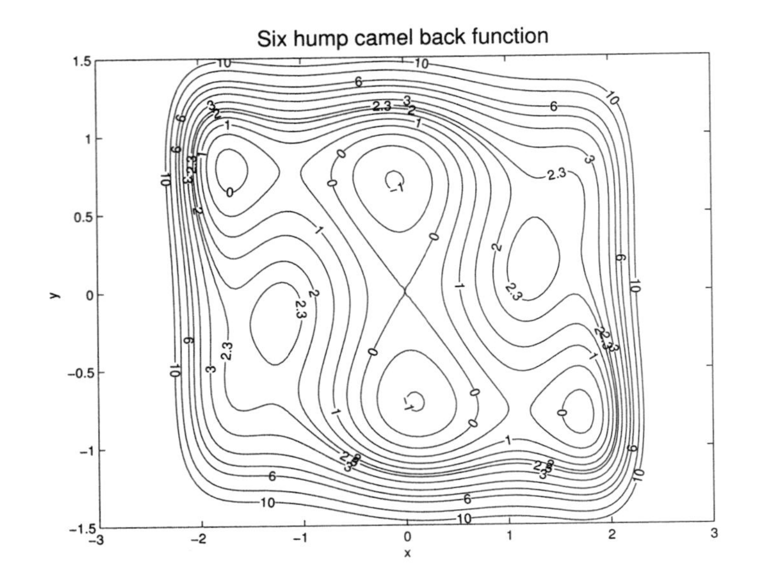

Figure 8.15. *Contour plot labelled using* `clabel`.

```
x = 0:.1:pi; y = 0:.1:pi;
[X,Y] = meshgrid(x,y);
Z = sin(Y.^2+X)-cos(Y-X.^2);

subplot(221)
mesh(Z)

subplot(222)
meshc(Z)

subplot(223)
mesh(x,y,Z)
axis([0 pi 0 pi -5 5])

subplot(224)
mesh(Z)
hidden off
```

The function `surf` differs from `mesh` in that it produces a solid filled surface plot, and `surfc` adds a contour plot below. In the next example we call MATLAB's `membrane`, which returns the first eigenfunction of an L-shaped membrane. The pictures in the first row of Figure 8.17 show the effect of `surf` and `surfc`. The (1,2) plot displays a color scale using `colorbar`. The color map for the current figure can be set using `colormap`; see `doc colormap`. The (2,1) plot uses the `shading` function with the `flat` option to remove the grid lines on the surface; another option is `interp`, which varies the color over each segment by interpolation. The (2,2) plot uses the related function `waterfall`, which is similar to `mesh` with the wireframes in the column direction removed.

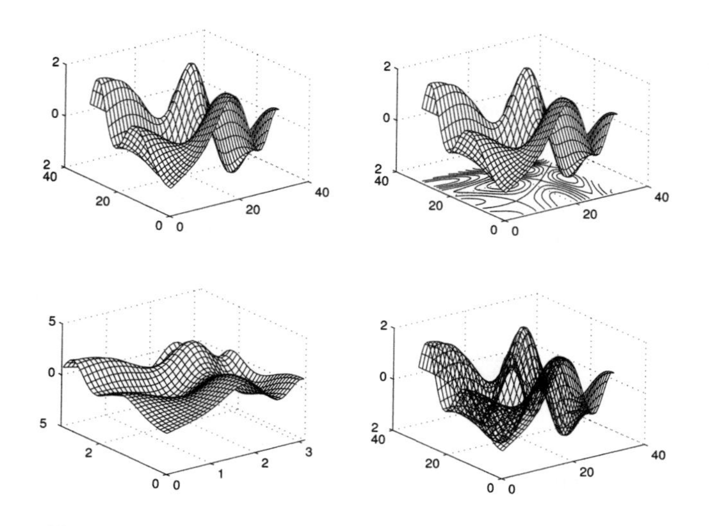

Figure 8.16. *Surface plots with* `mesh` *and* `meshc`.

```
Z = membrane; FS = 'FontSize';
subplot(221), surf(Z), title('\bf{surf}',FS,14)
subplot(222), surfc(Z), title('\bf{surfc}',FS,14), colorbar
subplot(223), surf(Z), shading flat
              title('\bf{surf} shading flat',FS,14)
subplot(224), waterfall(Z), title('\bf{waterfall}',FS,14)
```

Ordinarily, the color of a surface represents the height above the x-y plane. However, **`mesh`**, **`surf`**, and related functions may also be used in the form **`mesh(x,y,Z,W)`**, which bases the colors on the array `W`; this form can be used to display other features of the surface, or to impose an independent coloring pattern.

The 3D pictures in Figures 8.13, 8.16, and 8.17 use MATLAB's default viewing angle. This can be overridden with the function **`view`**. Typing **`view(a,b)`** sets the counterclockwise rotation about the z-axis to `a` degrees and the vertical elevation to `b` degrees. The default is `view(-37.5,30)`, while `view(2)` is equivalent to `view(0,90)` and gives a 2D view of a surface looking down from above. The **`rotate 3D`** tool on the toolbar of the figure window enables the mouse to be used to change the angle of view by clicking and dragging within the axis area.

It is possible to view a 2D plot as a 3D one, by using the `view` command to specify a viewing angle, or simply by typing `view(3)`. Figure 8.18 shows the result of typing

```
plot(fft(eye(17))); view(3); grid
```

In the next example we generate a fractal landscape with the recursive function **`land`** shown in Listing 8.1, which uses a variant of the random midpoint displacement algorithm [90, Sec. 7.6]; see Figure 8.19. Recursion is discussed further in Section 10.9. The basic step taken by **`land`** is to update an `N`-by-`N` matrix with nonzeros only in each corner by filling in the entries in positions `(1,d)`, `(d,1)`, `(d,d)`, `(d,N)`, and

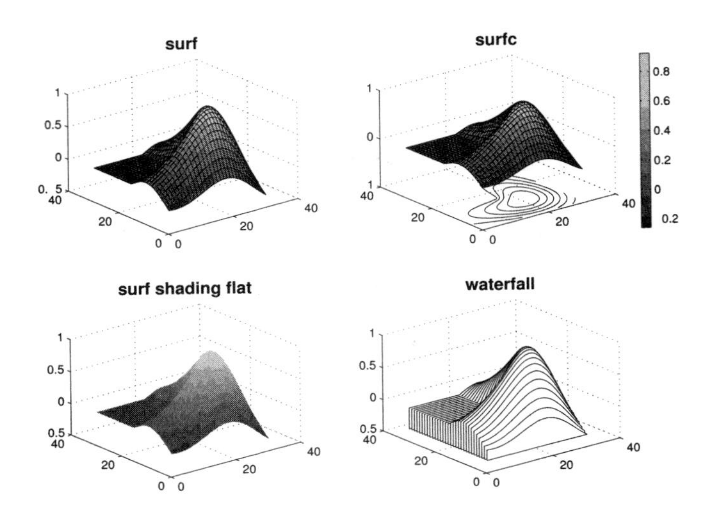

Figure 8.17. *Surface plots with* `surf`, `surfc`, *and* `waterfall`.

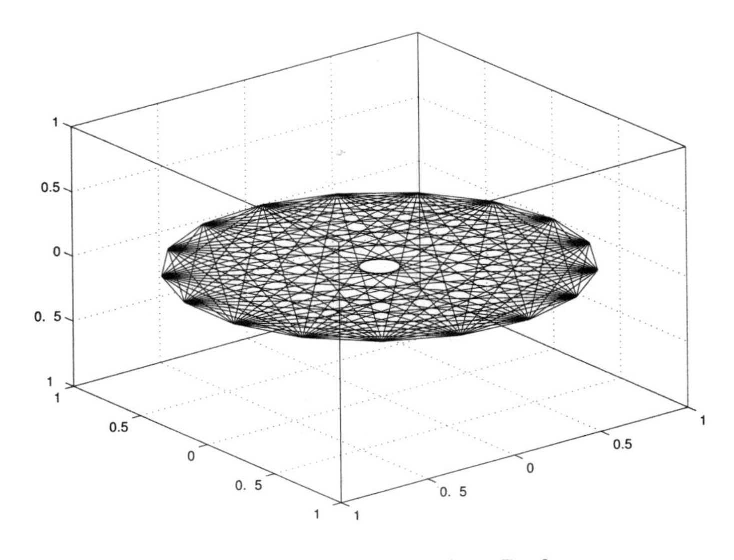

Figure 8.18. *3D view of a 2D plot.*

Listing 8.1. *Function* land.

```
function  B = land(A)
%LAND     Fractal landscape.
%         B = LAND(A) generates a random fractal landscape
%         represented by B, where A is a square matrix of
%         dimension N = 2^n + 1 whose four corner elements
%         are used as input parameters.

N = size(A,1);
d = (N+1)/2;
level = log2(N-1);
scalef = 0.05*(2^(0.9*level));

B = A;

B(d,d) = mean([A(1,1),A(1,N),A(N,1),A(N,N)]) + scalef*randn;
B(1,d) = mean([A(1,1),A(1,N)]) + scalef*randn;
B(d,1) = mean([A(1,1),A(N,1)]) + scalef*randn;
B(d,N) = mean([A(1,N),A(N,N)]) + scalef*randn;
B(N,d) = mean([A(N,1),A(N,N)]) + scalef*randn;

if N > 3
   B(1:d,1:d) = land(B(1:d,1:d));
   B(1:d,d:N) = land(B(1:d,d:N));
   B(d:N,1:d) = land(B(d:N,1:d));
   B(d:N,d:N) = land(B(d:N,d:N));
end
```

(N,d), where d = (N+1)/2, in the following manner:

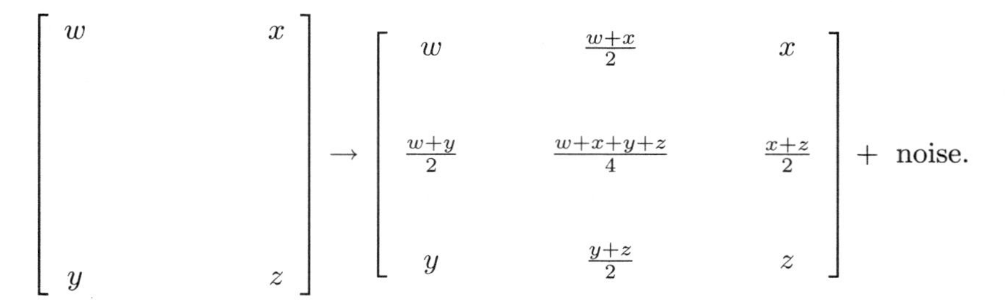

$$\begin{bmatrix} w & & x \\ & & \\ & & \\ y & & z \end{bmatrix} \to \begin{bmatrix} w & \frac{w+x}{2} & x \\ \frac{w+y}{2} & \frac{w+x+y+z}{4} & \frac{x+z}{2} \\ y & \frac{y+z}{2} & z \end{bmatrix} + \text{noise}.$$

The noise is introduced by adding a multiple of **randn** to each new nonzero element. The process is repeated recursively on the four square submatrices whose corners are defined by the nonzero elements, until the whole matrix is filled. The scaling factor for the noise is reduced by $2^{0.9}$ at each level of recursion. Note that the input argument A in **land**(A) must be a square matrix with dimension of the form $2^n + 1$, and only the corner elements of A have any effect on the result.

In the example below that produces Figure 8.19 we use **land** to set up a height matrix, B. For the surface plots, we use **meshz**, which works like **mesh** but hangs a vertical curtain around the edges of the surface. The first subplot shows the default view of B. For the second subplot we impose a "sea level" by raising all heights that are below the average value. This resulting data matrix, `Bisland`, is also plotted with the default view. The third and fourth subplots use `view([-75 40])` and `view([240 65])`, respectively. For these two subplots we also control the axis limits.

```
randn('state',10);
k = 2^5+1;
A = zeros(k);
A([1 k], [1 k]) = [1   1.25
                   1.1  2.0];
B = land(A);

subplot(221), meshz(B)
FS = 'FontSize'; title('Default view',FS,12)

Bisland = max(B,mean(mean(B)));
Bmin = min(min(Bisland));
Bmax = max(max(Bisland));
subplot(222), meshz(Bisland)
title('Default view',FS,12)

subplot(223), meshz(Bisland)
view([-75 40])
axis([0 k 0 k Bmin Bmax])
title('view([-75 40])',FS,12)

subplot(224), meshz(Bisland)
view([240 65])
axis([0 k 0 k Bmin Bmax])
title('view([240 65])',FS,12)
```

Table 8.6 summarizes the most popular 3D plotting functions. As the table indicates, several of the functions have "easy-to-use" alternative versions with names beginning `ez`. Section 8.3 discusses some of these functions.

A feature common to all graphics functions is that NaNs are interpreted as "missing data" and are not plotted. For example,

```
plot([1 2 NaN 3 4])
```

draws two disjoint lines and does not connect "2" to "3", while

```
A = peaks(80); A(28:52,28:52) = NaN; surfc(A)
```

produces the `surfc` plot with a hole in the middle shown in Figure 8.20. (The function `peaks` generates a matrix of height values corresponding to a particular function of two variables and is useful for demonstrating 3D plots.)

MATLAB contains in its `demos` directory several functions with names beginning `cplx` for visualizing functions of a complex variable (type `what demos`). Figure 8.21 shows the plot produced by `cplxroot(3)`. In general, `cplxroot(n)` plots the Riemann surface for the function $z^{1/n}$.

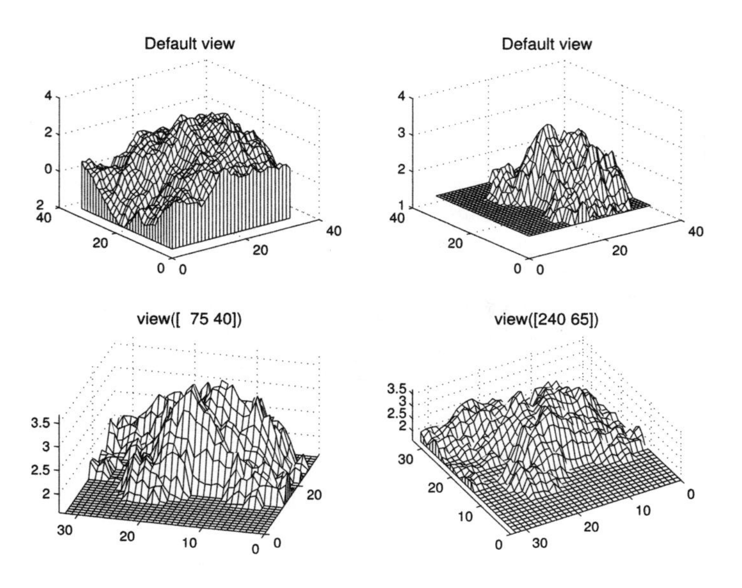

Figure 8.19. *Fractal landscape views.*

Table 8.6. *3D plotting functions.*

`plot3`*	Simple x-y-z plot
`contour`*	Contour plot
`contourf`*	Filled contour plot
`contour3`	3D contour plot
`mesh`*	Wireframe surface
`meshc`*	Wireframe surface plus contours
`meshz`	Wireframe surface with curtain
`surf`*	Solid surface
`surfc`*	Solid surface plus contours
`waterfall`	Unidirectional wireframe
`bar3`	3D bar graph
`bar3h`	3D horizontal bar graph
`pie3`	3D pie chart
`fill3`	Polygon fill
`comet3`	3D animated, comet-like plot
`scatter3`	3D scatter plot
`stem3`	Stem plot

* These functions `fun` have `ezfun` counterparts, too.

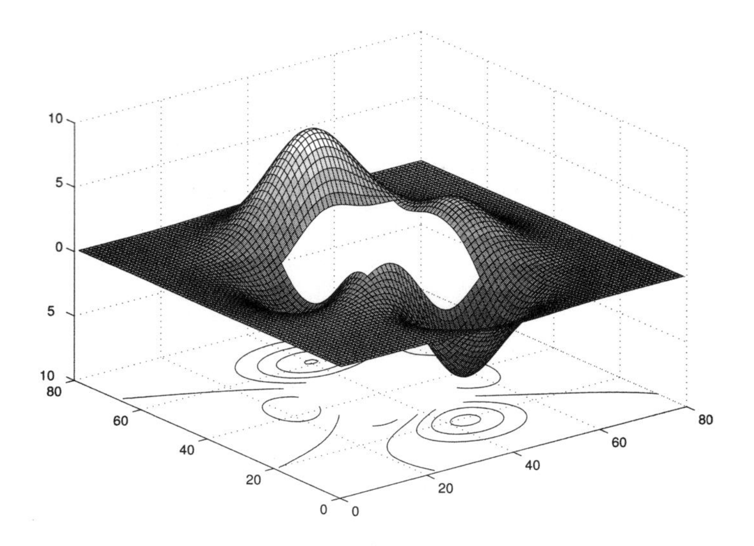

Figure 8.20. `surfc` *plot of matrix containing NaNs.*

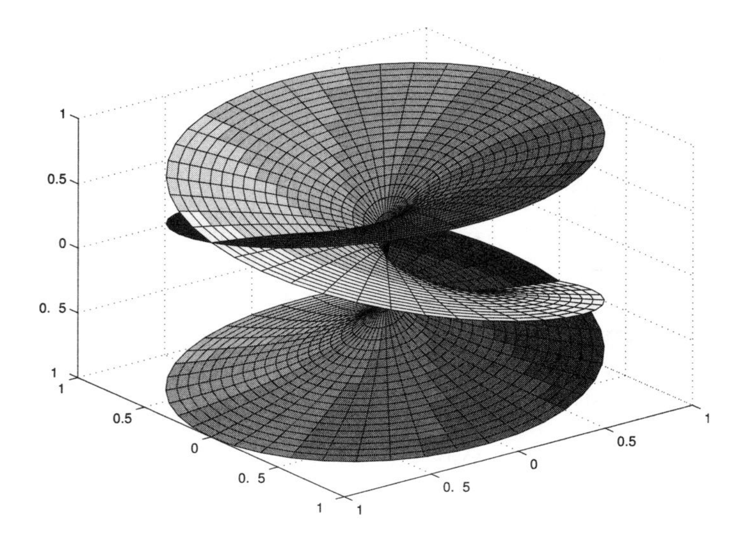

Figure 8.21. *Riemann surface for* $z^{1/3}$.

8.3. Specialized Graphs for Displaying Data

In this section we describe some additional functions from Tables 8.5 and 8.6 that are useful for displaying data (as opposed to plotting mathematical functions).

MATLAB has four functions for plotting bar graphs, covering 2D and 3D vertical or horizontal bar graphs, with options to stack or group the bars. The simplest usage of the bar plot functions is with a single `m`-by-`n` matrix input argument. For 2D bar plots elements in a row are clustered together, either in a group of `n` bars with the default `'grouped'` argument, or in one bar apportioned among the `n` row entries with the `'stacked'` argument.

The following code uses `bar` and `barh` to produce Figure 8.22:

```
Y = [7 6 5
     6 8 1
     4 5 9
     2 3 4
     9 7 2];

subplot(2,2,1)
bar(Y)
title('bar(...,''grouped'')')

subplot(2,2,2)
bar(0:5:20,Y)
title('bar(...,''grouped'')')

subplot(2,2,3)
bar(Y,'stacked')
title('bar(...,''stacked'')')

subplot(2,2,4)
barh(Y)
title('barh')
```

Note that in the two-argument form `bar(x,Y)` the vector `x` provides the x-axis locations for the bars.

For 3D bar graphs the default arrangement is `'detached'`, with the bars for the elements in each column distributed along the y-axis. The arguments `'grouped'` and `'stacked'` give 3D views of the corresponding 2D bar plots with the same arguments. With the same data matrix, `Y`, Figure 8.23 is produced by

```
subplot(2,2,1)
bar3(Y)
title('bar3(...,''detached'')')

subplot(2,2,2)
bar3(Y,'grouped')
title('bar3(...,''grouped'')')

subplot(2,2,3)
bar3(Y,'stacked')
```

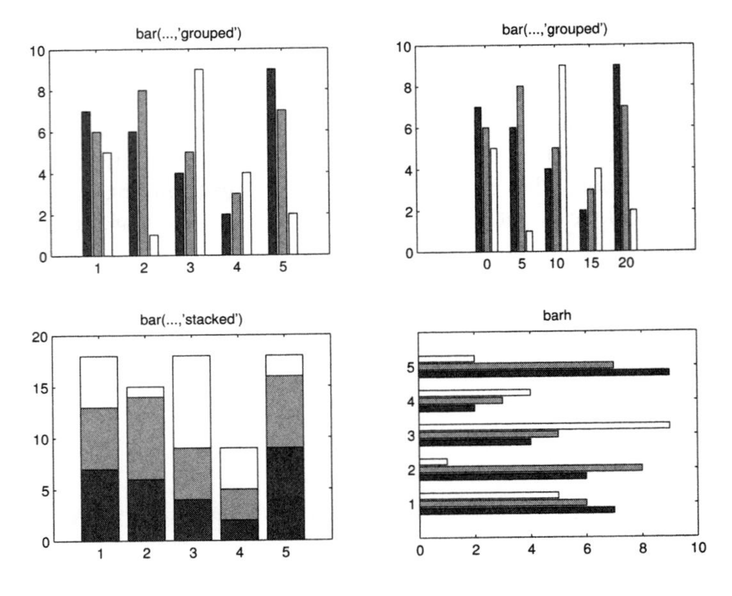

Figure 8.22. *2D bar plots.*

```
title('bar3(...,''stacked'')')

subplot(2,2,4)
bar3h(Y)
title('bar3h')
```

Note that with the default **'detached'** arrangement some bars are hidden behind others. A satisfactory solution to this problem can sometimes be found by rotating the plot using `view` or the mouse.

Histograms, which show the distribution of data by intervals, are produced by the **hist** function. The first argument, `y`, to **hist** is the data vector and the second is either a scalar specifying the number of bins or a vector defining the midpoints of the bins; if only `y` is supplied then 10 bins are used. If no output arguments are specified then `hist` plots a bar graph; otherwise it returns the frequency counts in the first output argument and the bin locations in the second output argument. If the input `y` is a matrix then bins are created for each column and a grouped bar graph is produced. The following code generates Figure 8.24:

```
randn('state',1)
y = exp(randn(1000,1)/3);
subplot(2,2,1)
hist(y)
title('1000-by-1 data vector, 10 bins')

subplot(2,2,2)
hist(y,25)
title('25 bins')
```

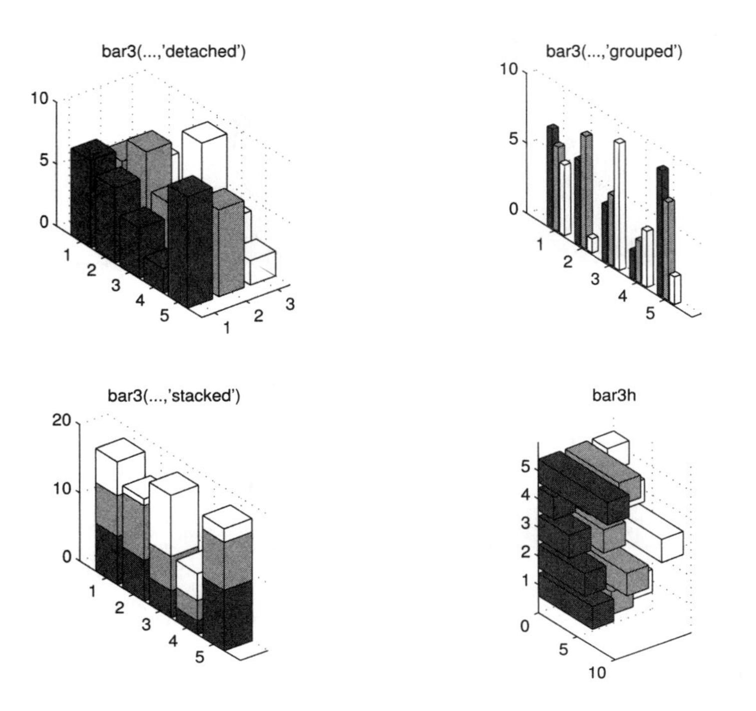

Figure 8.23. *3D bar plots.*

```
subplot(2,2,3)
hist(y,min(y):.1:max(y))
title('Bin width 0.1')

Y = exp(randn(1000,3)/3);
subplot(2,2,4)
hist(Y)
title('1000-by-3 data matrix')
```

Pie charts can be produced with `pie` and `pie3`. They take a vector argument, `x`, and corresponding to each element `x(i)` they draw a slice with area proportional to `x(i)`. A second argument `explode` can be given, which is a 0-1 vector with a 1 in positions corresponding to slices that are to be offset from the chart. By default, the slices are labelled with the percentage of the total area that they occupy; replacement labels can be specified in a cell array of strings (see Section 18.3). The following code produces Figure 8.25.

```
x = [1.5 3.4 4.2];

subplot(2,2,1)
pie(x)

subplot(2,2,2)
pie(x,[0 0 1])
```

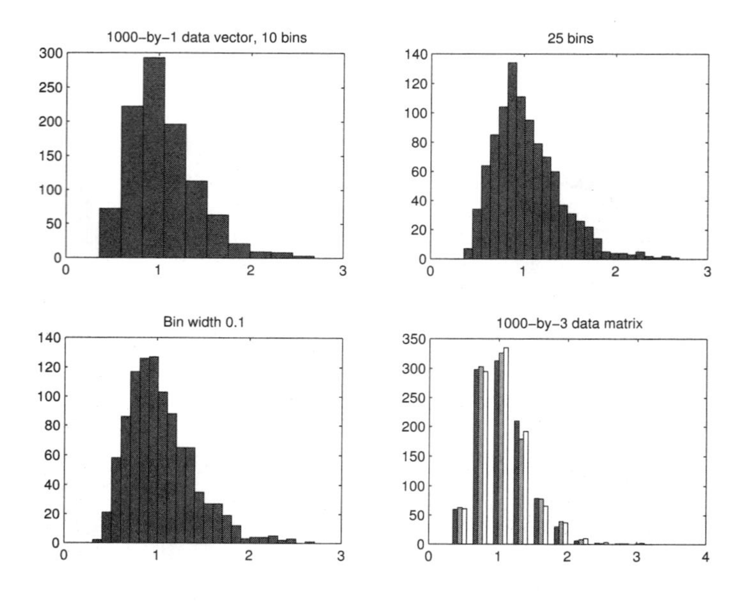

Figure 8.24. *Histograms produced with* `hist`.

```
subplot(2,2,3)
pie(x,{'Slice 1','Slice 2','Slice 3'})

subplot(2,2,4)
pie3(x,[0 1 0])
```

The **area** function produces a stacked area plot. With vector arguments, **area** is similar to **plot** except that the area between the y-values and 0 (or the level specified by the optional second argument) is filled; for matrix arguments the plots of the columns are stacked, showing the sum at each x-value. The following code produces Figure 8.26.

```
randn('state',1)
x = [1:12 11:-1:8 10:15]; Y = [x' x'];

subplot(2,1,1)
area(Y+randn(size(Y)))

subplot(2,1,2)
Y = Y + 5*randn(size(Y));
   (Y,min(min(Y)))
```

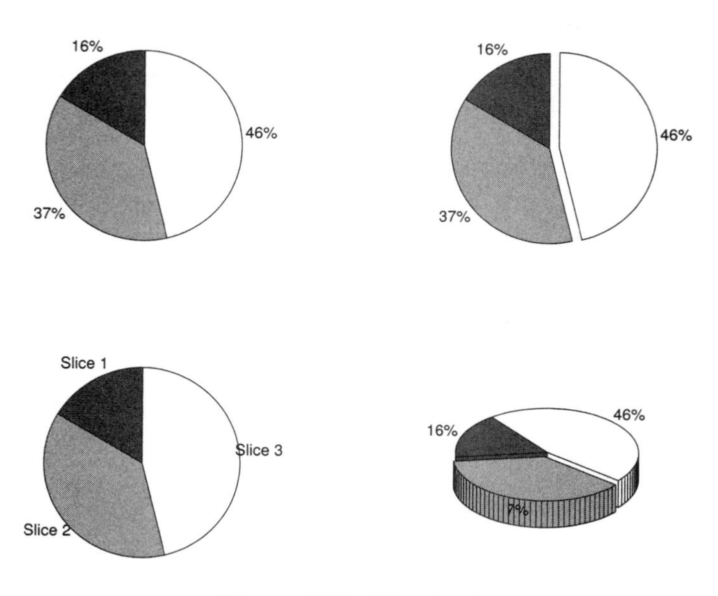

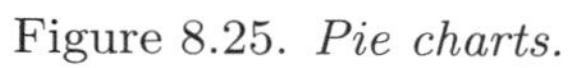
Figure 8.25. *Pie charts.*

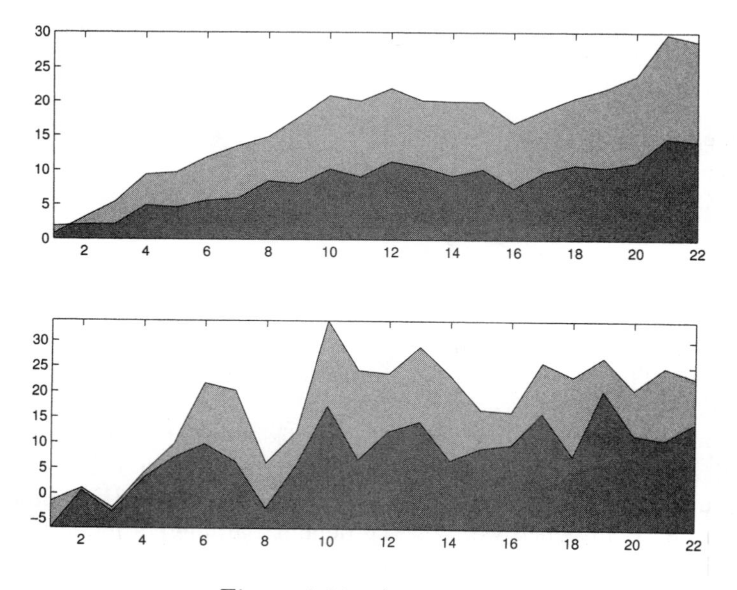

Figure 8.26. *Area graphs.*

8.4. Saving and Printing Figures

If your default printer has been set appropriately, simply typing **print** will send the contents of the current figure window to your printer. An alternative is to use the **print** command to save the figure as a file. For example,

```
print -deps2 myfig.eps
```

creates an encapsulated level 2 black and white PostScript file **myfig.eps** that can subsequently be printed on a PostScript printer or included in a document. This file can be incorporated into a LaTeX document, as in the following outline:

```
\documentclass{article}
\usepackage[dvips]{graphicx} % Assumes use of dvips dvi driver.
...
\begin{document}
...
\begin{center}
\includegraphics[width=8cm]{myfig.eps}
\end{center}
...
\end{document}
```

See [33], [61], or [64] for more about LaTeX.

The many options of the **print** command can be seen with **help print**. The **print** command also has a functional form, illustrated by

```
print('-deps2','myfig.eps')
```

(an example of command/function duality—see Section 7.5). To illustrate the utility of the functional form, the next example generates a sequence of five figures and saves them to files **fig1.eps**, ..., **fig5.eps**:

```
x = linspace(0,2*pi,50);
for i=1:5
    plot(x,sin(i*x))
    print('-deps2',['fig' int2str(i) '.eps'])
end
```

The second argument to the **print** command is formed by string concatenation (see Section 18.1), making use of the function **int2str**, which converts its integer argument to a string. Thus when **i=1**, for example, the **print** statement is equivalent to **print('-deps2','fig1.eps')**.

It is important to realize that graphics that look good on the screen may not ...od in print. In particular, if you accept the default values of the various ...ters your printed figures could be hard to read. Compare the two ... which we used the default parameters in the left-hand plot ...we increased **LineWidth**, **FontSize**, and **MarkerSize** and ...engths. To produce visually attractive, readable printed ...y to increase parameters such as these from their default ... three ways:

...rs such as **'FontSize',12** to the relevant commands.

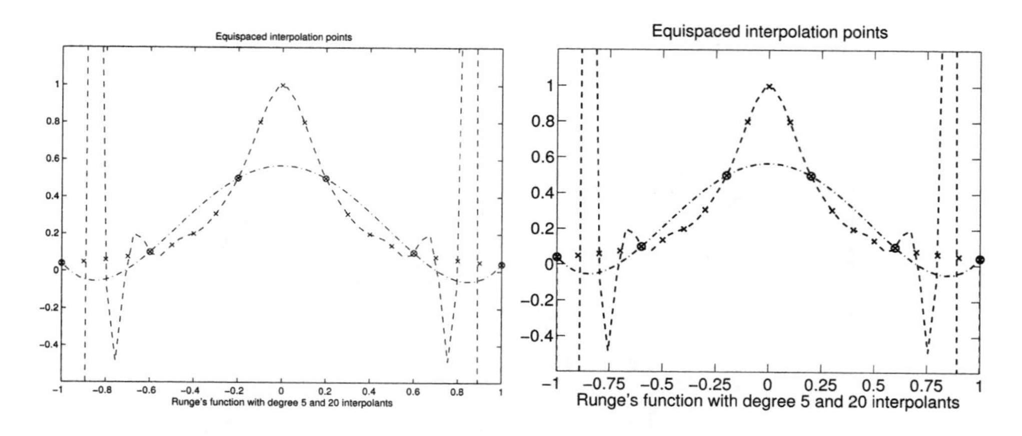

Figure 8.27. *Same plot with default parameters (left) and parameters tuned for the printed page (right).*

2. By resetting the default values for properties, prior to creating the MATLAB figure. See Section 17.2 for details of this approach.

3. By using the function `exportfig`, which is not part of MATLAB but is available from the MATLAB Central File Exchange. This function writes a figure to a file, changing many of the parameters that control the appearance of the figure according to the specified arguments. For example,

```
exportfig(gcf,'myfig.eps','FontSize',12,'LineWidth',0.8,...
          'Color','bw')
```

   creates a black and white encapsulated PostScript file `myfig.eps` from the current figure, setting all text and lines to the specified size and width, respectively.

The `saveas` command saves a figure to a file in a form that can be reloaded into MATLAB. For example,

```
saveas(gcf,'myfig','fig')
```

saves the current figure as a binary FIG-file, which can be reloaded into MATLAB with `open('myfig.fig')`.

It is also possible to save and print figures from the pulldown `File` menu in the figure window.

8.5. On Things Not Treated

We have restricted our treatment in this chapter to high-level graphics functions that deal with common 2D and 3D visualization tasks. MATLAB's graphics capabilities extend far beyond what is described here. On the one hand, MATLAB provides access to lighting, transparency control, solid model building, texture mapping, and the construction of graphical user interfaces. On the other hand, it is possible to control low-level details such as the tick labels and the position and size of the axes, and to produce animation; how to do this is described in Chapter 17 on Handle

Figure 8.28. *From the* 1964 *Gatlinburg Conference on Numerical Algebra. From left to right: J. H. Wilkinson, W. J. Givens, G. E. Forsythe, A. S. Householder, P. Henrici, and F. L. Bauer.* (*Source of photograph: Oak Ridge National Laboratory.*)

Graphics. A good place to find out more about MATLAB graphics is [76]. You can also learn by exploring the demonstrations in the `matlab\demos` directory. Try **`help demos`**, but note that not all files in this directory are documented in the help information.

Another area of MATLAB that we have not discussed is image handling and manipulation. If you type `what demos`, you will find that the `demos` directory contains a selection of MAT-files, most of which contain image data. These can be loaded and displayed as in the following example, which produces the image shown in Figure 8.28:

```
>> load gatlin, image(X); colormap(map), axis off
```

This picture was taken at a meeting in Gatlinburg, Tennessee, in 1964, and shows six major figures in the development of numerical linear algebra and scientific computing (you can find some of their names in Table 5.3).

Before coding graphs in MATLAB you should think carefully about the design, aiming for a result that is uncluttered and conveys clearly the intended message. Good references on graphical design are [11, Chaps. 10, 11], [115], [116], [117].

Finally, for many more examples of creative MATLAB graphics, see the books by D. J. Higham [39] and Trefethen [113].

"What is the use of a book," thought Alice,
"without pictures or conversation?"

— LEWIS CARROLL, *Alice's Adventures in Wonderland* (1865)

The close *command closes the current figure window.*
If there is no open figure window MATLAB opens one and then closes it.

— CLEVE B. MOLER

A picture is worth a thousand words.

— ANONYMOUS

Given their low data-density and
failure to order numbers along a visual dimension,
pie charts should never be used.

— EDWARD R. TUFTE, *The Visual Display of Quantitative Information* (1983)

It's kind of scandalous that the world's calculus books,
up until recent years, have never had a good picture[4] of a cardioid...
Nobody ever knew what a cardioid looked like, when I took calculus,
because the illustrations were done by graphic artists
who were trying to imitate drawings by previous artists,
without seeing the real thing.

— DONALD E. KNUTH, *Digital Typography* (1999)

[4] `ezpolar('1+cos(t)')`

Chapter 9
Linear Algebra

MATLAB was originally designed for linear algebra computations, so it not surprising that it has a rich set of functions for solving linear equation and eigenvalue problems. Many of the linear algebra functions are based on routines from the LAPACK [3] Fortran library.

Most of the linear algebra functions work for both real and complex matrices. We write A^* for the conjugate transpose of A. Recall that a square matrix A is Hermitian if $A^* = A$ and unitary if $A^*A = I$, where I is the identity matrix. To avoid clutter, we use the appropriate adjectives for complex matrices. Thus, when the matrix is real, "Hermitian" can be read as "symmetric" and "unitary" can be read as "orthogonal". For background on numerical linear algebra see [20], [31], [110], [114], [123], and, particularly for the algorithmic aspects, [8], [107], [108].

9.1. Norms and Condition Numbers

A norm is a scalar measure of the size of a vector or matrix. The p-norm of an n-vector x is defined by

$$\|x\|_p = \left(\sum_{i=1}^{n} |x_i|^p \right)^{1/p}, \qquad 1 \le p < \infty.$$

For $p = \infty$ the norm is defined by

$$\|x\|_\infty = \max_{1 \le i \le n} |x_i|.$$

The `norm` function can compute any p-norm and is invoked as `norm(x,p)`, with default `p` $= 2$. As a special case, for `p` $= -$`inf` the quantity $\min_i |x_i|$ is computed. Example:

```
>> x = 1:4;
>> [norm(x,1) norm(x,2) norm(x,inf) norm(x,-inf)]
ans =
   10.0000    5.4772    4.0000    1.0000
```

The p-norm of a matrix is defined by

$$\|A\|_p = \max_{x \ne 0} \frac{\|Ax\|_p}{\|x\|_p}.$$

The 1- and ∞-norms of an m-by-n matrix A can be characterized as

$$\|A\|_1 = \max_{1 \le j \le n} \sum_{i=1}^{m} |a_{ij}|, \qquad \text{"max column sum"},$$

$$\|A\|_\infty = \max_{1\le i\le m} \sum_{j=1}^{n} |a_{ij}|, \qquad \text{“max row sum”}.$$

The 2-norm of A can be expressed as the largest singular value of A, `max(svd(A))` (singular values and the `svd` function are described in Section 9.6). For matrices the `norm` function is invoked as `norm(A,p)` and supports `p = 1,2,inf` and `p = 'fro'`, the Frobenius norm

$$\|A\|_F = \left(\sum_{i=1}^{m}\sum_{j=1}^{n} |a_{ij}|^2\right)^{1/2}.$$

(This is a an example of a function with an argument that can vary in type: `p` can be a `double` or a string.) Example:

```
>> A = [1 2 3; 4 5 6; 7 8 9]
A =
      1     2     3
      4     5     6
      7     8     9
>> [norm(A,1) norm(A,2) norm(A,inf) norm(A,'fro')]
ans =
   18.0000   16.8481   24.0000   16.8819
```

For cases in which computation of the 2-norm of a matrix is too expensive the function `normest` can be used to obtain an estimate. The call `normest(A,tol)` uses the power method on A^*A to estimate $\|A\|_2$ to within a relative error `tol`; the default is `tol` = `1e-6`.

For a nonsingular square matrix A, $\kappa(A) = \|A\|\,\|A^{-1}\| \ge 1$ is the condition number with respect to inversion. It measures the sensitivity of the solution of a linear system $Ax = b$ to perturbations in A and b. The matrix A is said to be well conditioned or ill conditioned according as $\kappa(A)$ is small or large, where the meaning of “small” and “large” depends on the context. A condition number of the order of the reciprocal of `eps` is certainly regarded as large, because it implies that A is within distance about `eps` of a singular matrix. The condition number is computed by the `cond` function as `cond(A,p)`. The p-norm choices `p = 1,2,inf,'fro'` are supported, with default `p = 2`. For `p = 2`, rectangular matrices are allowed, in which case the condition number is defined by $\kappa_2(A) = \|A\|_2\|A^+\|_2$, where A^+ is the pseudo-inverse (see Section 9.3).

Computing the exact condition number is expensive, so MATLAB provides two functions for estimating the 1-norm condition number of a square matrix A, `rcond` and `condest`. Both functions produce estimates usually of the correct order of magnitude at about one third the cost of explicitly computing A^{-1}. Function `rcond` uses the LAPACK condition estimator to estimate the reciprocal of $\kappa_1(A)$, producing a result between 0 and 1, with 0 signalling exact singularity. Function `condest` estimates $\kappa_1(A)$ and also returns an approximate null vector, which is required in some applications. The command `[c,v] = condest(A)` produces a scalar `c` and vector `v` so that `c` $\le \kappa_1$`(A)` and `norm(A*v,1)` = `norm(A,1)*norm(v,1)/c`. Example:

```
>> A = gallery('grcar',8);

>> [cond(A,1) 1/rcond(A) condest(A)]
ans =
```

```
      7.7778    5.3704    7.7778

>> [cond(A,1) 1/rcond(A) condest(A)]
ans =
      7.7778    5.3704    7.2222
```

As this example illustrates, `condest` does not necessarily return the same result on each invocation, as it makes use of `rand`.

9.2. Linear Equations

The fundamental tool for solving a linear system of equations is the backslash operator, \. It handles three types of linear system $Ax = b$, where the matrix A and the vector b are given. The three possible shapes for A lead to square, overdetermined, and underdetermined systems, as described below. More generally, the \ operator can be used to solve $AX = B$, where B is a matrix with p columns; in this case MATLAB solves $AX(:,j) = B(:,j)$ for $j = 1:p$.

9.2.1. Square System

If A is an n-by-n nonsingular matrix then `A\b` is the solution `x` to $Ax = b$, computed by LU factorization with partial pivoting. During the solution process MATLAB computes `rcond(A)`, and it prints a warning message if the result is smaller than about `eps`:

```
>> x = hilb(15)\ones(15,1);
Warning: Matrix is close to singular or badly scaled.
         Results may be inaccurate. RCOND = 1.543404e-018.
```

These warning messages can be turned off using `warning off`; see Section 14.2.

MATLAB recognizes three special forms of square systems and takes advantage of them to reduce the computation.

- Triangular matrix, or permutation of a triangular matrix. (The square matrix A is upper triangular if $a_{ij} = 0$ for $i > j$, and lower triangular if $a_{ij} = 0$ for $j > i$.) The system is solved by substitution.

- Upper Hessenberg matrix. (The square matrix A is upper Hessenberg if $a_{ij} = 0$ for $i > j + 1$.) The system is solved by LU factorization with partial pivoting, taking advantage of the Hessenberg form.

- Hermitian positive definite matrix. (The Hermitian matrix A is positive definite if $x^*Ax > 0$ for all nonzero vectors x, or, equivalently, if all the eigenvalues are real and positive.) Cholesky factorization is used instead of LU factorization. How does MATLAB know the matrix is definite? When \ is called with a Hermitian matrix that has positive diagonal elements MATLAB attempts to Cholesky factorize the matrix. If the Cholesky factorization succeeds it is used to solve the system; otherwise an LU factorization with partial pivoting is carried out.

When efficiency is important and A is structured, it may be appropriate to use the `linsolve` function in place of the backslash operator. The syntax is `X =`

`linsolve(A,B,opts)`, which solves $AX = B$, or $A^*X = B$ if `opts.TRANSA` has the value `true`. The structure `opts` (see Section 18.3 for details of structures) specifies one of several possible matrix properties and `linsolve` gains efficiency over backslash by not attempting to verify the stated property. Table 9.1 shows how to set `opts` in order to use `linsolve` to solve some particular types of structured system. The usage is illustrated by the following M-file:

```
n = 8; rand('seed',1)
B = rand(n,2);

% Upper triangular A.
A = triu(rand(n));
opts = struct('UT',true);
X = linsolve(A,B,opts);
res = norm(B-A*X)

% Give LINSOLVE incorrect opts structure.
opts = struct('LT',true);
X = linsolve(A,B,opts);
res = norm(B-A*X)

% Upper triangular A, solve transposed system.
opts = struct('UT',true,'TRANSA',true);
X = linsolve(A,B,opts);
res = norm(B-A'*X)

% Symmetric positive definite A.
A = rand(n); A = A*A';
opts = struct('SYM',true,'POSDEF',true);
X = linsolve(A,B,opts);
res = norm(B-A*X)
```

The output is

```
res =
  1.2613e-015
res =
   10.7664
res =
  1.9675e-016
res =
  4.6226e-015
```

The residuals in this example should be of order `eps` for a correct solution. The second call to `linsolve` wrongly specifies the matrix A as lower triangular. Since `linsolve` performs no checks on the matrix properties it gives an incorrect answer without any warning or error. Hence care is required in the use of this function. The execution time saved by using `linsolve` instead of backslash is highly dependent on the matrix property and the dimension of the matrix. Rectangular systems, as described in the next two subsections, can also be solved by `linsolve`; see the online help for details.

Table 9.1. *Some examples of how to set* `opts` *structure in* `linsolve`.

Matrix property	opts structure
Lower triangular	`opts = struct('LT',true)`
Upper triangular	`opts = struct('UT',true)`
Upper Hessenberg	`opts = struct('UHESS',true)`
Symmetric/Hermitian and positive definite	`opts = struct('SYM',true,'POSDEF',true)`
Use (conjugate) transpose of matrix	`opts = struct('TRANSA',true)`

9.2.2. Overdetermined System

If A has dimension m-by-n with $m > n$ then $Ax = b$ is an overdetermined system: there are more equations than unknowns. In general, there is no x satisfying the system. MATLAB's `A\b` gives a least squares solution to the system, that is, it minimizes `norm(A*x-b)` (the 2-norm of the residual) over all vectors `x`. If A has full rank n there is a unique least squares solution. If A has rank k less than n then `A\b` is a basic solution—one with at most k nonzero elements (k is determined, and `x` computed, using the QR factorization with column pivoting). In the latter case MATLAB displays a warning message.

A least squares solution to $Ax = b$ can also be computed as `x_min = pinv(A)*b`, where the function `pinv` computes the pseudo-inverse; see Section 9.3. In the case where A is rank-deficient `x_min` is the unique solution of minimal 2-norm.

A vector that minimizes the 2-norm of $Ax - b$ over all nonnegative vectors x, for real A and b, is computed by `lsqnonneg`. The simplest usage is `x = lsqnonneg(A,b)`, and several other input and output arguments can be specified, including a starting vector for the iterative algorithm that is used. Example:

```
>> A = gallery('lauchli',3,0.25), b = [1 2 4 8]';
A =
    1.0000    1.0000    1.0000
    0.2500         0         0
         0    0.2500         0
         0         0    0.2500

>> x = A\b;                % Least squares solution.
>> xn = lsqnonneg(A,b);   % Nonnegative least squares solution.

>> [x xn], [norm(A*x-b) norm(A*xn-b)]
ans =
   -9.9592         0
   -1.9592         0
   14.0408    2.8235
ans =
    7.8571    8.7481
```

9.2.3. Underdetermined System

If A has dimension m-by-n with $m < n$ then $Ax = b$ is an underdetermined system: there are fewer equations than unknowns. The system has either no solution or infinitely many. In the latter case `A\b` produces a basic solution, one with at most k nonzero elements, where k is the rank of A. This solution is generally not the solution of minimal 2-norm, which can be computed as `pinv(A)*b`. If the system has no solution (that is, it is inconsistent) then `A\b` is a least squares solution. Here is an example that illustrates the difference between the \ and `pinv` solutions:

```
>> A = [1 1 1; 1 1 -1], b = [3; 1]
A =
     1     1     1
     1     1    -1
b =
     3
     1

>> x = A\b; y = pinv(A)*b;
>> [x y]
ans =
    2.0000    1.0000
         0    1.0000
    1.0000    1.0000

>> [norm(x) norm(y)]
ans =
    2.2361    1.7321
```

9.3. Inverse, Pseudo-Inverse, and Determinant

The inverse of an n-by-n matrix A is a matrix X satisfying $AX = XA = I$, where I is the identity matrix (`eye(n)`). A matrix without an inverse is called singular. A singular matrix can be characterized in several ways: in particular, its determinant is zero and it has a nonzero null vector, that is, there exists a nonzero vector v such that $Av = 0$.

The matrix inverse is computed by the function `inv`. For example:

```
>> A = pascal(3), X = inv(A)
A =
     1     1     1
     1     2     3
     1     3     6
X =
     3    -3     1
    -3     5    -2
     1    -2     1

>> norm(A*X-eye(3))
ans =
     0
```

The inverse is formed using LU factorization with partial pivoting and the reciprocal condition estimate `rcond` is computed. A warning message is produced if exact singularity is detected or if `rcond` is smaller than about `eps`.

Note that it is rarely necessary to compute the inverse of a matrix. For example, solving a square linear system $Ax = b$ by `A\b` is 2–3 times faster than by `inv(A)*b` and often produces a smaller residual. It is usually possible to reformulate computations involving a matrix inverse in terms of linear system solving, so that explicit inversion is avoided.

The determinant of a square matrix is computed by the function `det`. It is calculated from the LU factors. Although the computation is affected by rounding errors in general, `det(A)` returns an integer when `A` has integer entries:

```
>> A = vander(1:5)
A =
     1     1     1     1     1
    16     8     4     2     1
    81    27     9     3     1
   256    64    16     4     1
   625   125    25     5     1

>> det(A)
ans =
   288
```

It is not recommended to test for nearness to singularity using `det`. Instead, `cond`, `rcond`, or `condest` should be used.

The (Moore–Penrose) pseudo-inverse generalizes the notion of inverse to rectangular and rank-deficient matrices A and is written A^+. It is computed with `pinv(A)`. The pseudo-inverse A^+ of A can be characterized as the unique matrix $X = A^+$ satisfying the four conditions $AXA = A$, $XAX = X$, $(XA)^* = XA$, and $(AX)^* = AX$. It can also be written explicitly in terms of the singular value decomposition (SVD): if the SVD of A is given by (9.1) on p. 130 then $A^+ = V\Sigma^+U^*$, where Σ^+ is n-by-m diagonal with (i,i) entry $1/\sigma_i$ if $\sigma_i > 0$ and otherwise 0. To illustrate,

```
>> pinv(ones(3))
ans =
    0.1111    0.1111    0.1111
    0.1111    0.1111    0.1111
    0.1111    0.1111    0.1111
```

and if

```
A =
     0     0     0     0
     0     1     0     0
     0     0     2     0
```

then

```
>> pinv(A)
ans =
         0         0         0
```

```
         0    1.0000         0
         0         0    0.5000
         0         0         0
```

9.4. LU and Cholesky Factorizations

An LU factorization of a square matrix A is a factorization $A = LU$ in which L is unit lower triangular (that is, lower triangular with 1s on the diagonal) and U is upper triangular. Not every matrix can be factorized in this way, but when row interchanges are incorporated the factorization always exists. The `lu` function computes an LU factorization with partial pivoting $PA = LU$, where P is a permutation matrix. The call `[L,U,P] = lu(A)` returns the triangular factors and the permutation matrix. With just two output arguments, `[L,U] = lu(A)` returns `L` $= P^T L$, so `L` is a triangular matrix with its rows permuted. Example:

```
>> format short g
>> A = gallery('fiedler',3), [L,U] = lu(A)
A =
     0     1     2
     1     0     1
     2     1     0
L =
             0             1             0
           0.5          -0.5             1
             1             0             0
U =
     2     1     0
     0     1     2
     0     0     2
```

The `lu` function also works for rectangular matrices. If A is m-by-n then `[L,U] = lu(A)` produces an m-by-n `L` and n-by-n `U` if $m \geq n$ and an m-by-m `L` and m-by-n `U` if $m < n$.

Using `x` = `A\b` to solve a linear system `Ax` = `b` with a square `A` is equivalent to LU factorizing the matrix and then solving with the factors:

```
[L,U] = lu(A); x = U\(L\b);
```

As noted in Section 9.2.1, MATLAB takes advantage of the fact that `L` is a permuted triangular matrix when forming `L\b`. An advantage of this two-step approach is that if further linear systems involving `A` are to be solved then the `LU` factors can be reused, with a saving in computation.

Any Hermitian positive definite matrix has a Cholesky factorization $A = R^*R$, where R is upper triangular with real, positive diagonal elements. The Cholesky factor is computed by `R = chol(A)`. For example:

```
>> A = pascal(4)
A =
     1     1     1     1
     1     2     3     4
     1     3     6    10
```

```
     1     4    10    20

>> R = chol(A)
R =
     1     1     1     1
     0     1     2     3
     0     0     1     3
     0     0     0     1
```

Note that `chol` looks only at the elements in the upper triangle of `A` (including the diagonal)—it factorizes the Hermitian matrix agreeing with the upper triangle of `A`. An error is produced if `A` is not positive definite. The `chol` function can be used to test whether a matrix is positive definite (indeed, this is as good a test as any) using the call `[R,p] = chol(A)`, where the integer `p` will be zero if the factorization succeeds and positive otherwise; see `help chol` for more details about `p`.

Function `cholupdate` modifies the Cholesky factorization when the original matrix is subjected to a rank 1 perturbation (either an update, $+xx^*$, or a downdate, $-xx^*$).

9.5. QR Factorization

A QR factorization of an m-by-n matrix A is a factorization $A = QR$, where Q is m-by-m unitary and R is m-by-n upper triangular. This factorization is very useful for the solution of least squares problems and for constructing an orthonormal basis for the columns of A. The command `[Q,R] = qr(A)` computes the factorization, while when $m > n$ `[Q,R] = qr(A,0)` produces an "economy size" version in which `Q` has only n columns and `R` is n-by-n. Example:

```
>> format short e, A
A =
     1     0     1
     1    -1     1
     2     0     0

>> [Q,R] = qr(A)
Q =
  -4.0825e-001  1.8257e-001 -8.9443e-001
  -4.0825e-001 -9.1287e-001 -6.1745e-017
  -8.1650e-001  3.6515e-001  4.4721e-001
R =
  -2.4495e+000  4.0825e-001 -8.1650e-001
             0  9.1287e-001 -7.3030e-001
             0            0 -8.9443e-001
```

A QR factorization with column pivoting has the form $AP = QR$, where P is a permutation matrix. The permutation strategy that is used produces a factor R whose diagonal elements are nonincreasing: $|r_{11}| \geq |r_{22}| \geq \cdots \geq |r_{nn}|$. Column pivoting is particularly appropriate when A is suspected of being rank-deficient, as it helps to reveal near rank-deficiency. Roughly speaking, if A is near a matrix of rank $r < n$ then the last $n - r$ diagonal elements of R will be of order `eps*norm(A)`. A third output argument forces function `qr` to use column pivoting and return the permutation matrix: `[Q,R,P] = qr(A)`. If the economy size factorization with column pivoting is requested, via `[Q,R,p] = qr(A,0)`, then `p` is a permutation vector and $A(:,p) = QR$. Continuing the previous example, we make `A` nearly singular and see how column pivoting reveals the near singularity in the last diagonal element of `R`:

```
>> A(2,2) = eps
A =
   1.0000e+000             0   1.0000e+000
   1.0000e+000   2.2204e-016   1.0000e+000
   2.0000e+000             0             0

>> [Q,R,P] = qr(A); R, P
R =
  -2.4495e+000  -8.1650e-001  -9.0649e-017
             0  -1.1547e+000  -1.2820e-016
             0             0   1.5701e-016
P =
       1       0       0
       0       0       1
       0       1       0
```

Functions `qrdelete`, `qrinsert`, and `qrupdate` modify the QR factorization when a column of the original matrix is deleted or inserted and when a rank 1 perturbation is added.

9.6. Singular Value Decomposition

The SVD of an m-by-n matrix A has the form

$$A = U\Sigma V^*, \tag{9.1}$$

where U is an m-by-m unitary matrix, V is an n-by-n unitary matrix, and Σ is a real m-by-n diagonal matrix with (i,i) entry σ_i. The singular values σ_i satisfy $\sigma_1 \geq \sigma_2 \geq \cdots \geq \sigma_{\min(m,n)} \geq 0$. The SVD is an extremely useful tool [31]. For example, the rank of A is the number of nonzero singular values and the smallest singular value is the 2-norm distance to the nearest rank-deficient matrix. The complete SVD is computed using `[U,S,V] = svd(A)`; if only one output argument is specified then a vector of singular values is returned. Example:

```
>> A = reshape(1:9,3,3); format short e
>> svd(A)'
ans =
   1.6848e+001   1.0684e+000   5.5431e-016
```

Here, the matrix is singular. The smallest computed singular value is at the level of the unit roundoff rather than zero because of rounding errors.

When $m > n$ the command `[U,S,V] = svd(A,0)` produces an "economy size" SVD in which `U` is m-by-n with orthonormal columns and `S` is n-by-n. The call `[U,S,V] = svd(X,'econ')` produces the same result as `svd(X,0)` when $m \geq n$, but if $m < n$ it returns an n-by-m `V` with orthonormal columns and an m-by-m `S`. Example:

```
>> B = gallery('triw',[2 4],-2)
B =
     1    -2    -2    -2
     0     1    -2    -2

>> [U,S,V] = svd(B,'econ')
U =
  -8.1124e-001 -5.8471e-001
  -5.8471e-001  8.1124e-001
S =
   4.1623e+000            0
             0  2.1623e+000
V =
  -1.9490e-001 -2.7041e-001
   2.4933e-001  9.1601e-001
   6.7076e-001 -2.0953e-001
   6.7076e-001 -2.0953e-001
```

Functions **rank**, **null**, and **orth** compute, respectively, the rank, an orthonormal basis for the null space, and an orthonormal basis for the range of their matrix argument. All three base their computation on the SVD, using a tolerance proportional to **eps** to decide when a computed singular value can be regarded as zero. For example, using the previous matrix:

```
>> format
>> rank(A)
ans =
     2

>> null(A)
ans =
    0.4082
   -0.8165
    0.4082

>> orth(A)
ans =
   -0.4797    0.7767
   -0.5724    0.0757
   -0.6651   -0.6253
```

Another function connected with rank computations is **rref**, which computes the reduced row echelon form. Since the computation of this form is very sensitive to rounding errors, this function is mainly of pedagogical interest.

The generalized singular value decomposition of an m-by-p matrix A and an n-by-p matrix B can be written

$$A = UCX^*, \quad B = VSX^*, \qquad C^*C + S^*S = I,$$

where U and V are unitary, X is nonsingular, and C and S are real diagonal matrices with nonnegative diagonal elements. The numbers $C(i,i)/S(i,i)$ are the generalized singular values. This decomposition is computed by `[U,V,X,C,S] = gsvd(A,B)`. See `help gsvd` for more details about the dimensions of the factors.

9.7. Eigenvalue Problems

Algebraic eigenvalue problems are straightforward to define, but their efficient and reliable numerical solution is a complicated subject. MATLAB's `eig` function simplifies the solution process by recognizing and taking advantage of the number of input matrices, as well as their structure and the output requested. It automatically chooses among 16 different algorithms or algorithmic variants, corresponding to

- standard (`eig(A)`) or generalized (`eig(A,B)`) problem,
- real or complex matrices `A` and `B`,
- symmetric/Hermitian `A` and `B` with `B` positive definite, or not,
- eigenvectors requested or not.

9.7.1. Eigenvalues

The scalar λ and nonzero vector x are an eigenvalue and corresponding eigenvector of the n-by-n matrix A if $Ax = \lambda x$. The eigenvalues are the n roots of the degree n characteristic polynomial $\det(\lambda I - A)$. The $n + 1$ coefficients of this polynomial are computed by `p = poly(A)`:

$$\det(\lambda I - A) = p_1\lambda^n + p_2\lambda^{n-1} + \cdots + p_n\lambda + p_{n+1}.$$

The eigenvalues of `A` are computed with the `eig` function: `e = eig(A)` assigns the eigenvalues to the vector `e`. More generally, `[V,D] = eig(A)` computes an n-by-n diagonal matrix `D` and an n-by-n matrix `V` such that `A*V` = `V*D`. Thus `D` contains eigenvalues on the diagonal and the columns of `V` are eigenvectors. Not every matrix has n linearly independent eigenvectors, so the matrix `V` returned by `eig` may be singular (or, because of roundoff, nonsingular but very ill conditioned). The matrix in the following example has two eigenvalues 1 and only one eigenvector:

```
>> [V,D] = eig([2 -1; 1  0])
V =
     0.7071    0.7071
     0.7071    0.7071
D =
     1     0
     0     1
```

The scaling of eigenvectors is arbitrary (if x is an eigenvector then so is any nonzero multiple of x). As the last example illustrates, MATLAB normalizes so that each column of V has unit 2-norm. Note that eigenvalues and eigenvectors can be complex, even for a real (non-Hermitian) matrix.

A Hermitian matrix has real eigenvalues and its eigenvectors can be taken to be mutually orthogonal. For Hermitian matrices MATLAB returns eigenvalues sorted in increasing order and the matrix of eigenvectors is unitary to working precision:

```
>> [V,D] = eig([2 -1; -1  1])
V =
   -0.5257   -0.8507
   -0.8507    0.5257
```

```
D =
    0.3820         0
         0    2.6180

>> norm(V'*V-eye(2))
ans =
  2.2204e-016
```

In the following example `eig` is applied to the (non-Hermitian) Frank matrix:

```
>> F = gallery('frank',5)
F =
     5     4     3     2     1
     4     4     3     2     1
     0     3     3     2     1
     0     0     2     2     1
     0     0     0     1     1

>> e = eig(F)'
e =
   10.0629    3.5566    1.0000    0.0994    0.2812
```

This matrix has some special properties, one of which we can see by looking at the reciprocals of the eigenvalues:

```
>> 1./e
ans =
    0.0994    0.2812    1.0000   10.0629    3.5566
```

Thus if λ is an eigenvalue then so is $1/\lambda$. The reason is that the characteristic polynomial is anti-palindromic:

```
>> poly(F)
ans =
    1.0000  -15.0000   55.0000  -55.0000   15.0000   -1.0000
```

Thus $\det(F - \lambda I) = -\lambda^5 \det(F - \lambda^{-1} I)$.

Function `condeig` computes condition numbers for the eigenvalues: a large condition number indicates an eigenvalue that is sensitive to perturbations in the matrix. The following example displays eigenvalues in the first row and condition numbers in the second:

```
>> A = gallery('frank',6);
>> [V,D,s] = condeig(A);
>> [diag(D)'; s']
ans =
   12.9736    5.3832    1.8355    0.5448    0.0771    0.1858
    1.3059    1.3561    2.0412   15.3255   43.5212   56.6954
```

For this matrix the small eigenvalues are slightly ill conditioned.

9.7.2. More about Eigenvalue Computations

The function `eig` works in several stages. First, when A is nonsymmetric, it balances the matrix, that is, it carries out a similarity transformation $A \leftarrow Y^{-1}AY$, where Y is a permutation of a diagonal matrix chosen to give A rows and columns of approximately equal norm. The motivation for balancing is that it can lead to a more accurate computed eigensystem. However, balancing can worsen rather than improve the accuracy (see `doc eig` for an example), so it may be necessary to turn balancing off with `eig(A,'nobalance')`.

After balancing, `eig` reduces A to Hessenberg form, then uses the QR algorithm to reach Schur form, after which eigenvectors are computed by substitution if required. The Hessenberg factorization takes the form $A = QHQ^*$, where H is upper Hessenberg and Q is unitary. If A is Hermitian then H is Hermitian and tridiagonal. The Hessenberg factorization is computed by `H = hess(A)` or `[Q,H] = hess(A)`. The real Schur decomposition of a real A has the form $A = QTQ^T$, where T is upper quasi-triangular, that is, block triangular with 1-by-1 and 2-by-2 diagonal blocks, and Q is orthogonal. The (complex) Schur decomposition has the form $A = QTQ^*$, where T is upper triangular and Q is unitary. If A is real then `T = schur(A)` and `[Q,T] = schur(A)` produce the real Schur decomposition. If A is complex then `schur` produces the complex Schur form. The complex Schur form can be obtained for a real matrix with `schur(A,'complex')` (it differs from the real form only when `A` has one or more nonreal eigenvalues).

The Schur decomposition can be reordered (i.e., the eigenvalues placed in a different order on the diagonal blocks of T) using the `ordschur` function. The function `ordeig` returns the eigenvalues of a quasi-triangular matrix in the order that they appear along the (block) diagonal; for such matrices it is more efficient than applying `eig`.

If A is real and symmetric (complex Hermitian), `[V,D] = eig(A)` reduces initially to symmetric (Hermitian) tridiagonal form then iterates to produce a diagonal Schur form, resulting in an orthogonal (unitary) `V` and a real, diagonal `D`.

9.7.3. Generalized Eigenvalues

The generalized eigenvalue problem is defined in terms of two n-by-n matrices A and B: λ is an eigenvalue and $x \neq 0$ an eigenvector if $Ax = \lambda Bx$. The generalized eigenvalues are computed by `e = eig(A,B)`, while `[V,D] = eig(A,B)` computes an n-by-n diagonal matrix `D` and an n-by-n matrix `V` of eigenvectors such that `A*V` = `B*V*D`. The theory of the generalized eigenproblem is more complicated than that of the standard eigenproblem, with the possibility of zero, finitely many, or infinitely many eigenvalues and of eigenvalues that are infinitely large. When B is singular `eig` may return computed eigenvalues containing NaNs. To illustrate the computation of generalized eigenvalues:

```
>> A = gallery('triw',3), B = magic(3)
A =
     1    -1    -1
     0     1    -1
     0     0     1

B =
     8     1     6
     3     5     7
```

```
     4     9     2

>> [V,D] = eig(A,B); V, eivals = diag(D)'
V =
   -1.0000   -1.0000    0.3526
    0.4844   -0.4574    0.3867
    0.2199   -0.2516   -1.0000
eivals =
    0.2751    0.0292   -0.3459
```

When A is Hermitian and B is Hermitian positive definite (the Hermitian definite generalized eigenproblem) the eigenvalues are real and A and B are simultaneously diagonalizable. In this case `eig` returns real computed eigenvalues sorted in increasing order, with the eigenvectors normalized (up to roundoff) so that `V'*B*V = eye(n)`; moreover, `V'*A*V` is diagonal. The method that `eig` uses (Cholesky factorization of B, followed by reduction to a standard eigenproblem and solution by the QR algorithm) can be numerically unstable when B is ill conditioned. You can force `eig` to ignore the structure and solve the problem in the same way as for general A and B by invoking it as `eig(A,B,'qz')`; the QZ algorithm (see below) is then used, which has guaranteed numerical stability but does not guarantee real computed eigenvalues. Example:

```
>> A = gallery('fiedler',3), B = gallery('moler',3)
A =
     0     1     2
     1     0     1
     2     1     0
B =
     1    -1    -1
    -1     2     0
    -1     0     3

>> format short g
>> [V,D] = eig(A,B); V, eivals = diag(D)'
V =
       0.55335       0.23393        2.3747
       0.15552      -0.57301        1.2835
      -0.36921       0.19163       0.90938
eivals =
      -0.75993      -0.30839        17.068

>> V'*A*V
ans =
      -0.75993 -2.1748e-016 -2.6845e-015
  -2.1982e-016     -0.30839  1.8584e-015
  -2.6665e-015  1.7522e-015       17.068

>> V'*B*V
ans =
             1 -8.4568e-018 -2.9295e-016
   2.1549e-018            1 -7.7954e-017
  -2.4248e-016 -7.7927e-017            1
```

The function `hess` computes the Hessenberg form of A and B: $QAZ = H$, $QBZ = T$, where H is upper Hessenberg, T is upper triangular, and Q and Z are unitary. The syntax is `[H,T,Q,Z] = hess(A,B)`.

The generalized Schur decomposition of a pair of matrices A and B has the form

$$QAZ = T, \quad QBZ = S,$$

where Q and Z are unitary and T and S are upper triangular. The generalized eigenvalues are the ratios `T(i,i)/S(i,i)` of the diagonal elements of T and S. The generalized real Schur decomposition of real `A` and `B` has the same form with `Q` and `Z` orthogonal and `T` and `S` upper quasi-triangular. These decompositions are computed by the `qz` function with the command `[T,S,Q,Z,V,W] = qz(A,B)`, where the output arguments `V` and `W` are matrices of generalized right eigenvectors and left eigenvectors, respectively. The function is named after the QZ algorithm that it implements. By default the (possibly) complex form with upper triangular T and S is produced. For real matrices, `qz(A,B,'real')` produces the real form and `qz(A,B,'complex')` the default complex form. The generalized Schur decomposition can be reordered using the `ordqz` function.

Function `polyeig` solves the polynomial eigenvalue problem $(\lambda^p A_p + \lambda^{p-1} A_{p-1} + \cdots + \lambda A_1 + A_0)x = 0$, where the A_i are given square coefficient matrices. The generalized eigenproblem is obtained for $p = 1$, with $A_0 = I$ then giving the standard eigenproblem. The quadratic eigenproblem $(\lambda^2 A + \lambda B + C)x = 0$ corresponds to $p = 2$. If A_p is n-by-n and nonsingular then there are pn eigenvalues. MATLAB's syntax is `e = polyeig(A0,A1,..,Ap)` or `[X,e] = polyeig(A0,A1,..,Ap)`, with `e` a pn-vector of eigenvalues and `X` an n-by-pn matrix whose columns are the corresponding eigenvectors. Example:

```
>> A = eye(2); B = [20 -10; -10 20]; C = [15 -5; -5 15];

>> [X,e] = polyeig(C,B,A)
X =
    0.7071   -0.7071   -0.7071    0.7071
   -0.7071    0.7071   -0.7071    0.7071
e =
  -29.3178
   -0.6822
   -1.1270
   -8.8730
```

An optional third output argument of `polyeig` returns condition numbers for the eigenvalues.

9.8. Iterative Linear Equation and Eigenproblem Solvers

In this section we describe functions that are based on iterative methods and primarily intended for large, possibly sparse problems, for which solution by one of the methods described earlier in the chapter could be prohibitively expensive. Sparse matrices are discussed further in Chapter 15.

Several functions implement iterative methods for solving square linear systems $Ax = b$; see Table 9.2. All apply to general A except `minres` and `symmlq`, which require A to be Hermitian, and `pcg`, which requires A to be Hermitian positive definite. All the methods employ matrix–vector products Ax and possibly A^*x and do not require explicit access to the elements of A. The functions have identical calling sequences, apart from `gmres` (see below). The simplest usage is `x = solver(A,b)` (where `solver` is one of the functions in Table 9.2). Alternatively, `x = solver(A,b,tol)` specifies a convergence tolerance `tol`, which defaults to `1e-6`. Convergence is declared when an iterate `x` satisfies `norm(b-A*x) <= tol*norm(b)`. The argument `A` can be a full or sparse matrix, or the name of a function `afun` such that `afun(x)` returns `A*x` and, in the case of `bicg` and `qmr`, such that `afun(x,'transp')` returns `A'*x`.

These iterative methods usually need preconditioning if they are to be efficient. All accept further arguments M_1 and M_2 or $M = M_1 M_2$ and effectively solve the preconditioned system

$$M_1^{-1} A M_2^{-1} \cdot M_2 x = M_1^{-1} b \quad \text{or} \quad M^{-1} A x = M^{-1} b.$$

The aim is to choose M_1 and M_2 so that $M_1^{-1} A M_2^{-1}$ or $M^{-1}A$ is in some sense close to the identity matrix. Choosing a good preconditioner is a difficult task that usually requires knowledge of the application from which the linear system came. The functions `luinc` and `cholinc` compute incomplete factorizations that provide one way of constructing preconditioners; see `doc luinc`, `doc cholinc`, and `doc bicg`. For background on iterative linear equation solvers and preconditioning see [9], [32], [50], [96].

To illustrate the usage of the iterative solvers we give an example involving `pcg`, which implements the preconditioned conjugate gradient method. For A we take a symmetric positive definite finite element matrix called the Wathen matrix, which has a fixed sparsity pattern and random entries.

```
>> A = gallery('wathen',12,12); n = length(A)
n =
   481
>> b = ones(n,1);

>> x = pcg(A,b);
pcg stopped at iteration 20 without converging to the desired
tolerance 1e-006 because the maximum number of iterations was
reached.
The iterate returned (number 19) has relative residual 0.12

>> x = pcg(A,b,1e-6,100);
pcg converged at iteration 93 to a solution with relative residual
9e-007
```

The bare minimum of arguments to `pcg` is the matrix and the right-hand side. The conjugate gradient method did not converge to the default tolerance (10^{-6}) within the default of 20 iterations, so we tried again with the same tolerance and a new limit of 100 iterations; convergence was then achieved. For this matrix it can be shown that `M = diag(diag(A))` is a good preconditioner. Supplying this preconditioner as a fifth argument leads to a useful reduction in the number of iterations:

```
>> [x,flag,relres,iter] = pcg(A,b,1e-6,100,diag(diag(A)));
>> flag, relres, iter
```

Table 9.2. *Iterative linear equation solvers.*

Function	Matrix type	Method
`bicg`	General	BiConjugate gradient method
`bicgstab`	General	BiConjugate gradient stabilized method
`cgs`	General	Conjugate gradient squared method
`gmres`	General	Generalized minimum residual method
`lsqr`	General	Conjugate gradients on normal equations
`minres`	Hermitian	Minimum residual method
`pcg`	Hermitian pos. def.	Preconditioned conjugate gradient method
`qmr`	General	Quasi-minimal residual method
`symmlq`	Hermitian	Symmetric LQ method

```
flag =
     0
relres =
  8.2661e-007
iter =
    28
```

Notice that when more than one output argument is requested the messages are suppressed. A zero value of `flag` denotes convergence with relative residual `relres = norm(b-A*x)/norm(b)` after `iter` iterations.

All the other functions in Table 9.2 have the same calling sequence as `pcg` with the exception of `gmres`, which has an extra argument `restart` in the third position that specifies at which iteration to restart the method.

Function `eigs` computes a few selected eigenvalues and eigenvectors for the standard eigenvalue problem or for the symmetric definite generalized eigenvalue problem $Ax = \lambda Bx$ with B real and symmetric positive definite. This is in contrast to `eig`, which always computes the full eigensystem. Like the iterative linear equation solvers, `eigs` needs just the ability to form matrix–vector products, so A can be given either as an explicit matrix or as a function that performs matrix–vector products. In its simplest form, `eigs` can be called in the same way as `eig`, with `[V,D] = eigs(A)`, when it computes the six eigenvalues of largest magnitude and the corresponding eigenvectors. See `doc eigs` for more details and examples of usage. This function is an interface to the ARPACK package [65]. As an example, we form a sparse symmetric matrix and compute its five algebraically largest eigenvalues using `eigs`. For comparison, we also apply `eig`, which requires that the matrix first be converted to a full matrix and always computes all the eigenvalues:

```
>> A = delsq(numgrid('N',40));
>> n = length(A)
n =
        1444

>> nnz(A)/n^2 % Percentage of nonzeros
ans =
    0.0034
```

```
>> tic, e_all = eig(full(A))'; toc
Elapsed time is 5.093000 seconds.
>> e_all(n:-1:n-4)
ans =
    7.9870    7.9676    7.9676    7.9482    7.9354

>> options.disp = 0; % Turn off intermediate output.
>> tic, e_big = eigs(A,5,'LA',options)'; toc % LA = largest algebraic
Elapsed time is 0.406000 seconds.
>> e_big
e_big =
    7.9870    7.9676    7.9676    7.9482    7.9354
```

The `tic` and `toc` functions provide an easy way of timing (in seconds) the code that they surround. Clearly, `eigs` is much faster than `eig` in this example, and it also uses much less storage.

A corresponding function `svds` computes a few singular values and singular vectors of an m-by-n matrix A. It does so by applying `eigs` to the Hermitian matrix

$$\begin{bmatrix} 0_m & A \\ A^* & 0_n \end{bmatrix}.$$

9.9. Functions of a Matrix

As mentioned in Section 5.3, some of the elementary functions defined for arrays have counterparts defined in the matrix sense, implemented in functions whose names end in `m`. The three main examples are `expm`, `logm`, and `sqrtm`. The exponential of a square matrix A can be defined by

$$e^A = I + A + \frac{A^2}{2!} + \frac{A^3}{3!} + \cdots$$

It is computed by `expm`. The logarithm of a matrix is an inverse to the exponential. A nonsingular matrix has infinitely many logarithms. Function `logm` computes the principal logarithm, which, for a nonsingular matrix with no negative real eigenvalues, is the logarithm whose eigenvalues have imaginary parts lying strictly between $-\pi$ and π.

A square root of a square matrix A is a matrix X for which $X^2 = A$. Every nonsingular matrix has at least two square roots. Function `sqrtm` computes the principal square root, which, for a nonsingular matrix with no negative real eigenvalues, is the square root with eigenvalues having positive real part.

We give some examples. The matrix

```
A =
    17     8     1     0
     8    18     8     1
     1     8    18     8
     0     1     8    17
```

has a tridiagonal square root:

```
>> format short g, sqrtm(A)
ans =
             4             1  4.4409e-016 -7.7716e-016
             1             4            1  1.1102e-015
   2.2204e-016             1            4            1
  -4.4409e-016   8.8818e-016            1            4
```

The Jordan block

```
>> A = gallery('jordbloc',4,1)
A =
     1     1     0     0
     0     1     1     0
     0     0     1     1
     0     0     0     1
```

has exponential

```
>> X = expm(A)
X =
         2.7183        2.7183        1.3591       0.45305
              0        2.7183        2.7183        1.3591
              0             0        2.7183        2.7183
              0             0             0        2.7183
```

and we can recover the original matrix using `logm`:

```
>> logm(X)
ans =
              1             1    -4.12e-018 -9.8256e-020
              0             1             1    -4.12e-018
              0             0             1            1
              0             0             0            1
```

The function `funm` computes general matrix functions, using a method based on the Schur decomposition (MATLAB 7 has a new version of this function that is more reliable than earlier versions). For the functions exp, log, cos, sin, cosh, and sinh, `funm` can be used as in the following example, which computes $\cos(A)$ and $\sin(A)$ and checks that $\cos(A)^2 + \sin(A)^2 = I$:

```
>> format short, A = gallery('frank',3)
A =
     3     2     1
     2     2     1
     0     1     1

>> X = funm(A,@cos), Y = funm(A,@sin)
X =
    0.3362   -0.2983   -0.1021
   -0.3926    0.6267   -0.1963
    0.1885   -0.3847    0.6345
Y =
   -0.2455   -0.8062   -0.5435
```

```
   -0.5255   -0.2635   -0.2628
   -0.5615    0.2987    0.5607

>> residual = norm(X^2+Y^2 - eye(3),1)
residual =
  1.5751e-015
```

To compute other functions, say $f(A)$ in general, it is necessary to write a separate function of the form

```
function fd = fun(x,k)
```

that accepts a vector `x` and integer `k` and returns the `k`th derivative of f evaluated at `x`. (**`funm`** has such functions built in for the cases listed above.) The derivatives are needed when the matrix has repeated (and, in finite precision, close) eigenvalues.

Nichols: "Transparent aluminum?"
Scott: "That's the ticket, laddie."
Nichols: "It'd take years just to figure out the dynamics of this matrix."
McCoy: "Yes, but you would be rich beyond the dreams of avarice!"

— *Star Trek IV: The Voyage Home* (Stardate 8390)

We share a philosophy about linear algebra:
we think basis-free,
we write basis-free,
but when the chips are down we close the office door and
compute with matrices like fury.

— IRVING KAPLANSKY, *Reminiscences [of Paul Halmos]* (1991)

The matrix of that equation system is negative *definite—which is a*
positive definite system that has been multiplied through by -1.
For all practical geometries the common finite difference
Laplacian operator gives rise to these,
the best of all possible matrices.
Just about any standard solution method will succeed,
and many theorems are available for your pleasure.

— FORMAN S. ACTON, *Numerical Methods That Work* (1970)

Chapter 10
More on Functions

10.1. Function Handles

Many problems tackled with MATLAB require one function to be passed as an argument to another. The usual mechanism for doing this is through a function handle. A function handle is a MATLAB data type that contains all the information necessary to evaluate a function. It can be created by putting the `@` character before a function name.

We illustrate using `ezplot`, which plots a function $f(x)$ over a default range of $[-2\pi, 2\pi]$. If `fun` is a function M-file of the form required by `ezplot` then we can write

```
ezplot(@fun)
```

As well as being an M-file, `fun` can be the name of a built-in function:

```
ezplot(@sin)
```

Function handles were introduced in MATLAB 6 and they supersede an earlier syntax in which a function name was passed in a string, as in `ezplot('exp')`.

A function that accepts another function as an argument will need to evaluate the passed function. Evaluation is achieved simply by treating the function handle as if it were a function name and appending a list of arguments to it. If the function being called takes no input arguments then empty parentheses are required after the function handle name. Consider the function `fd_deriv` in Listing 10.1. This function evaluates the finite difference approximation

$$\frac{f(x+h)-f(x)}{h} \approx f'(x)$$

to the function passed as its first argument. When we type

```
>> fd_deriv(@sqrt,0.1)
ans =
    1.5811
```

the first `f` call in `fd_deriv` is equivalent to `sqrt(x+h)`. We can use our N
root function `sqrtn` (Listing 7.5) instead of the built-in square roo

```
>> fd_deriv(@sqrtn,0.1)
 k              x_k              rel. change
 1:  5.5000000745058064e-001   8.18e-001
% Remaining output from sqrtn omitted.
ans =
    1.5811
```

The direct evaluation of function handles is new to MATLAB 7. In earlier versions of MATLAB it was necessary to use `feval` to evaluate the passed function. The syntax is `feval(fun,x1,x2,...,xn)`, where `fun` is the function handle (or a function name) and `x1, x2, ..., xn` are its arguments. The function `fd_deriv2` in Listing 10.1 is a modified version of `fd_deriv` that uses `feval`.

One difference between direct evaluation of function handles and `feval` is that the latter works for function names passed as strings, while the former does not:

```
>> fd_deriv2('sqrt',0.1)
ans =
    1.5811

>> fd_deriv('sqrt',0.1)
??? Subscript indices must either be real positive integers or
    logicals.

Error in ==> fd_deriv.m
On line 8  ==> y = (f(x+h) - f(x))/h;
```

One way to modify `fd_deriv` so that it accepts function names passed as strings is to insert

```
f = fcnchk(f);
```

at the start of the function. `fcnchk` is a helper function for some of MATLAB's functions in the directory `matlab\funfun` and it can convert the string name of a function into the corresponding function handle. If the function `fcnchk` is given a final argument `'vectorized'`, as in `fcnchk(f,'vectorized')`, then it vectorizes the string `f`: it converts multiplication, exponentiation, and division into array operations, so that vector and matrix arguments can be used.

10.2. Anonymous Functions

Anonymous functions, introduced in MATLAB 7, provide a way of creating a "one line" function without writing an M-file. For example:

```
>> f = @(x) exp(x)-1
f =
    @(x) exp(x)-1
>> f(2)
ans =
    6.3891
```

Here, `f` is a function handle to the anonymous function. The `@` character, which constructs the function handle, is followed by a list of input arguments to the function in parentheses, and then by a single MATLAB expression. Being a function handle, n anonymous function can be passed to other functions.

The expression that evaluates an anonymous function can contain variables not he argument list. The values of such variables are captured when the function is ted and they are held constant throughout the life of the function, as the next ple of a three-argument function illustrates:

Listing 10.1. *Functions* fd_deriv *and* fd_deriv2 *(for MATLAB 6.5 and earlier).*

```
function y = fd_deriv(f,x,h)
%FD_DERIV   Finite difference approximation to derivative.
%           FD_DERIV(F,X,H) is a finite difference approximation
%           to the derivative of function F at X with difference
%           parameter H.  H defaults to SQRT(EPS).

if nargin < 3, h = sqrt(eps); end
y = (f(x+h) - f(x))/h;
```

```
function y = fd_deriv2(f,x,h)
%FD_DERIV2  Finite difference approximation to derivative.
%           FD_DERIV2(F,X,H) is a finite difference approximation
%           to the derivative of function F at X with difference
%           parameter H.  H defaults to SQRT(EPS).

if nargin < 3, h = sqrt(eps); end
y = (feval(f,x+h) - feval(f,x))/h;
```

```
>> alpha = 1;
>> g = @(x,y,z) x^2+y^2-alpha*z^2;
>> g(1,2,3)
ans =
    -4

>> alpha = 0;
>> g(1,2,3)
ans =
    -4
```

In order for the changed value of alpha to be reflected in g, the anonymous function must be reconstructed.

One of the advantages of anonymous functions can be seen in connection with the example from the previous section in which we invoked fd_deriv on the function sqrtn. Recall from Listing 7.5 that sqrtn has a second input argument that specifies a convergence tolerance. Suppose we wish to use a different tolerance (say 1e-14) when sqrtn is called by fd_deriv. This does not seem possible, because fd_deriv invokes its input function with only one argument. A way round this difficulty is to set up an anonymous function that calls sqrtn with the required convergence tolerance and pass this one-argument function to fd_deriv:

```
>> fd_deriv(@(x) sqrtn(x,1e-14), 0.1)
 k           x_k             rel. change
 1:  5.5000000745058064e-001  8.18e-001
% Remaining output from sqrtn omitted.
ans =
    1.5811
```

Note that here we set up the anonymous function within the call to **fd_deriv**. This technique is very useful when calling the MATLAB "function-function" routines that are described in the next two chapters.

10.3. Inline Objects

A mathematical function can be passed to a MATLAB function in another way besides via the function handle of an appropriate function: as an inline object. Inline objects were introduced in MATLAB 6 and are largely superseded by the more versatile and efficient anonymous functions introduced in MATLAB 7. We describe them here because the reader is likely to meet them in existing MATLAB codes, but we use anonymous functions in preference throughout the book. For an indication of the greater efficiency of anonymous functions over inline functions, see the Fresnel integrals example in Section 12.1.

An inline object is essentially a "one line" function defined by a string and it can be assigned to a variable and then evaluated or passed to another function:

```
>> f = inline('exp(x)-1')
f =
     Inline function:
     f(x) = exp(x)-1
>> f(2)
ans =
    6.3891

>> ezplot(f)
```

Note that `f` here is not a function handle.

MATLAB automatically determines and orders the arguments to an inline function. If its choice is not suitable then the arguments can be explicitly defined and ordered via extra arguments to `inline`:

```
>> f = inline('log(a*x)/(1+y^2)')
f =
     Inline function:
     f(a,x,y) = log(a*x)/(1+y^2)

>> f = inline('log(a*x)/(1+y^2)','x','y','a')
f =
     Inline function:
     f(x,y,a) = log(a*x)/(1+y^2)
```

We can pass an inline object to `fd_deriv` in Listing 10.1:

```
>> f = inline('exp(-x)/(1+x^2)');
>> fd_deriv(f,pi)
ans =
   -0.0063
```

It is sometimes necessary to vectorize an inline object or a string expression. This can be done with the `vectorize` function:

```
>> f = inline('log(a*x)/(1+y^2)');
>> f = vectorize(f)
f =

     Inline function:
     f(a,x,y) = log(a.*x)./(1+y.^2)
```

10.4. Subfunctions

A function M-file may contain other functions, called subfunctions, which appear in any order after the main (or primary) function. Subfunctions are visible only to the main function and to any other subfunctions. They typically carry out a task that needs to be separated from the main function but that is unlikely to be needed in other M-files, or they may override existing functions of the same names (since subfunctions take precedence). The use of subfunctions helps to avoid proliferation of M-files.

For an example of a subfunction see `poly1err`[5] in Listing 10.2, which approximates the maximum error in the linear interpolating polynomial to subfunction `f` on $[0, 1]$ based on `n` sample points on the interval:

```
>> poly1err(5)
ans =
    0.0587

>> poly1err(50)
ans =
    0.0600
```

Alternative ways to code `poly1err` are to define `f` as an anonymous function rather than a subfunction, or to make `f` an input argument.

Another example is shown in Listing 10.3, some graphical output from which is displayed in Figure 10.1. It could be argued that the subfunction `spiro` in this example is unnecessary and that the subfunction should be "inlined". Our reason for retaining the subfunction is to remind us that the formulae underlying `rosy` are a special case of more complicated formulae—a fact that would be less clear after inlining.

Help for a subfunction is displayed by specifying the main function name followed by ">" and the subfunction name. Thus help for subfunction `f` of `poly1err` is listed by

```
>> help poly1err>f

  F       Function to be interpolated, F(X).
```

A subfunction can be passed to another function as a function handle. Thus, for example, in the main body of `poly1err` we can write `ezplot(@f)` in order to plot the subfunction `f`.

For less trivial examples of subfunctions see Chapter 12.

[5]This function is readily vectorized; see Section 20.1.

Listing 10.2. *Function* poly1err.

```
function max_err = poly1err(n)
%POLY1ERR   Error in linear interpolating polynomial.
%           POLY1ERR(N) is an approximation based on N sample points
%           to the maximum difference between subfunction F and its
%           linear interpolating polynomial at 0 and 1.

max_err = 0;
f0 = f(0); f1 = f(1);
for x = linspace(0,1,n)
    p = x*f1 + (x-1)*f0;
    err = abs(f(x)-p);
    max_err = max(max_err,err);
end

% Subfunction.
function y = f(x)
%F       Function to be interpolated, F(X).
y = sin(x);
```

Listing 10.3. *Function* rosy.

```
function rosy(a, b)
%ROSY    "Rose" figures.
%         ROSY(A, B) plots the curve
%            X = R*COS(A*theta), Y = R*SIN(A*theta), where
%            R = SIN(A*B*theta) and 0 <= theta <= 2*PI (360 values).
%         Suggestions: ROSY(97, 5); ROSY(43, 4); ROSY(79, n9), n a digit.

%         P. M. Maurer, A rose is a rose..., Amer. Math. Monthly, 94 (1987),
%         pp. 631-645.

if nargin < 2, b = 1; end
if nargin < 1, a = 1; end

c = 0; d = 1; p = a*b;
[x, y] = spiro(a, a, c, d, p, .5);

plot(x,y)
axis square, axis off

% Subfunction.
function [x, y] = spiro(a, b, c, d, p, k)
h = k*2*pi/180;
t = (0:h:2*pi)';
r = c + d*sin(t*p);
x = r.*cos(a*t);
y = r.*sin(b*t);
```

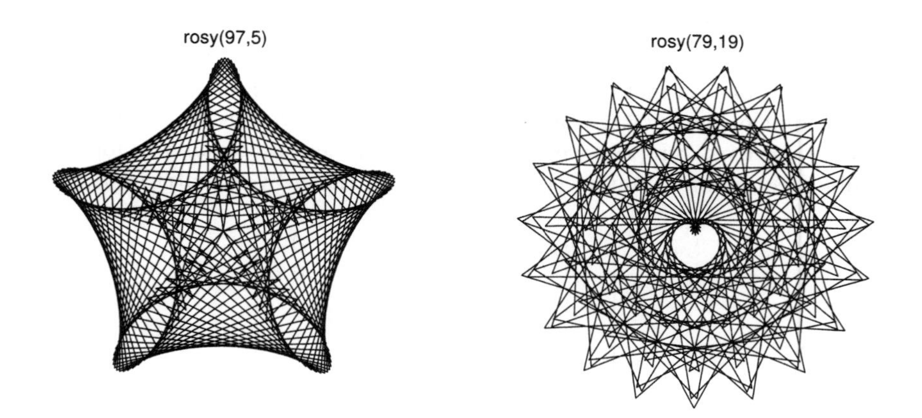

Figure 10.1. *Sample output from* `rosy`.

10.5. Default Input Arguments

As we saw in Section 7.1, by using **nargin** a function can be written so that only the first few of its input arguments need be supplied on a particular call, the rest being set to default values within the function. When a function is designed, its input arguments should therefore be ordered starting with those that must be supplied and continuing with those that can or can not be specified, in decreasing order of importance. To allow full flexibility a function can be written so that the user can request default values to be defined for arguments that occur before the last argument specified. The idea is for the empty matrix `[]` to be supplied as input argument and for **isempty** to be used to detect an empty input argument.

Consider the following snippet from a function:

```
function f = fun_ex(A, npts, nangles, step)
if nargin < 2 | isempty(npts), npts = length(A); end
if nargin < 3 | isempty(nangles), nangles = 10; end
if nargin < 4, step = 0.1; end
```

Here, the array `A` must be supplied, but the other arguments are all optional. Example invocations of the function are:

```
funarg_ex(rand(2),10,20,1e-3)
funarg_ex(rand(2),10)
funarg_ex(rand(2),[],50)
funarg_ex(rand(2),[],[],0.5)
```

When the empty matrix `[]` is passed as the second or third input argument the **isempty** part of the corresponding **if** test ensures that the default value is assigned to the relevant variable.

This technique is used by numerous MATLAB functions (for example, `lsqr` and **quad**). An alternative to a long list of input arguments, all of which are equally likely to be required to take default values, is to use a structure to specify some of the arguments. This is done by several of MATLAB's numerical methods functions; see Chapters 11 and 12.

10.6. Variable Numbers of Arguments

In certain situations a function must accept or return a variable, possibly unlimited, number of input or output arguments. This can be achieved using the **varargin** and **varargout** functions. Suppose we wish to write a function **companb** to form the mn-by-mn block companion matrix corresponding to the n-by-n matrices A_1, A_2, ..., A_m:

$$C = \begin{bmatrix} -A_1 & -A_2 & \dots & \dots & -A_m \\ I & 0 & & & 0 \\ & I & \ddots & & \vdots \\ & & \ddots & 0 & \vdots \\ & & & I & 0 \end{bmatrix}.$$

We could use a standard function definition such as

```
function C = companb(A_1,A_2,A_3,A_4,A_5)
```

but m is then limited to 5 and handling the different values of m between 1 and 5 is tedious. The solution is to use **varargin**, as shown in Listing 10.4. When **varargin** is specified as the input argument list, the input arguments supplied are copied into a cell array called **varargin**. Cell arrays, described in Section 18.3, are a special kind of array in which each element can hold a different type and size of data. The elements of a cell array are accessed using curly braces. Consider the call

```
>> X = ones(2); C = companb(X, 2*X, 3*X)
C =
    -1    -1    -2    -2    -3    -3
    -1    -1    -2    -2    -3    -3
     1     0     0     0     0     0
     0     1     0     0     0     0
     0     0     1     0     0     0
     0     0     0     1     0     0
```

If we insert the line

```
varargin
```

at the beginning of **companb** then the above call produces

```
varargin =
    [2x2 double]    [2x2 double]    [2x2 double]
```

Thus **varargin** is a 1-by-3 cell array whose elements are the 2-by-2 matrices passed as arguments to **companb**, and **varargin{j}** is the jth input matrix, A_j.

It is not necessary for **varargin** to be the only input argument but it must be the last one, appearing after any named input arguments.

An example using the analogous statement **varargout** for output arguments is shown in Listing 10.5. Here we use **nargout** to determine how many output arguments have been requested and then create a **varargout** cell array containing the required output. (It is the curly brackets on the right-hand side of the assignment statement that make **varargout** a cell array.) To illustrate:

Listing 10.4. *Function* companb.

```
function C = companb(varargin)
%COMPANB     Block companion matrix.
%            C = COMPANB(A_1,A_2,...,A_m) is the block companion matrix
%            corresponding to the n-by-n matrices A_1,A_2,...,A_m.

m = nargin;
n = length(varargin{1});

C = diag(ones(n*(m-1),1),-n);
for j = 1:m
    Aj = varargin{j};
    C(1:n,(j-1)*n+1:j*n) = -Aj;
end
```

Listing 10.5. *Function* moments.

```
function varargout = moments(x)
%MOMENTS  Moments of a vector.
%         [m1,m2,...,mk] = MOMENTS(X) returns the first, second, ...,
%         k'th moments of the vector X, where the j'th moment
%         is SUM(X.^j)/LENGTH(X).

for j = 1:nargout, varargout(j) = {sum(x.^j)/length(x)}; end
```

```
>> m1 = moments(1:4)
m1 =
    2.5000

>> [m1,m2,m3] = moments(1:4)
m1 =
    2.5000
m2 =
    7.5000
m3 =
    25
```

10.7. Nested Functions

MATLAB allows one or more functions to be nested wholly inside another. To define the nesting, **end** statements must be placed at the end of the nested functions and the main function (otherwise the functions are subfunctions), and it is good style to indent the nested functions.

Nested functions have two key properties:

- A nested function has access to the workspaces of all functions within which it is nested.

- A function handle for a nested function stores the information needed to access the nested function *and* the values of any variables in functions containing the nested function ("externally scoped" variables) that are needed to evaluate it.

An example of a nested function is given in `rational_ex` in Listing 10.6, which makes use of `fd_deriv` in Listing 10.1. The example illustrates how a function depending on parameters can be passed to another function. In the body of `rational_ex` a function handle to the nested function `rational` is passed to `fd_deriv`. The variables `a`, `b`, `c`, and `d` from the main function's workspace are available inside `rational` and their values are encapsulated in the function handle that is passed to `fd_deriv`. Even though these four variables are not in the scope of `fd_deriv`, this function can correctly evaluate `rational` in terms of the values of the variables at the time the function handle was created:

```
>> rational_ex(2)
ans =
    3.0000
```

This example could be rewritten using an anonymous function, as in the example involving `sqrtn` in Section 10.2. However, the anonymous function approach is applicable only when the function is given by a single expression, which is quite limiting. Nested functions have the advantage that they enable a single function (containing nested functions) to be written to solve a complete problem involving a parametrized function; we will use them extensively for this purpose in Chapters 12 and 22.

Nested functions are new to MATLAB 7, and many existing functions can profitably be converted to use them. For example, a function with many subfunctions will usually need to pass data to the subfunctions when it calls them. If much of this data consists of variables (rather than expressions) then it is advantageous to convert the subfunctions into nested functions so that the variables are automatically available inside the subfunctions and need not be passed.

The precise scoping rules of nested functions, can be found in the online MATLAB documentation.

For further examples of nested functions, see Chapters 12 and 22.

10.8. Private Functions

A typical MATLAB installation contains hundreds of M-files on the user's path, all accessible just by typing the name of the M-file. While this ease of accessing M-files is an advantage, it can lead to clutter and clashes of names, not least due to the presence of "helper functions" that are used by other functions but not intended to be called directly by the user. Private functions provide an elegant way to avoid these problems. Any functions residing in a directory called `private` are visible only to functions in the parent directory. They can therefore have the same names as functions in other directories. When MATLAB looks for a function it searches subfunctions, then private functions (relative to the directory in which the function making the call is located), then the current directory and the path. Hence if a private function has the same name as a nonprivate function (even a built-in function), the private function will be found first.

Listing 10.6. *Function* rational_ex.

```
function rational_ex(x)
%RATIONAL_EX   Illustration of nested function.

a = 1; b = 2; c = 1; d = -1;
fd_deriv(@rational,x)

   function r = rational(x)
   % Rational function.
   r = (a+b*x)/(c+d*x);
   end

end
```

Good use of private functions is made by the **gallery** function, which lives in the matlab\elmat directory and provides a collection of test matrices (see Section 5.1). The 50 or so matrix-generating functions invoked by **gallery** live in **matlab\elmat\private**, and their names could be chosen without fear of clashing with an existing function (perhaps one in a toolbox).

Private directories should not be put on the path.

Help for a private function **fun** can be accessed using **help private\fun**.

10.9. Recursive Functions

Functions can be recursive, that is, they can call themselves, as we have seen with function gasket in Listing 1.7 and function land in Listing 8.1. Recursion is a powerful tool, but not all computations that are described recursively are best programmed this way.

The function koch in Listing 10.7 uses recursion to draw a Koch curve [90, Sec. 2.4]. The basic construction in koch is to replace a line by four shorter lines. The upper left-hand picture in Figure 10.2 shows the four lines that result from applying this construction to a horizontal line. The upper right-hand picture then shows what happens when each of these four lines is processed. The lower left- and right-hand pictures show the next two levels of recursion.

We see that koch has three input arguments. The first two, pl and pr, give the (x, y) coordinates of the current line and the third, level, indicates the level of recursion required. If level $= 0$ then a line is drawn; otherwise koch calls itself four times with level one less and with endpoints that define the four shorter lines.

Figure 10.2 was produced by the following code:

```
pl = [0;0]; % Left endpoint
pr = [1;0]; % Right endpoint

for k = 1:4
   subplot(2,2,k)
   koch(pl,pr,k)
   axis('equal')
```

Listing 10.7. *Function* koch.

```
function koch(pl,pr,level)
%KOCH   Recursively generated Koch curve.
%       KOCH(PL, PR, LEVEL) recursively generates a Koch curve,
%       where PL and PR are the current left and right endpoints and
%       LEVEL is the level of recursion.

if level == 0
  plot([pl(1),pr(1)],[pl(2),pr(2)]); % Join pl and pr.
  hold on
else
  A = (sqrt(3)/6)*[0 1; -1 0];       % Rotate/scale matrix.

  pmidl = (2*pl + pr)/3;
  koch(pl,pmidl,level-1)             % Left branch.

  ptop = (pl + pr)/2 + A*(pl-pr);
  koch(pmidl,ptop,level-1)           % Left mid branch.

  pmidr = (pl + 2*pr)/3;
  koch(ptop,pmidr,level-1)           % Right mid branch.

  koch(pmidr,pr,level-1)             % Right branch.

end
```

```
   title(['Koch curve: level = ' num2str(k)],'FontSize',16)
end
hold off
```

To produce Figure 10.3 we called **koch** with pairs of endpoints equally spaced around the unit circle, so that each edge of the snowflake is a copy of the same Koch curve. The relevant code is

```
level = 4; edges = 7;

for k = 1:edges
    pl = [cos(2*k*pi/edges); sin(2*k*pi/edges)];
    pr = [cos(2*(k+1)*pi/edges); sin(2*(k+1)*pi/edges)];
    koch(pl,pr,level)
end
axis('equal')
title('Koch snowflake','FontSize',16,'FontAngle','italic')
hold off
```

For other examples of recursion, see the function lsys in Listing 22.6 and quad and quadl described in Section 12.1.

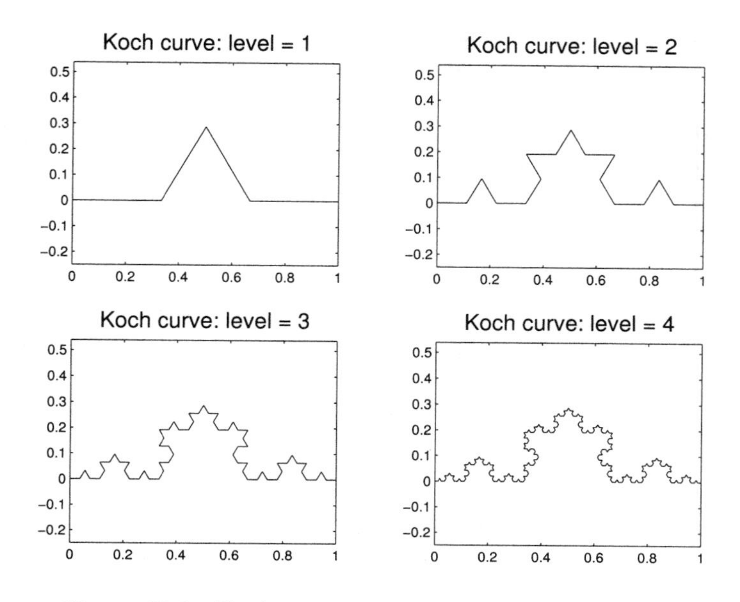

Figure 10.2. *Koch curves created with function* `koch`.

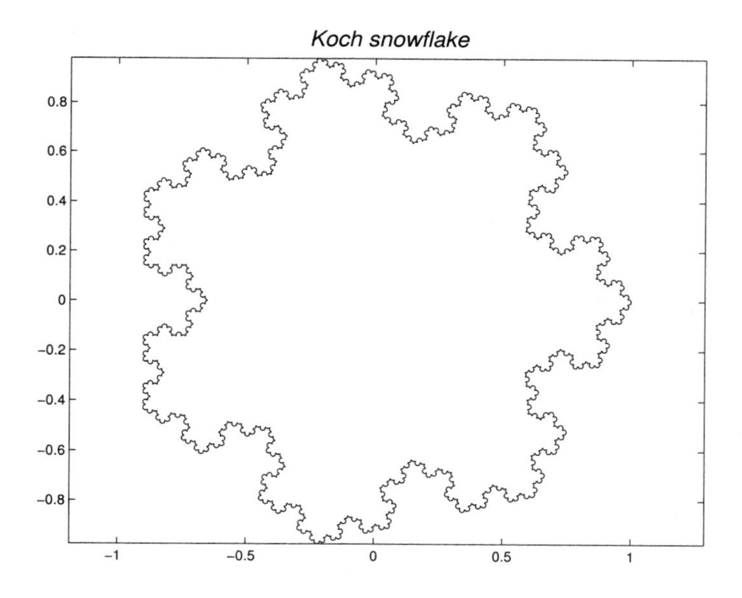

Figure 10.3. *Koch snowflake created with function* `koch`.

10.10. Global Variables

Variables within a function are local to that function's workspace. Occasionally it is convenient to create variables that exist in more than one workspace including, possibly, the main workspace. This can be done using the `global` statement. As an example, suppose we wish to study plots of the function $f(x) = 1/(a + (x - b)^2)$ on $[0, 1]$ for various a and b. We can define a function

```
function f = myfun(x)
global A B
f = 1/(A + (x-B)^2);
```

At the command line we type

```
>> global A B
>> A = 0.01; B = 0.5;
>> fplot(@myfun,[0 1])
```

The values of `A` and `B` set at the command line are available within `myfun`. New values for `A` and `B` can be assigned and `fplot` called again without editing `myfun.m`. Note that it is possible to avoid the use of `global` in this example by passing the parameters a and b through the argument list of `fplot`, as in the example on p. 99.

Within a function, the `global` declaration should appear before the first occurrence of the relevant variables, ideally at the top of the file. By convention the names of global variables are comprised of capital letters, and ideally the names are long in order to reduce the chance of clashes with other variables.

The use of global variables is not recommended. Anonymous and nested functions provide one way to avoid them.

10.11. Exemplary Functions in MATLAB

Perhaps the best way to learn how to write functions is by studying well-written examples. An excellent source of examples is MATLAB itself, since all functions that are not built into the interpreter are M-files that can be examined. We list below some M-files that illustrate particular aspects of MATLAB programming. The source can be viewed with `type function_name`, by loading the file into the Editor/Debugger with `edit function_name` (the editor searches the path for the function, so the pathname need not be given), or by loading the file into your favorite editor (in which case you will need to know the path, which we indicate).

- `datafun/cov`: use of `varargin`.
- `datafun/var`: argument checking.
- `elmat/hadamard`: matrix building.
- `elmat/why`: `switch` construct, subfunctions.
- `funfun/fminbnd`: argument checking, loop constructs.
- `funfun/quad`, `funfun/quadl`: recursive functions.
- `matfun/gsvd`: subfunctions.

- `funfun/ode45`: use of `varargin` and `varargout`.
- `sparfun/pcg`: sophisticated argument handling and error checking.

In this example, ALEVIL is a function name being passed to ROOT and MONEY is the ROOT of ALEVIL.

— ROGER EMANUEL KAUFMAN, *A FORTRAN Coloring Book* (1978)

Use recursive procedures for recursively-defined data structures.

— BRIAN W. KERNIGHAN and P. J. PLAUGER,
The Elements of Programming Style (1978)

Great fleas have little fleas upon their backs to bite 'em,
And little fleas have lesser fleas and so ad infinitum.
And the great fleas themselves, in turn, have greater fleas to go on;
While these again have greater still, and greater still, and so on.

— AUGUSTUS DEMORGAN

Chapter 11
Numerical Methods: Part I

This chapter describes MATLAB's functions for solving problems involving polynomials, nonlinear equations, optimization, and the fast Fourier transform. In many cases a function `fun` must be passed as an argument. As described in Section 10.1, `fun` can be a function handle (including an anonymous function) or a string expression. The MATLAB functions described in this chapter place various demands on the function that is to be passed, but most require it to return a vector of values when given a vector of inputs.

When a function `fun` is passed to and evaluated by another MATLAB function, it is sometimes necessary to communicate problem parameters to `fun`. In versions of MATLAB prior to MATLAB 7 this was done through the argument lists. Thanks to the anonymous functions and nested functions introduced in MATLAB 7, passing parameters through arguments lists is no longer necessary, and indeed it is now not recommended. All our examples involving problem parameters make use of anonymous or nested functions.

Several functions described in this chapter make use of structures, which are one of MATLAB's data types: see Section 18.3 for details of structures.

For mathematical background on the methods described in this and the next chapter suitable textbooks are [6], [7], [17], [27], [47], [83], [101], [120].

11.1. Polynomials and Data Fitting

MATLAB represents a polynomial

$$p(x) = p_1 x^n + p_2 x^{n-1} + \cdots + p_n x + p_{n+1}$$

by a row vector `p = [p(1) p(2) ... p(n+1)]` of the coefficients. (Note that compared with the representation $\sum_{i=0}^{n} p_i x^i$ used in many textbooks, MATLAB's vector is reversed and its subscripts are increased by 1.)

Here are three problems related to polynomials:

Evaluation: Given the coefficients evaluate the polynomial at one or more points.

Root finding: Given the coefficients find the roots (the points at which the polynomial evaluates to zero).

Data fitting: Given a set of data $\{x_i, y_i\}_{i=1}^{m}$, find a polynomial that "fits" the data.

The standard technique for evaluating $p(x)$ is Horner's method, which corresponds to the nested representation

$$p(x) = \Big(\cdots\Big((p_1 x + p_2)x + p_3\Big)x + \cdots + p_n\Big)x + p_{n+1}.$$

Function polyval carries out Horner's method: y = polyval(p,x). In this command x can be a matrix, in which case the polynomial is evaluated at each element of the matrix (that is, in the array sense). Evaluation of the polynomial p in the matrix (as opposed to array) sense is defined for a square matrix argument X by

$$p(X) = p_1 X^n + p_2 X^{n-1} + \cdots + p_n X + p_{n+1} I.$$

The command Y = polyvalm(p,X) carries out this evaluation.

The roots (or zeros) of p are obtained with z = roots(p). Of course, some of the roots may be complex even if p is a real polynomial. The function poly carries out the converse operation: given a set of roots it constructs a polynomial. Thus if z is an n-vector then p = poly(z) gives the coefficients of the polynomial

$$p_1 x^n + p_2 x^{n-1} + \cdots + p_n x + p_{n+1} = (x - z_1)(x - z_2)\ldots(x - z_n).$$

(The normalization $p_1 = 1$ is always used.) The function poly also accepts a matrix argument: as explained in Section 9.7, for a square matrix A, p = poly(A) returns the coefficients of the characteristic polynomial $\det(xI - \mathtt{A})$.

Function polyder computes the coefficients of the derivative of a polynomial, but it does not evaluate the polynomial.

As an example, consider the quadratic $p(x) = x^2 - x - 1$. First, we find the roots:

```
>> format short g, p = [1 -1 -1]; z = roots(p)
z =
          1.618
       -0.61803
```

The next command verifies that these are roots, up to roundoff:

```
>> polyval(p,z)
ans =
   2.2204e-016
  -1.1102e-016
```

Next, we observe that a certain 2-by-2 matrix has p as its characteristic polynomial:

```
>> A = [0 1; 1 1]; cp = poly(A)
cp =
             1            -1            -1
```

The Cayley–Hamilton theorem says that every matrix satisfies its own characteristic polynomial. This is confirmed, modulo roundoff, for our matrix:

```
>> polyvalm(cp, A)
ans =
  -2.2204e-016             0
             0 -2.2204e-016
```

Polynomials can be multiplied and divided using conv and deconv, respectively. When a polynomial g is divided by a polynomial h there is a quotient q and a remainder r: $g(x) = h(x)q(x) + r(x)$, where the degree of r is less than that of h. The syntax for deconv is [q,r] = deconv(g,h). In the following example we divide $x^3 - 6x^2 + 12x - 8$ by $x - 2$, obtaining quotient $x^2 - 4x + 4$ and zero remainder. Then we reproduce the original polynomial using conv.

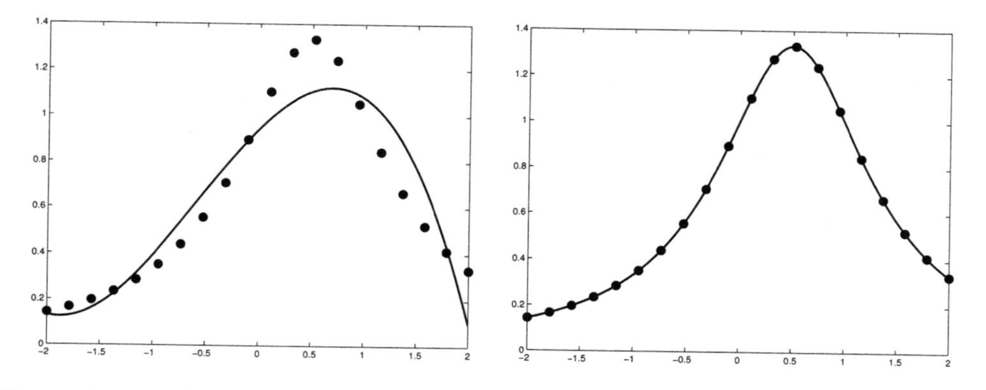

Figure 11.1. *Left: least squares polynomial fit of degree 3. Right: cubic spline. Data is from* $1/(x + (1 - x)^2)$.

```
>> g = [1 -6 12 -8]; h = [1 -2];

>> [q,r] = deconv(g,h)
q =
     1    -4     4
r =
     0     0     0     0

>> conv(h,q)
ans =
     1    -6    12    -8
```

The data fitting problem can be addressed with `polyfit`. Suppose the data $\{x_i, y_i\}_{i=1}^m$ has distinct x_i values, and we wish to find a polynomial p of degree at most n such that $p(x_i) \approx y_i$, $i = 1\colon m$. The `polyfit` function computes the least squares polynomial fit, that is, it determines p so that $\sum_{i=1}^m (p(x_i) - y_i)^2$ is minimized. The syntax is `p = polyfit(x,y,n)`. Specifying the degree n so that $n \geq m$ produces an interpolating polynomial, that is, $p(x_i) = y_i$, $i = 1\colon m$, so the polynomial fits the data exactly. However, high-degree polynomials can be extremely oscillatory, so small values of n are generally preferred. The following example computes and plots a least squares polynomial fit of degree 3. The data comprises the function $1/(x+(1-x)^2)$ evaluated at 20 equally spaced points on the interval $[-2, 2]$, generated by `linspace`. The resulting plot is on the left-hand side of Figure 11.1.

```
x = linspace(-2,2,20);
y = 1./(x+(1-x).^2);
p = polyfit(x,y,3);
xx = linspace(-2,2,100);
plot(x,y,'.',xx,polyval(p,xx),'-','MarkerSize',30,'LineWidth',2)
```

The `spline` function can be used if exact data interpolation is required. It fits a cubic spline, $sp(x)$, to the data $\{x_i, y_i\}_{i=1}^m$: $sp(x)$ has the following properties:

- it is a cubic polynomial between each pair of successive points x_i and x_{i+1} (i.e., it is a piecewise cubic polynomial),

- $sp(x_i) = y_i$, $i = 1\colon m$ (i.e., sp interpolates the data),
- it has continuous first and second derivatives at the points x_i (i.e., the cubic pieces join up smoothly).

In addition, the extra freedom in the spline is used up by enforcing the not-a-knot end conditions, the meaning of which can be found in the textbooks cited at the start of this section.

Given data vectors `x` and `y`, the command `yy = spline(x,y,xx)` returns in the vector `yy` the value of the spline at the points given by `xx`. The following code fits a cubic spline to the data in the polynomial example above. The resulting curve is on the right-hand side of Figure 11.1.

```
yy = spline(x,y,xx);
plot(x,y,'.',xx,yy,'-','MarkerSize',30,'LineWidth',2)
```

It is also possible to work with the coefficients of the spline curve. The command `pp = spline(x,y)` stores the coefficients in a structure that is interpreted by the `ppval` function, so `plot(x,y,'.',xx,ppval(pp,xx),'--')` would then produce the same plot as in the example above. Low-level manipulation of splines is possible with the functions `mkpp` and `unmkpp`.

MATLAB has another function for piecewise polynomial interpolation, `pchip`. This function produces a piecewise cubic $p(x)$ whose second derivative is generally not continuous. However, `pchip` has one very special property: it maintains both the shape and the monotonicity of the data. This means that on intervals where the data is monotonic, so is p, and at points where the data has a local extremum, so does p. To illustrate, consider this script:

```
x = [-12:4:12];
y = atan(x);
t = [-12:.1:12];
p = pchip(x,y,t);
s = spline(x,y,t);
plot(x,y,'o',t,p,'-',t,s,'-.','LineWidth',1.25)
xlim([-12 12])
legend('data','pchip','spline','Location','NW')
```

The script produces Figure 11.2. The spline is smooth, but oscillates between the first three and last three data points; `pchip` has no oscillations, at the expense of a slight loss of smoothness at the data points.

MATLAB has functions for interpolation in one, two, and more dimensions. For one-dimensional interpolation, the function `interp1` accepts `x(i),y(i)` data pairs and a further vector `xi`. It fits an interpolant to the data and then returns the values of the interpolant at the points in `xi`:

```
yi = interp1(x,y,xi)
```

The vector `x` must have monotonically increasing elements. Several types of interpolant are supported, as specified by a fourth input parameter, the main choices for which are

`'nearest'`	nearest neighbor interpolation
`'linear'`	linear interpolation (default)
`'spline'`	cubic spline interpolation
`'pchip'`	piecewise cubic Hermite interpolation

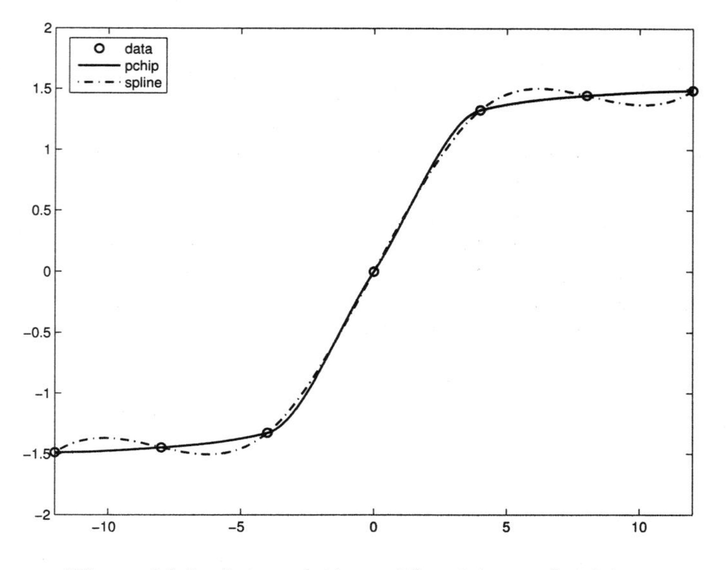

Figure 11.2. *Interpolation with* `pchip` *and* `spline`.

Linear interpolation puts a line between adjacent data pairs, while nearest neighbor interpolation reproduces the y-value of the nearest x point. The following example illustrates `interp1`.

```
x = [0 pi/4 3*pi/8 3*pi/4 pi]; y = sin(x);
xi = linspace(0,pi,40)';
yn = interp1(x,y,xi,'nearest');
yl = interp1(x,y,xi,'linear');
ys = interp1(x,y,xi,'spline');
yp = interp1(x,y,xi,'pchip');
xx = linspace(0,pi,50);
plot(xi,yn,'*', xi,yl,'+', xi,ys,'v', xi,yp,'o')
legend('nearest','linear','spline','pchip')
hold on
plot(xx,sin(xx),'-',x,y,'.k','MarkerSize',30)
set(gca,'XTick',x), set(gca,'XTickLabel','0|pi/4|3pi/8|3pi/4|pi')
set(gca,'XGrid','on')
axis([-0.25 3.5 -0.1 1.1])
hold off
```

This code samples 5 points from a sine curve on $[0, \pi]$, computes interpolants using the four methods above, and evaluates the interpolants at 40 points on the interval. In Figure 11.3 the solid circles plot the `x(i),y(i)` data pairs and the symbols plot the interpolants. The graphics commands are discussed in Chapters 8 and 17.

MATLAB has two functions for two-dimensional interpolation: `griddata` and `interp2`. The syntax for `griddata` is

```
ZI = griddata(x,y,z,XI,YI)
```

Here, the vectors `x`, `y`, and `z` are the data and `ZI` is a matrix of interpolated values

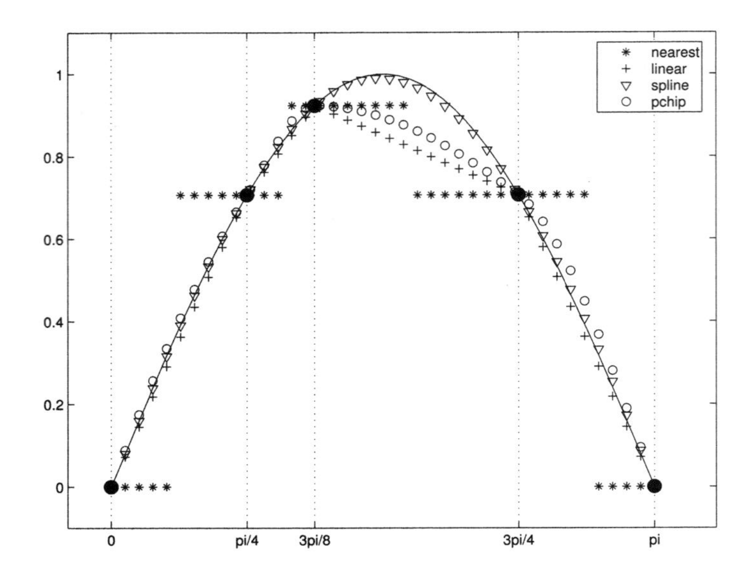

Figure 11.3. *Interpolating a sine curve at five points using* `interp1`.

corresponding to the matrices `XI` and `YI`, which are usually produced with `meshgrid`. A sixth string argument specifies the method:

`'linear'`	Triangle-based linear interpolation (default)
`'cubic'`	Triangle-based cubic interpolation
`'nearest'`	Nearest neighbor interpolation

Function `interp2` has a similar argument list, but it requires `x` and `y` to be monotonic matrices in the form produced by `meshgrid`. Here is an example in which we use `griddata` to interpolate values on a surface.

```
x = rand(100,1)*4-2; y = rand(100,1)*4-2;
z = x.*exp(-x.^2-y.^2);
hi = -2:.1:2;
[XI,YI] = meshgrid(hi);
ZI = griddata(x,y,z,XI,YI);
mesh(XI,YI,ZI), hold
plot3(x,y,z,'o'), hold off
```

The result is shown in Figure 11.4, which plots the original data points as circles and the interpolated surface as a mesh.

Other interpolation functions include `interp3`, `griddata3`, `interpn`, and `griddatan`, for three- and n-dimensional interpolation.

11.2. Nonlinear Equations

MATLAB has routines for finding a zero of a function of one variable (`fzero`) and for minimizing a function of one variable (`fminbnd`) or of n variables (`fminsearch`). In all

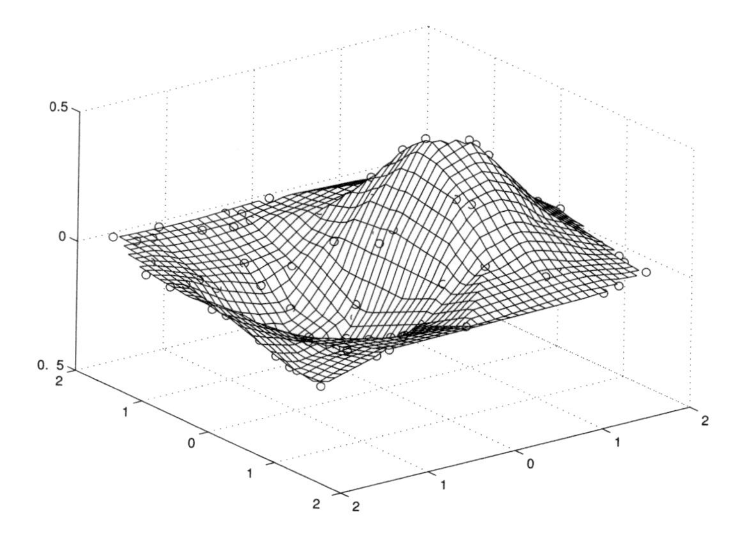

Figure 11.4. *Interpolation with* `griddata`.

cases the function must be real-valued and have real arguments. Unfortunately, there is no provision for directly solving a system of n nonlinear equations in n unknowns.[6]

The simplest invocation of `fzero` is `x = fzero(fun,x0)`, with `x0` a scalar, which attempts to find a zero of `fun` near `x0`. The function `fun` should be passed in one of two ways:

1. as the handle to an anonymous function: `fzero(@(x)cos(x)-x,x0)`,
2. as the handle to a function M-file: `fzero(@myfun,x0)`, where

```
function f = myfun(x)
f = cos(x)-x;
```

Thus to find a zero near 0, using an anonymous function:

```
>> fzero(@(x)cos(x)-x,0)
ans =
    0.7391
```

More precisely, `fzero` looks for a point where `fun` changes sign, and will not find zeros of even multiplicity. An initial search is carried out starting from `x0` to find an interval on which `fun` changes sign. The function `fun` must return a real scalar when passed a real scalar argument. Failure of `fzero` is signalled by the return of a NaN.

If, instead of being a scalar, `x0` is a 2-vector such that `fun(x0(1))` and `fun(x0(2))` have opposite sign, then `fzero` works on the interval defined by `x0`. Providing a starting interval in this way can be important when the function has a singularity. Consider the example

[6]However, an attempt at solving such a system could be made by minimizing the sum of squares of the residual. The Optimization Toolbox contains a nonlinear equation solver. See also the MATLAB codes provided with [52].

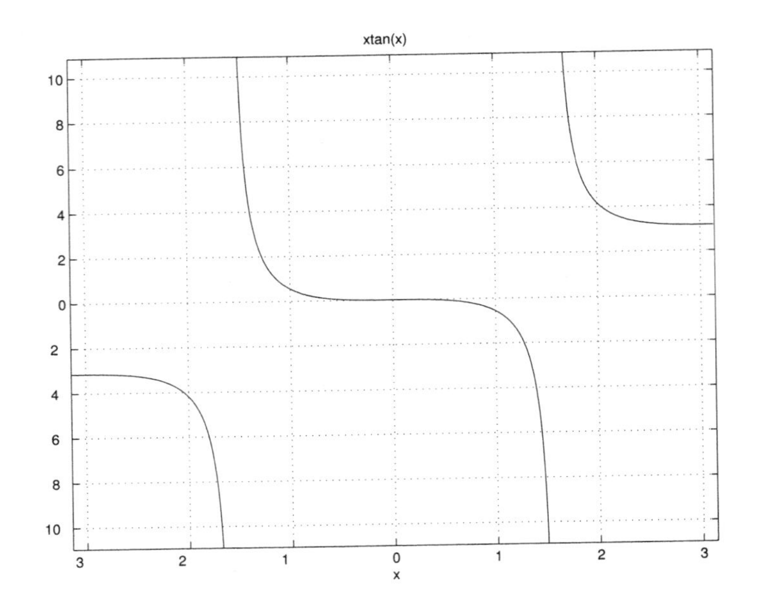

Figure 11.5. *Plot produced by* `ezplot(@(x)x-tan(x),[-pi,pi]), grid`.

```
>> [x, fval] = fzero(@(x)x-tan(x),1)
x =
    1.5708
fval =
  1.2093e+015
```

The second output argument is the function value at `x`, the purported zero. Clearly, in this example `x` is not a zero but an approximation to the point $\pi/2$ at which the function has a singularity; see Figure 11.5. To force `fzero` to keep away from singularities we can give it a starting interval that encloses a zero but not a singularity:

```
>> [x, fval] = fzero(@(x)x-tan(x),[-1 1])
x =
     0
fval =
     0
```

The convergence tolerance and the display of output in `fzero` are controlled by a third argument, the structure `options`, which is best defined using the `optimset` function. Four of the fields of the `options` structure are used: `Display` specifies the level of reporting, with values `off` for no output, `iter` for output at each iteration, `final` for just the final output, and `notify` for output only when the iteration fails to converge; `TolX` is a convergence tolerance; `FunValCheck` determines whether function values are checked for complex or NaN values; and `OutputFcn` specifies a user-defined function that is called at each iteration. Example uses are

```
fzero(fun,x0,optimset('Display','iter'))
fzero(fun,x0,optimset('TolX',1e-4))
```

The default corresponds to

```
optimset('Display','notify','TolX',eps,'FunValCheck','off',...
        'OutputFcn',[])
```

Note that the field names passed to `optimset` can be any combination of upper and lower case, and it is sufficient to type just enough characters of the field name to uniquely identify it.

Suppose now that we wish to find a zero of the function $f(x) = a\sin(x) + be^{-x^2/2}$, where a and b are parameters that we wish to vary. Our function is called from within `fzero` with just one argument, `x`, so how do we communicate the values of `a` and `b`? As we discussed in Section 10.2, this can be achieved using an anonymous function:

```
>> a = 1; b = 2;
>> fzero(@(x)a*sin(x) + b*exp(-x^2/2),0)
ans =
   -1.2274

>> a = 3; b = -1;
>> fzero(@(x)a*sin(x) + b*exp(-x^2/2),0)
ans =
    0.3220
```

Note the importance of reconstructing the anonymous function each time `a` and `b` change. Compare this example with

```
>> a = 1; b = 2;
>> f = @(x)a*sin(x) + b*exp(-x^2/2)
>> fzero(f,0)
ans =
   -1.2274

>> a = 3; b = -1;
>> fzero(f,0)
ans =
   -1.2274
```

where the second invocation of `fzero` produces the same result as the first, since `f` always uses the values of `a` and `b` current at the time the anonymous function was constructed. Another approach is to construct the basic function once and for all with

```
>> fun = @(x,a,b)a*sin(x)+b*exp(-x^2/2);
```

and then construct a wrapping anonymous function from it each time:

```
>> a = 1; b = 2;
>> fzero(@(x)fun(x,a,b),0)
ans =
   -1.2274

>> a = 3; b = -1;
>> fzero(@(x)fun(x,a,b),0)
ans =
    0.3220
```

The algorithm used by fzero, a combination of the bisection method, the secant method, and inverse quadratic interpolation, is described in [27, Chap. 7].

11.3. Optimization

The command x = fminbnd(fun,x1,x2) attempts to find a local minimizer x of the function of one variable specified by fun over the interval [x1, x2]. A point x is a local minimizer of f if it minimizes f in an interval around x. In general, a function can have many local minimizers. MATLAB does not provide a function for the difficult problem of computing a global minimizer (one that minimizes $f(x)$ over all x). Example:

```
>> [x,fval] = fminbnd(@(x)sin(x)-cos(x),-pi,pi)
x =
   -0.7854
fval =
   -1.4142
```

As for fzero, options can be specified using a structure options set via the optimset function. In addition to the fields used by fzero, fminbnd uses MaxFunEvals (the maximum number of function evaluations allowed) and MaxIter (the maximum number of iterations allowed). The defaults correspond to

```
optimset('Display','notify','MaxFunEvals',500,'MaxIter',500,...
         'TolX',1e-4,'FunValCheck','off','OutputFcn',[])
```

The algorithm used by fminbnd, a combination of golden section search and parabolic interpolation, is described in [27, Chap. 8].

If you wish to maximize a function f rather than minimize it you can minimize $-f$, since $\max_x f(x) = -\min_x(-f(x))$.

Function fminsearch searches for a local minimum of a real function of n real variables. The syntax is similar to fminbnd except that a starting vector rather than an interval is supplied: x = fminsearch(fun,x0,options). The fields in options are those supported by fminbnd plus TolFun, a termination tolerance on the function value. Both TolX and TolFun default to 1e-4. To illustrate the use of fminsearch we consider the quadratic function

$$F(x) = x_1^2 + x_2^2 - x_1 x_2,$$

which has a minimum at $x = [0\ 0]^T$. Given the function

```
function f = fquad(x)
f = x(1)^2 + x(2)^2 - x(1)*x(2);
```

we can type

```
>> [x,fval] = fminsearch(@fquad,ones(2,1),optimset('Disp','final'))

Optimization terminated:
 the current x satisfies the termination criteria using
 OPTIONS.TolX of 1.000000e-004
 and F(X) satisfies the convergence criteria using
```

```
 OPTIONS.TolFun of 1.000000e-004

x =
  1.0e-004 *
   -0.4582
   -0.4717
fval =
  2.1635e-009
```

Alternatively, we can define F using an anonymous function:

```
[x,fval] = fminsearch(@(x) x(1)^2+x(2)^2-x(1)*x(2),ones(2,1))
```

Function **fminsearch** is based on the Nelder–Mead simplex algorithm [51, Sec. 8.1], [93, Sec. 10.4], a direct search method that uses function values but not derivatives. The method can be very slow to converge, or may fail to converge to a local minimum. However, it has the advantage of being insensitive to discontinuities in the function. More sophisticated minimization functions can be found in the Optimization Toolbox.

11.4. The Fast Fourier Transform

The discrete Fourier transform of an n-vector x is the vector $y = F_n x$, where F_n is an n-by-n unitary matrix made up of roots of unity and illustrated by

$$F_4 = \begin{bmatrix} 1 & 1 & 1 & 1 \\ 1 & \omega & \omega^2 & \omega^3 \\ 1 & \omega^2 & \omega^4 & \omega^6 \\ 1 & \omega^3 & \omega^6 & \omega^9 \end{bmatrix}, \qquad \omega = e^{-2\pi i/4}.$$

The fast Fourier transform (FFT) is a more efficient way of forming y than the obvious matrix–vector multiplication. The `fft` function implements the FFT and is called as `y = fft(x)`. The efficiency of `fft` depends on the value of n; prime values are bad, highly composite numbers are better, and powers of 2 are best. A second argument can be given to `fft`: `y = fft(x,n)` causes `x` to be truncated or padded with zeros to make `x` of length `n` before the FFT algorithm is applied. The inverse FFT, $x = n^{-1} F_n^* y$, is carried out by the `ifft` function: `x = ifft(y)`. Example:

```
>> y = fft([1 1 -1 -1]')
y =
         0
   2.0000 - 2.0000i
         0
   2.0000 + 2.0000i

>> x = ifft(y)
x =
     1
     1
    -1
    -1
```

MATLAB also implements higher dimensional discrete Fourier transforms and their inverses: see functions `fft2`, `fftn`, `ifft2`, and `ifftn`.

To compute FFTs MATLAB uses a package called FFTW (the "Fastest Fourier Transform in the West") [28]. FFTW is an example of self-adapting numerical software [15], which tunes itself to obtain the best speed on the computational environment in which it is running. The function `fftw` provides control over the tuning process; see the online documentation for details.

Life as we know it would be very different without the FFT.
— CHARLES F. VAN LOAN, *Computational Frameworks for the Fast Fourier Transform* (1992)

Do you ever want to kick the computer?
Does it iterate endlessly on your newest algorithm
that should have converged in three iterations?
And does it finally come to a crashing halt
with the insulting message that you divided by zero?
These minor trauma are, in fact,
the ways the computer manages to kick you and,
unfortunately, you almost always deserve it!
For it is a sad fact that most of us
can more easily compute than think—
which might have given rise to that famous definition,
"Research is when you don't know what you're doing."
— FORMAN S. ACTON, *Numerical Methods That Work* (1970)

Chapter 12
Numerical Methods: Part II

We now move on to MATLAB's capabilities for evaluating integrals and solving ordinary and partial differential equations.

Most of the functions discussed in this chapter support mixed absolute/relative error tests, with tolerances `AbsTol` and `RelTol`, respectively. This means that they test whether an estimate `err` of some measure of the error in the vector `x` is small enough by testing whether, for all `i`,

```
err(i) <= max(AbsTol,RelTol*abs(x(i)))
```

If `AbsTol` is zero this is a pure relative error test and if `RelTol` is zero it is a pure absolute error test. Since we cannot expect to obtain an answer with more correct significant digits than the 16 or so to which MATLAB works, `RelTol` should be no smaller than about `eps`; and since `x` $= 0$ is a possibility we should also take `AbsTol` > 0. A rough way of interpreting the mixed error test above is that `err(i)` is acceptably small if `x(i)` has as many correct digits as specified by `RelTol` or is smaller than `AbsTol` in absolute value. `AbsTol` can be a vector of absolute tolerances, in which case the test is

```
err(i) <= max(AbsTol(i),RelTol*abs(x(i)))
```

Several of the functions described in this chapter employ structures as input and output arguments, in order to group several related pieces of information in one variable. See Section 18.3 for full details of structures.

The introduction of anonymous functions and nested functions in MATLAB 7 has had a big effect on the M-files in this chapter, because these types of function remove the need to pass problem parameters through argument lists. In all our examples we use the approach now recommended for handling problem parameters, but for `ode45` and the Rössler system we show both the new way (Listing 12.1) and the old, MATLAB 6.5 and earlier, way (Listing 12.2) of doing things; this should be helpful for those wishing to read or convert code written in the earlier style.

12.1. Quadrature

Quadrature is a synonym for numerical integration, the approximation of definite integrals $\int_a^b f(x)\,dx$. MATLAB has two main functions for quadrature, `quad` and `quadl`. Both require a and b to be finite and the integrand to have no singularities on $[a,b]$. For infinite integrals and integrals with singularities a variety of approaches can be used in order to produce an integral that can be handled by `quad` and `quadl`; these include change of variable, integration by parts, and analytic treatment of the integral over part of the range. See numerical analysis textbooks for details, for example, [6, Sec. 5.6], [17, Sec. 7.4.3], and [101, Sec. 5.4].

The basic usage is `q = quad(fun,a,b,tol)` (similarly, for `quadl`), where `fun` specifies the function to be integrated. The function `fun` must accept a vector argument and return a vector of function values. The argument `tol` is an absolute error tolerance, which defaults to 10^{-6}. To approximate $\int_2^4 x \log x \, dx$ we can type

```
>> quad(@(x)x.*log(x),2,4)
ans =
    6.7041
```

Note the use of array multiplication (`.*`) to make the anonymous function work for vector inputs.

The number of (scalar) function evaluations is returned in a second output argument:

```
[q,count] = quad(fun,a,b)
```

The `quad` routine is based on Simpson's rule, which is a Newton–Cotes 3-point rule (exact for polynomials of degree up to 3) and can also be interpreted as a 3-point Gauss–Lobatto rule; `quadl` employs a more accurate 4-point Gauss–Lobatto rule together with a 7-point Kronrod extension [29] (exact for polynomials of degrees up to 5 and 9, respectively). Both routines use adaptive quadrature. They break the range of integration into subintervals and apply the basic integration rule over each subinterval. They choose the subintervals according to the local behavior of the integrand, placing the smallest ones where the integrand is changing most rapidly. Warning messages are produced if the subintervals become very small or if an excessive number of function evaluations is used, either of which could indicate that the integrand has a singularity.

To illustrate how `quad` and `quadl` work, we consider the integral

$$\int_0^1 \left(\frac{1}{(x-0.3)^2+0.01} + \frac{1}{(x-0.9)^2+0.04} - 6 \right) dx = 29.858\ldots.$$

The integrand is the function `humps` provided with MATLAB, which has a large peak at 0.3 and a smaller one at 0.9. We applied `quad` to this integral, using a tolerance of `1e-4`. Figure 12.1 plots the integrand and shows with tick marks on the x-axis where the integrand was evaluated; circles show the corresponding values of the integrand. The figure shows that the subintervals are smallest where the integrand is most rapidly varying.

For another example we take the Fresnel integrals

$$x(t) = \int_0^t \cos(u^2)\, du, \qquad y(t) = \int_0^t \sin(u^2)\, du.$$

Plotting $x(t)$ against $y(t)$ produces a spiral [34, Sec. 2.6]. The following code plots the spiral by sampling at 2001 equally spaced points t on the interval $[-4\pi, 4\pi]$; the result is shown in Figure 12.2. For efficiency we exploit symmetry and avoid repeatedly integrating from 0 to t by integrating over each subinterval and then evaluating the cumulative sums using `cumsum`:

```
n = 1000;
x = zeros(1,n); y = x;
t = linspace(0,4*pi,n+1);
```

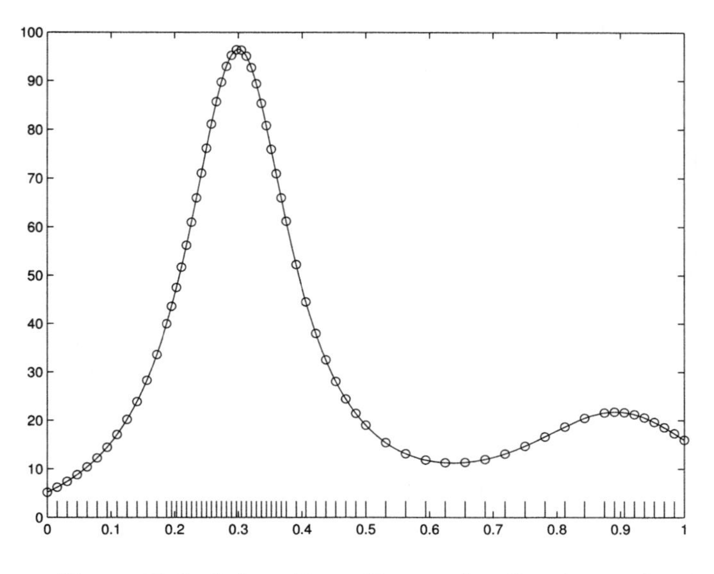

Figure 12.1. *Integration of* `humps` *function by* `quad`.

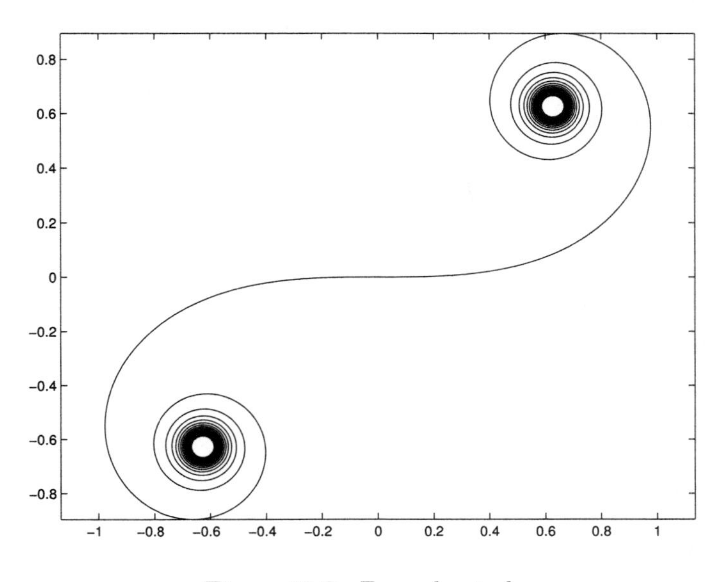

Figure 12.2. *Fresnel spiral.*

```
for i = 1:n
    x(i) = quadl(@(x)cos(x.^2),t(i),t(i+1),1e-3);
    y(i) = quadl(@(x)sin(x.^2),t(i),t(i+1),1e-3);
end
x = cumsum(x); y = cumsum(y);
plot([-x(end:-1:1) 0 x], [-y(end:-1:1) 0 y])
axis equal
```

Note the use of anonymous functions to specify the integrands. On the computer used to typeset this book, this calculation runs an order of magnitude faster than when we express the integrands using `inline` (e.g., `inline('cos(x.^2)')`).

Another quadrature function is `trapz`, which applies the repeated trapezium rule. It differs from `quad` and `quadl` in that its input comprises vectors of x_i and $f(x_i)$ values rather than a function representing the integrand f; therefore it is not adaptive. Example:

```
>> x = linspace(0,2*pi,10);
>> f = sin(x).^2./sqrt(1+cos(x).^2);
>> trapz(x,f)
ans =
    2.8478
```

In this example the error in the computed integral is of the order 10^{-7}, which is much smaller than the standard error expression for the repeated trapezium rule would suggest. The reason is that we are integrating a periodic function over a whole number of periods and the repeated trapezium rule is known to be highly accurate in this situation [6, Sec. 5.4], [101, p. 182]. In general, provided that a function is available to evaluate the integrand at arbitrary points, `quad` and `quadl` are preferable to `trapz`.

Double integrals can be evaluated with `dblquad`. To illustrate, suppose we wish to approximate the integral

$$\int_4^6 \int_0^1 \left(y^2 e^x + x\cos y\right) dx\, dy.$$

Using the function

```
function out = fxy(x,y)
      out = y^2*exp(x)+x*cos(y);
```

we type

```
>> dblquad(@fxy,0,1,4,6)
ans =
   87.2983
```

assed to `dblquad` must accept a vector `x` and a scalar `y` and return tput. Additional arguments to `dblquad` can be used to specify the the integrator (the default is `quad`).
ous function `triplequad` evaluates triple integrals.

12.2. Ordinary Differential Equations

MATLAB has a range of functions for solving initial value ordinary differential equations (ODEs). These mathematical problems have the form

$$\frac{d}{dt}y(t) = f(t, y(t)), \quad y(t_0) = y_0, \tag{12.1}$$

where t is a real scalar, $y(t)$ is an unknown m-vector, and the given function f of t and y is also an m-vector. To be concrete, we regard t as representing time. The function f defines the ODE, and the initial condition $y(t_0) = y_0$ then defines an initial value problem. The simplest way to solve such a problem is to write a function that evaluates f and then call one of MATLAB's ODE solvers. The minimum information that the solver must be given is the function name, the range of t values over which the solution is required, and the initial condition y_0. However, MATLAB's ODE solvers allow for extra (optional) input and output arguments that make it possible to specif more about the mathematical problem and how it is to be solved. Each of MATLA ODE solvers is designed to be efficient in specific circumstances, but all are essent interchangeable. In the next subsection we develop examples that illustrate the ode45. This function implements an adaptive Runge–Kutta algorithm and is t the most efficient solver for the classes of ODEs that concern MATLAB users range of ODE solving functions is discussed in Section 12.2.3 and listed in on p. 190. The functions follow a naming convention: all names begin followed by digits denoting the orders of the underlying integration fo final "s", "t", or "tb" denoting a function intended for stiff problems denoting a function intended for fully implicit systems.

12.2.1. Examples with ode45

In order to solve the scalar ($m = 1$) ODE

$$\frac{d}{dt}y(t) = -y(t) - 5e^{-t}\sin 5t, \quad y(0)$$

for $0 \le t \le 3$ with ode45, we create in the file myf.m t

```
function yprime = myf(t,y)
%MYF    ODE example function.
%       YPRIME = MYF(T,Y) evaluates d

yprime = -y - 5*exp(-t)*sin(5*t);
```

and then type

```
tspan = [0 3]; yzero = 1;
[t,y] = ode45(@myf,tspan,yze
plot(t,y,'*--')
xlabel t, ylabel y(t)
```

This produces the plot in Fig mand/function duality in se input arguments to ode45 a time interval, and the ini

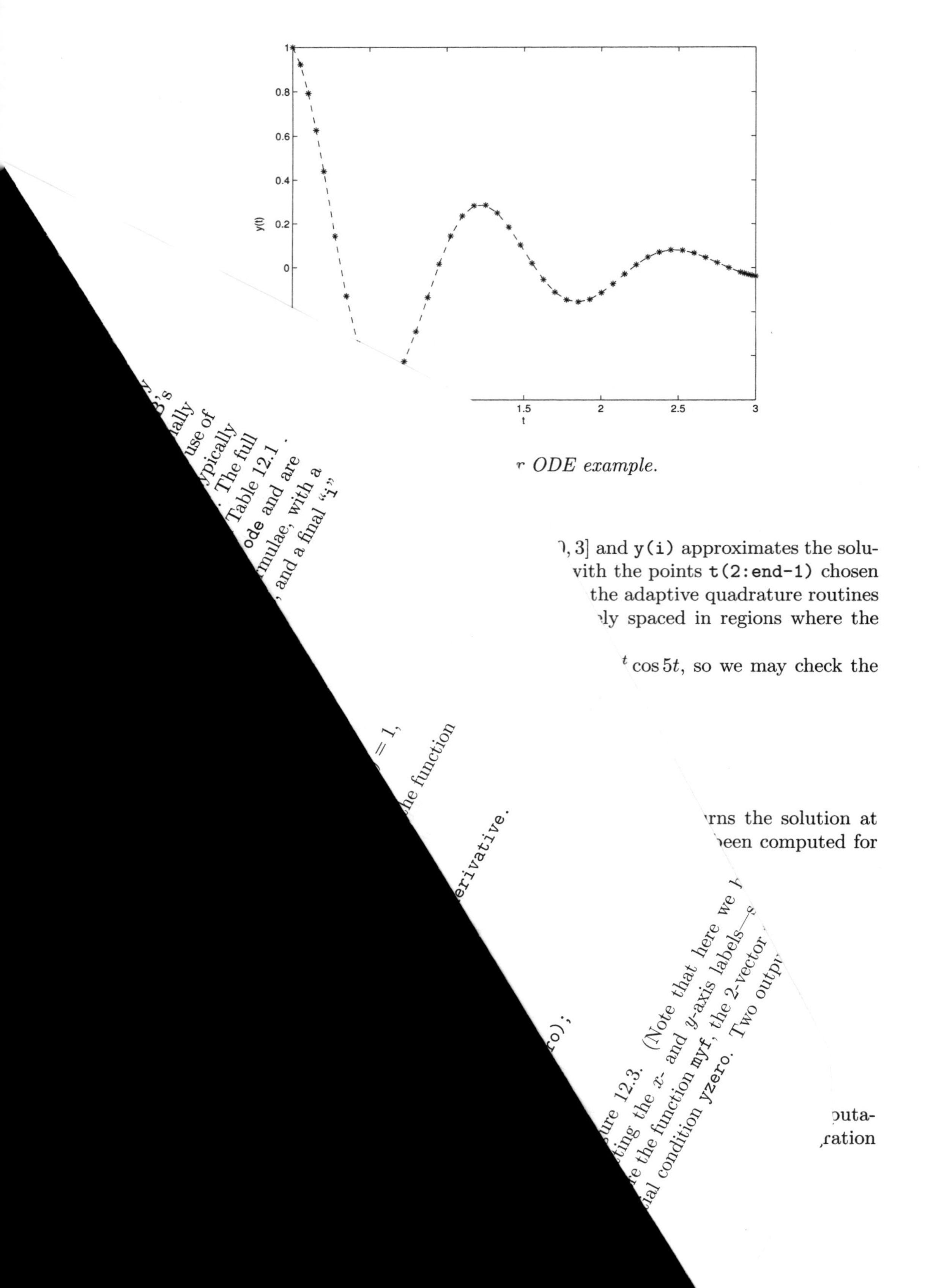

r ODE example.

], 3] and y(i) approximates the solu-
vith the points t(2:end-1) chosen
the adaptive quadrature routines
ly spaced in regions where the

$^t \cos 5t$, so we may check the

rns the solution at
een computed for

puta-
ration

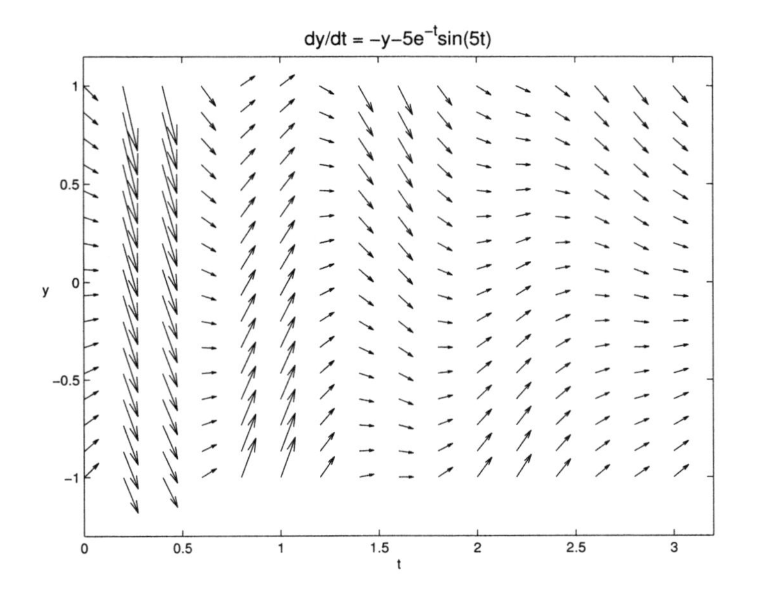

Figure 12.4. *Vector field for scalar ODE example.*

```
>> [t3,y3] = ode45(@myf,tspan3,yzero);
>> disp([t3 y3])
         0    1.0000
   -0.5000   -1.3209
   -1.0000    0.7711
```

At any point in the (t, y) plane the differential equation (12.1) gives the gradient of the solution $y(t)$. Therefore it defines a *vector field* (or *direction field*) through which all solutions "navigate". We can picture the vector field by plotting at each point (t, y) an arrow whose slope is $f(t, y)$. The following code uses the `quiver` function to plot the vector field for the differential equation at the start of this section:

```
n = 16;
tpts = linspace(0,3,n); ypts = linspace(-1,1,n);
[t,y] = meshgrid(tpts,ypts);
pt = ones(size(y));
py = -y-5*exp(-t).*sin(5*t);
quiver(t,y,pt,py,1.5);
title('dy/dt = -y-5e^{-t}sin(5t)','FontSize',14)
xlabel('t'), ylabel('y','Rotation',0)
xlim([0 3.2]), ylim([-1.3 1.15])  % Tune axis limits.
```

The command `quiver(x,y,u,v,scale)` plots arrows with components `(u,v)` at the locations `(x,y)`, producing arrows whose length is `scale` times the 2-norm of the `[u(i),v(i)]` vectors. Figure 12.4 shows the resulting graph.

Higher order ODEs can be solved if they are first rewritten as a larger system of first-order ODEs [98, Chap. 1]. For example, the simple pendulum equation [111,

Sec. 6.7] has the form

$$\frac{d^2}{dt^2}\theta(t) + \sin\theta(t) = 0.$$

Defining $y_1(t) = \theta(t)$ and $y_2(t) = d\theta(t)/dt$, we may rewrite this equation as the two first-order equations

$$\frac{d}{dt}y_1(t) = y_2(t),$$
$$\frac{d}{dt}y_2(t) = -\sin y_1(t).$$

This information can be encoded for use by `ode45` in the function `pend` as follows.

```
function yprime = pend(t,y)
%PEND   Simple pendulum.
%       YPRIME = PEND(T,Y).

yprime = [y(2); -sin(y(1))];
```

The following commands compute solutions over $0 \le t \le 10$ for three different initial conditions. Since we are solving a system of $m = 2$ equations, in the output `[t,y]` from `ode45` the `i`th row of the matrix `y` approximates $(y_1(t), y_2(t))$ at time $t =$ `t(i)`.

```
tspan = [0 10];
yazero = [1; 1]; ybzero = [-5; 2]; yczero = [5; -2];
[ta,ya] = ode45(@pend,tspan,yazero);
[tb,yb] = ode45(@pend,tspan,ybzero);
[tc,yc] = ode45(@pend,tspan,yczero);
```

To produce phase plane plots, that is, plots of $y_1(t)$ against $y_2(t)$, we simply plot the first column of the numerical solution against the second. The commands below generate phase plane plots of the solutions `ya`, `yb`, and `yc` computed above, and make use of `quiver` to superimpose a vector field. The resulting picture is shown in Figure 12.5.

```
[y1,y2] = meshgrid(-5:.5:5,-3:.5:3);
Dy1Dt = y2; Dy2Dt = -sin(y1);
quiver(y1,y2,Dy1Dt,Dy2Dt)
hold on
plot(ya(:,1),ya(:,2),yb(:,1),yb(:,2),yc(:,1),yc(:,2))
axis equal, axis([-5 5 -3 3])
xlabel y_1(t), ylabel y_2(t), hold off
```

The pendulum ODE preserves energy: any solution keeps $y_2(t)^2/2 - \cos y_1(t)$ constant for all t. We can check that this is approximately true for `yc` as follows.

```
>> Ec = .5*yc(:,2).^2 - cos(yc(:,1));
>> max(abs(Ec(1)-Ec))
ans =
    0.0263
```

The general form of a call to `ode45` is[7]

[7] In this argument list, and in those in the rest of the chapter, we assume that functions passed as arguments are specified by their handles (see Section 10.1), which is usually the case for the differential equation solvers.

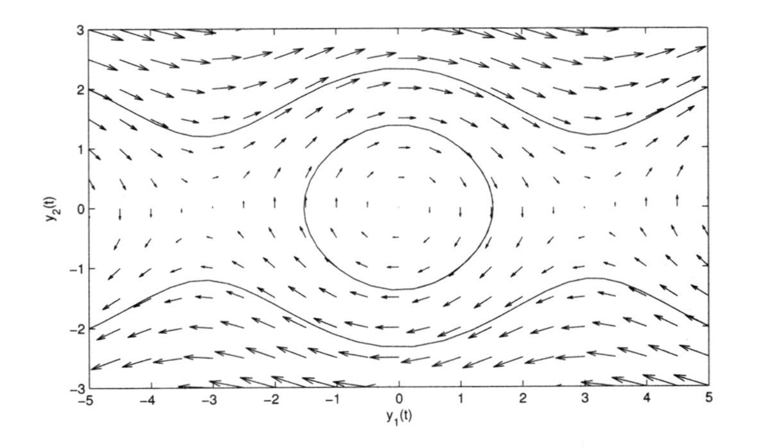

Figure 12.5. *Pendulum phase plane solutions.*

```
[t,y] = ode45(@fun,tspan,yzero,options,p1,p2,...);
```

The optional trailing arguments `p1`, `p2`, ... represent problem parameters that, if provided, are passed on to the function `fun`. However, in MATLAB 7 it is preferred to communicate problem parameters using anonymous functions or nested functions and the use of the arguments `p1`, `p2`, ... is no longer recommended. These two different approaches are illustrated in Listings 12.1 and 12.2. The optional argument `options` is a structure that controls many features of the solver and can be set via the `odeset` function. In our next example we create a structure `options` by the assignment

```
options = odeset('AbsTol',1e-7,'RelTol',1e-4);
```

Passing this structure as an input argument to `ode45` causes the absolute and relative error tolerances to be set to 10^{-7} and 10^{-4}, respectively. (The default values are 10^{-6} and 10^{-3}; see `help odeset` for the precise meaning of the tolerances.) These tolerances apply on a local, step-by-step, basis and it is not generally the case that the overall error is kept within these limits. However, under reasonable assumptions about the ODE, it can be shown that decreasing the tolerances by some factor, say 100, will decrease the overall error by a similar factor, so the error is usually roughly proportional to the tolerances. See [98, Chap. 7] for further details about error control in ODE solvers.

We next consider the Rössler system [111, Secs. 10.6, 12.3],

$$\begin{aligned}
\frac{d}{dt}y_1(t) &= -y_2(t) - y_3(t),\\
\frac{d}{dt}y_2(t) &= y_1(t) + ay_2(t),\\
\frac{d}{dt}y_3(t) &= b + y_3(t)\left(y_1(t) - c\right),
\end{aligned}$$

where a, b, and c are parameters. The function `rossler_ex1` in Listing 12.1 solves the Rössler system over $0 \le t \le 100$ with initial condition $y(0) = [1; 1; 1]$ for $(a, b, c) = (0.2, 0.2, 2.5)$ and $(a, b, c) = (0.2, 0.2, 5)$. The ODE is defined in the nested function

Listing 12.1. *Function* rossler_ex1.

```
function rossler_ex1
%ROSSLER_EX1       Run Rossler example.

tspan = [0 100]; yzero = [1;1;1];
options = odeset('AbsTol',1e-7,'RelTol',1e-4);

a = 0.2; b = 0.2; c = 2.5;
[t,y] = ode45(@rossler,tspan,yzero,options);
subplot(221), plot3(y(:,1),y(:,2),y(:,3)), mytitle, zlabel y_3(t), grid
subplot(223), plot(y(:,1),y(:,2)), mytitle

c = 5;
[t,y] = ode45(@rossler,tspan,yzero,options);
subplot(222), plot3(y(:,1),y(:,2),y(:,3)), mytitle, zlabel y_3(t), grid
subplot(224), plot(y(:,1),y(:,2)), mytitle

     % ----------------------- Nested functions ----------------------
     function yprime = rossler(t,y)
     %ROSSLER    Rossler system, parameterized.
     yprime = [-y(2)-y(3); y(1)+a*y(2); b+y(3)*(y(1)-c)];
     end

     function mytitle
     title(sprintf('c = %2.1f',c),'FontSize',14)
     xlabel y_1(t), ylabel y_2(t)
     end

end
```

rossler, in order that the parameters a, b, and c, defined in the main function, are captured in the function handle @rossler that is passed to ode45. The nested function mytitle is used to avoid repetition of title commands. For more on nested functions, see Section 10.7. Figure 12.6 shows the results. The 221 subplot gives the 3D phase space solution for $c = 2.5$ and the 223 subplot gives the 2D projection onto the y_1-y_2 plane. The 222 and 224 subplots give the corresponding pictures for $c = 5$. The function rossler_ex0 in Listing 12.2 shows how this example would be coded in MATLAB 6.5 and earlier versions, in which problem parameters had to be passed via the argument list.

Function rossler_ex1 illustrates how a complete problem specification and solution can be encapsulated in a single function (which does not need to have any input or output arguments) by making use of nested functions, subfunctions, and function handles. (Note that nested functions are needed only when the ODE is parameter-dependent.)

The ODE solvers may also be called with a single output argument. Specifying

```
sol = ode45(@fun,tspan,yzero,options);
```

causes the solution to be returned in a structure sol. The fields sol.x and sol.y

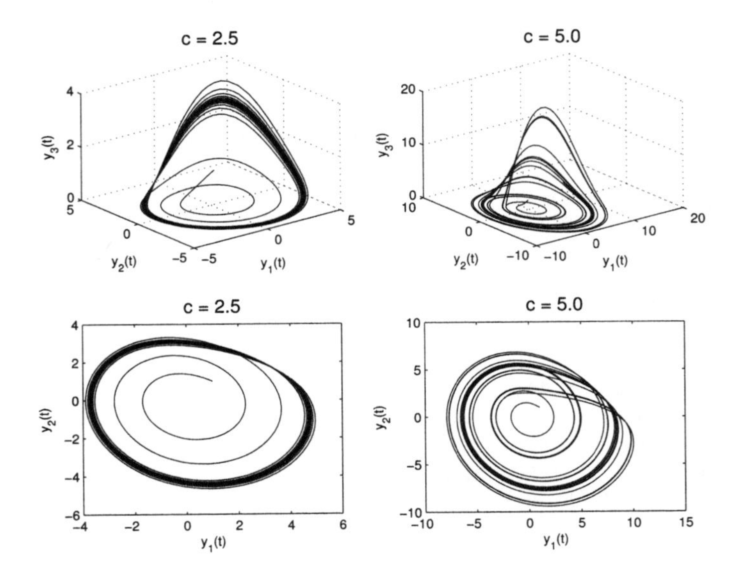

Figure 12.6. *Rössler system phase space solutions.*

Listing 12.2. *Function* rossler_ex0.

```
function rossler_ex0
%ROSSLER_EX0  Run Rossler example.
% This is the recommended approach for MATLAB 6.5 and earlier.
% ROSSLER_EX0 runs in MATLAB 7, but ROSSLER_EX1 illustrates the style of
% coding now recommended for MATLAB 7.

tspan = [0 100]; yzero = [1;1;1];
options = odeset('AbsTol',1e-7,'RelTol',1e-4);

a = 0.2; b = 0.2; c = 2.5;
[t,y] = ode45(@rossler,tspan,yzero,options,a,b,c);
subplot(221), plot3(y(:,1),y(:,2),y(:,3)), mytitle(c), zlabel y_3(t), grid
subplot(223), plot(y(:,1),y(:,2)), mytitle(c)

c = 5;
[t,y] = ode45(@rossler,tspan,yzero,options,a,b,c);
subplot(222), plot3(y(:,1),y(:,2),y(:,3)), mytitle(c), zlabel y_3(t), grid
subplot(224), plot(y(:,1),y(:,2)), mytitle(c)

function yprime = rossler(t,y,a,b,c)
%ROSSLER    Rossler system, parameterized.
yprime = [-y(2)-y(3); y(1)+a*y(2); b+y(3)*(y(1)-c)];

function mytitle(c)
title(sprintf('c = %2.1f',c),'FontSize',14)
xlabel y_1(t), ylabel y_2(t)
```

Listing 12.3. *Function* rossler_ex2.

```
function rossler_ex2
%ROSSLER_EX2     Attractor reconstruction for Rossler system.

tspan = [0 100]; yzero = [1;1;1];
options = odeset('AbsTol',1e-7,'RelTol',1e-4);

a = 0.2; b = 0.2; c = 2.5;
sol = ode45(@rossler,tspan,yzero,options);
tau = 1.5;
t = linspace(tau,100,1000);
y = deval(sol,t,1);
ylag = deval(sol,t-tau,1);
plot(y,ylag), title('\tau = 1.5','FontSize',14)
xlabel('y_1(t)','FontSize',14)
ylabel('y_1(t-\tau)','FontSize',14,'Rotation',0,...
       'HorizontalAlignment','right')

     function yprime = rossler(t,y)
     %ROSSLER    Rossler system, parametrized.
     yprime = [-y(2)-y(3); y(1)+a*y(2); b+y(3)*(y(1)-c)];
     end

end
```

are equivalent to the output arguments `t` and `y`, respectively, that arise from the call `[t,y] = ode45(@fun,tspan,yzero,options)`. Hence, the field `sol.x` is a row vector containing the t values chosen by `ode45` and the field `sol.y` is an array whose ith column `sol.y(:,i)` contains the solution at the points `sol.x(i)`. A utility function **`deval`** is available that, given `sol`, will evaluate the solution (and, optionally, the derivative) at any set of intermediate t values. So, if `trange` is a vector of points between `sol.x(1)` and `sol.x(end)`, then `ysol = deval(sol,trange)` will return an
...v `ysol` whose ith column corresponds to the solution at `trange(i)`. Adding a
...put argument, `ysol = deval(sol,trange,idx)`, restricts the output to solu-
...nents specified by the array `idx`. For example, `deval(sol,trange,[1,3])`
...rst and third solution components. We may add a second output argu-
...l] = deval(sol,trange,idx)` returns an array `ypsol` containing
...roximations to the first derivative of the solution.
...l.x`, rather than `sol.t`, for the array containing values of
...e because the structure output format was first intro-
...roblem solver, `bvp4c`; see Section 12.3.
...ook at the task of plotting $y_1(t)$ against $y_1(t-\tau)$,
...n *attractor reconstruction* [111, Sec. 12.4],
...mics in complete phase space from a single
... `ode45(....)` mode would be inconvenient
...ears in the array `t` it is not generally true that
...ssler_ex2 in Listing 12.3 uses two calls to `deval`

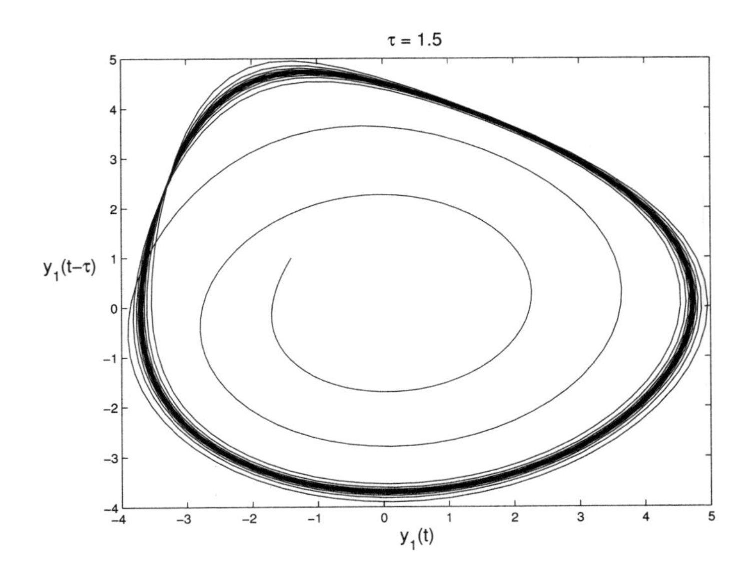

Figure 12.7. *Attractor reconstruction using* `deval`.

12.2.2. Case Study: Pursuit Problem with Event Location

Next we consider a pursuit problem [18, Chap. 5]. Suppose that a rabbit follows a predefined path $(r_1(t), r_2(t))$ in the plane, and that a fox chases the rabbit in such a way that (a) at each moment the tangent of the fox's path points towards the rabbit and (b) the speed of the fox is some constant k times the speed of the rabbit. Then the path $(y_1(t), y_2(t))$ of the fox is determined by the ODE

$$\frac{d}{dt}y_1(t) = s(t)\left(r_1(t) - y_1(t)\right),$$
$$\frac{d}{dt}y_2(t) = s(t)\left(r_2(t) - y_2(t)\right),$$

where

$$s(t) = \frac{k\sqrt{\left(\frac{d}{dt}r_1(t)\right)^2 + \left(\frac{d}{dt}r_2(t)\right)^2}}{\sqrt{\left(r_1(t) - y_1(t)\right)^2 + \left(r_2(t) - y_2(t)\right)^2}}.$$

Note that this ODE system becomes ill-defined if the fox approaches the rabbit. We let the rabbit follow an outward spiral,

$$\begin{bmatrix} r_1(t) \\ r_2(t) \end{bmatrix} = \sqrt{1+t}\begin{bmatrix} \cos t \\ \sin t \end{bmatrix},$$

and start the fox at $y_1(0) = 3, y_2(0) = 0$. The function `fox1` in Listing 12.4 implements the ODE, with k set to 0.75. The `error` function (see Section 14.1) has been used so that execution terminates with an error message if the denominator of $s(t)$ in the ODE becomes too small. The script below calls `fox1` to produce Figure 12.8. Initial conditions are denoted by circles and the dashed and solid lines show the phase plane paths of the rabbit and fox, respectively.

```
tspan = [0 10]; yzero = [3;0];
[tfox,yfox] = ode45(@fox1,tspan,yzero);
plot(yfox(:,1),yfox(:,2)), hold on
plot(sqrt(1+tfox).*cos(tfox),sqrt(1+tfox).*sin(tfox),'--')
plot([3 1],[0 0],'o');
axis equal, axis([-3.5 3.5 -2.5 3.1])
legend('Fox','Rabbit'), hold off
```

The implementation above is unsatisfactory for $k > 1$, that is, when the fox is faster than the rabbit. In this case, if the rabbit is caught within the specified time interval then no solution is displayed. It would be more natural to ask `ode45` to return with the computed solution if the fox and rabbit become close. Function `fox_rabbit` in Listing 12.5 does this by using the ODE solvers' event location facility, producing Figure 12.9. We have allowed `k` to be a parameter and set `k = 1.1`. The initial condition and the rabbit's path are as for Figure 12.8.

We use `odeset` to set the event location property to the handle of the subfunction `events`. This function has the three output arguments `value`, `isterminal`, and `direction`. It is the responsibility of `ode45` to use `events` to check whether any component passes through zero by monitoring the quantity returned in `value`. In our example `value` is a scalar, corresponding to the distance between the rabbit and fox, minus a threshold of 10^{-4}. Hence, `ode45` checks if the fox has approached within distance 10^{-4} of the rabbit. We set `direction = -1`, which signifies that `value` must be decreasing through zero in order for the event to be considered. The alternative choice `direction = 1` tells MATLAB to consider only crossings where `value` is increasing, and `direction = 0` allows for any type of zero. Since we set `isterminal = 1`, integration will cease when a suitable zero crossing is detected. With the other option, `isterminal = 0`, the event is recorded and the integration continues.

The output arguments from `ode45` are `[tfox,yfox,te,ye,ie]`. Here, `tfox` and `yfox` are the usual solution approximations, so `yfox(i,:)` approximates $y(t)$ at time $t =$ `tfox(i)`. The arguments `te` and `ye` record those t and y values at which the event(s) were recorded and, for vector-valued events, `ie` specifies which component of the event occurred each time. (If no events are detected then `te`, `ye`, and `ie` are returned as empty matrices.) In our example, we have

```
>> te, ye
te =
    5.0710
ye =
    0.8646   -2.3073
```

showing that the rabbit was captured after 5.07 time units at the point $(0.86, -2.31)$.

12.2.3. Stiff Problems, Differential-Algebraic Equations, and the Choice of Solver

The Robertson ODE system

$$\begin{aligned}
\frac{d}{dt}y_1(t) &= -0.04y_1(t) + 10^4 y_2(t)y_3(t),\\
\frac{d}{dt}y_2(t) &= 0.04y_1(t) - 10^4 y_2(t)y_3(t) - 3\times 10^7 y_2(t)^2,\\
\frac{d}{dt}y_3(t) &= 3\times 10^7 y_2(t)^2
\end{aligned}$$

Listing 12.4. *Function* **fox1**.

```
function yprime = fox1(t,y)
%FOX1   Fox-rabbit pursuit simulation.
%       YPRIME = FOX1(T,Y).

k = 0.75;
r = sqrt(1+t)*[cos(t); sin(t)];
r_p =(0.5/sqrt(1+t))*[cos(t)-2*(1+t)*sin(t);sin(t)+2*(1+t)*cos(t)];
dist = norm(r-y);
if dist > 1e-4
   factor = k*norm(r_p)/dist;
   yprime = factor*(r-y);
else
   error('ODE model ill-defined.')
end
```

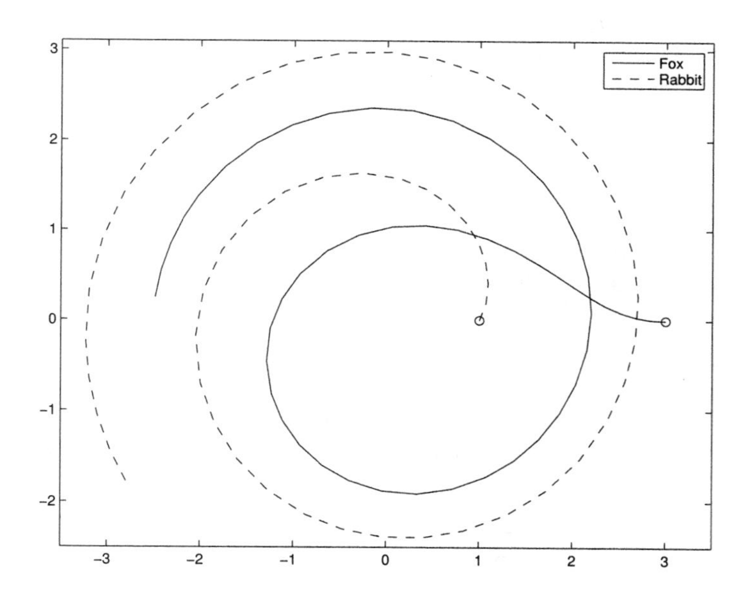

Figure 12.8. *Pursuit example.*

Listing 12.5. *Function* fox_rabbit.

```
function fox_rabbit
%FOX_RABBIT  Fox-rabbit pursuit simulation.
%            Uses relative speed parameter, K.

k = 1.1;
tspan = [0 10]; yzero = [3;0];
options = odeset('RelTol',1e-6,'AbsTol',1e-6,'Events',@events);
[tfox,yfox,te,ye,ie] = ode45(@fox2,tspan,yzero,options);
plot(yfox(:,1),yfox(:,2)), hold on
plot(sqrt(1+tfox).*cos(tfox),sqrt(1+tfox).*sin(tfox),'--')
plot([3 1],[0 0],'o'), plot(yfox(end,1),yfox(end,2),'*')
axis equal, axis([-3.5 3.5 -2.5 3.1])
legend('Fox','Rabbit'), hold off

    function yprime = fox2(t,y)
    %FOX2   Fox-rabbit pursuit simulation ODE.

    r = sqrt(1+t)*[cos(t); sin(t)];
    r_p = (0.5/sqrt(1+t)) * [cos(t)-2*(1+t)*sin(t); sin(t)+2*(1+t)*cos(t)];
    dist = max(norm(r-y),1e-6);
    factor = k*norm(r_p)/dist;
    yprime = factor*(r-y);

    end

end

function [value,isterminal,direction] = events(t,y)
%EVENTS     Events function for FOX2.
%           Locate when fox is close to rabbit.

r = sqrt(1+t)*[cos(t); sin(t)];
value = norm(r-y) - 1e-4;     % Fox close to rabbit.
isterminal = 1;               % Stop integration.
direction = -1;               % Value must be decreasing through zero.

end
```

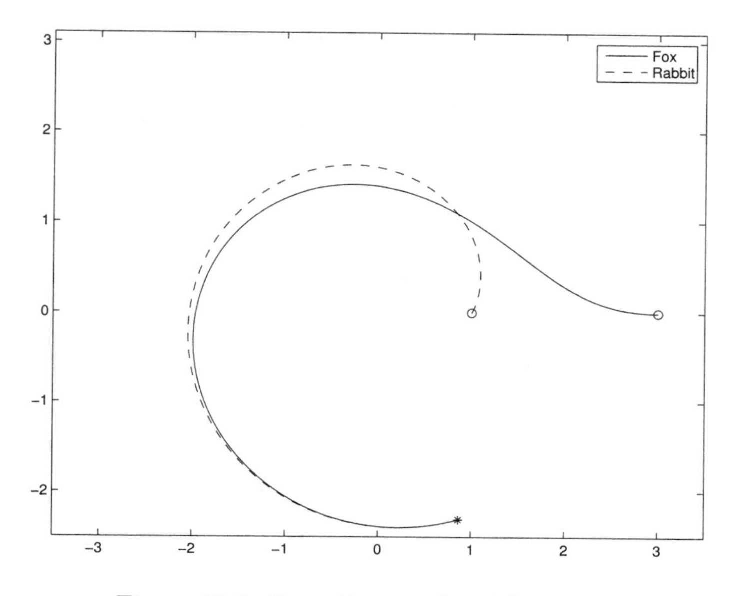

Figure 12.9. *Pursuit example, with capture.*

models a reaction between three chemicals [35, p. 3], [98, p. 418]. We set the system up as the function chem:

```
function yprime = chem(t,y)
%CHEM    Robertson's chemical reaction model.
%        YPRIME = CHEM(T,Y).

yprime = [-0.04*y(1) + 1e4*y(2)*y(3);
           0.04*y(1) - 1e4*y(2)*y(3) - 3e7*y(2)^2;
           3e7*y(2)^2];
```

The script file below solves this ODE for $0 \le t \le 3$ with initial condition $[1;0;0]$, first using ode45 and then using another solver, ode15s, which is based on implicit linear multistep methods. (Implicit means that a nonlinear algebraic equation must be solved at each step.) The results for $y_2(t)$ are plotted in Figure 12.10.

```
tspan = [0 3]; yzero = [1;0;0];
[ta,ya] = ode45(@chem,tspan,yzero);
subplot(121), plot(ta,ya(:,2),'-*')
ax = axis; ax(1) = -0.2; axis(ax) % Make initial transient clearer.
xlabel('t'), ylabel('y_2(t)'), title('ode45','FontSize',14)
[tb,yb] = ode15s(@chem,tspan,yzero);
subplot(122), plot(tb,yb(:,2),'-*'), axis(ax)
xlabel('t'), ylabel('y_2(t)'), title('ode15s','FontSize',14)
```

We see from Figure 12.10 that the solutions agree to within a small absolute tolerance (note the scale factor 10^{-5} for the y-axis labels). However, the left-hand solution from ode45 has been returned at many more time values than the right-hand solution from ode15s and seems to be less smooth. To emphasize these points, Figure 12.11 plots ode45's $y_2(t)$ for $2.0 \le t \le 2.1$. We see that the t values are densely

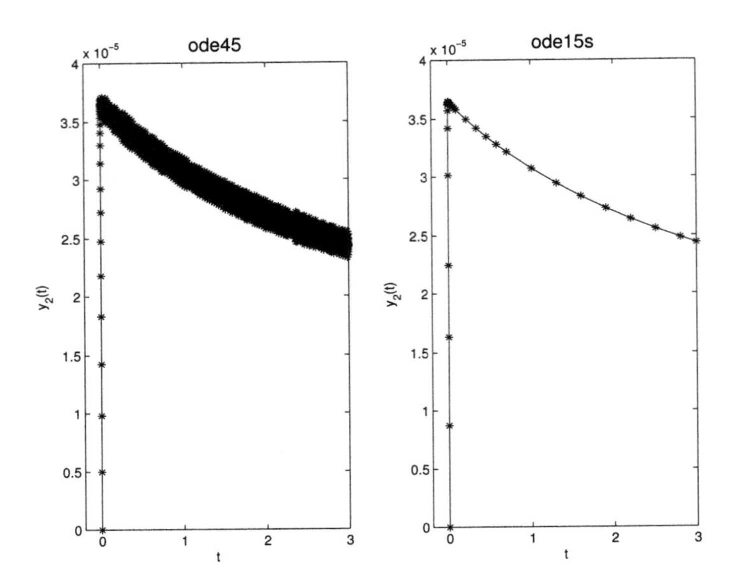

Figure 12.10. *Chemical reaction solutions. Left:* `ode45`. *Right:* `ode15s`.

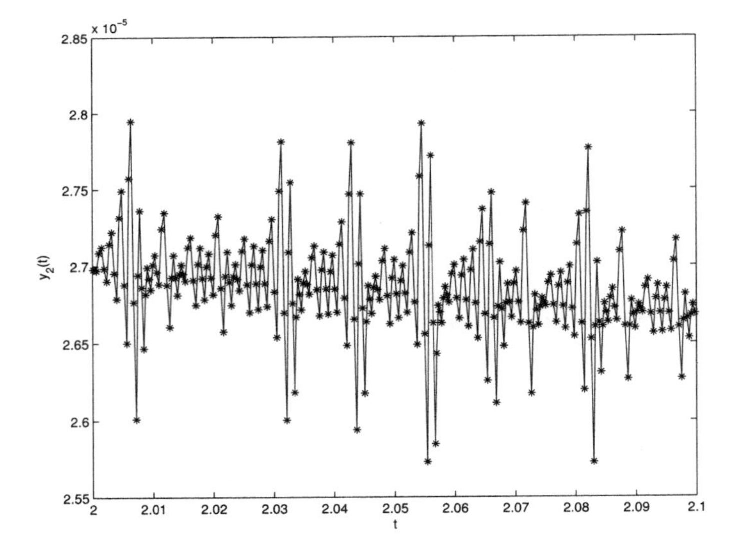

Figure 12.11. *Zoom of chemical reaction solution from* `ode45`.

packed and spurious oscillations are present at the level of the default absolute error tolerance, 10^{-6}. The Robertson problem is a classic example of a *stiff* ODE; see [35] or [98, Chap. 8] for full discussions about stiffness and its effects. Stiff ODEs arise in a number of application areas, including the modeling of chemical reactions and electrical circuits. Semi-discretized time-dependent partial differential equations are also a common source of stiffness (we give an example below). Many solvers behave inefficiently on stiff ODEs—they take an unnecessarily large number of intermediate steps in order to complete the integration and hence make an unnecessarily large number of calls to the ODE function (in this case, `chem`). We can obtain statistics on the computational cost of the integration by setting

```
options = odeset('Stats','on');
```

and providing `options` as an input argument:

```
[ta,ya] = ode45(@chem,tspan,yzero,options);
```

On completion of the run of `ode45`, the following statistics are then printed:

```
2052 successful steps
440 failed attempts
14953 function evaluations
```

Using the same `options` argument with `ode15s` gives

```
33 successful steps
5 failed attempts
73 function evaluations
2 partial derivatives
13 LU decompositions
63 solutions of linear systems
```

The behavior of `ode45` typifies what happens when an adaptive algorithm designed for nonstiff ODEs operates in the presence of stiffness. The solver does not break down or compute an inaccurate solution, but it does behave nonsmoothly and extremely inefficiently in comparison with solvers that are customized for stiff problems. This is one reason why MATLAB provides a suite of ODE solvers.

Note that in the computation above, we have

```
>> disp([length(ta), length(tb)])
        8205          34
```

showing that `ode45` returned output at almost 250 times as many points as `ode15s`. However, the statistics show that `ode45` took 2051 steps, only about 62 times as many as `ode15s`. The explanation is that by default `ode45` uses interpolation to return four solution values at equally spaced points over each "natural" step. The default interpolation level can be overridden via the `Refine` property with `odeset`.

A full list of MATLAB's ODE solvers is given in Table 12.1. The authors of these solvers, Shampine and Reichelt, discuss some of the theoretical and practical issues that arose during their development in [104]. The functions are designed to be interchangeable in basic use. So, for example, the illustrations in the previous subsection continue to work if `ode45` is replaced by any of the other solvers. The functions mainly differ in (a) their efficiency on different problem types and (b) their capacity for accepting information about the problem in connection with Jacobians and mass matrices. With regard to efficiency, Shampine and Reichelt write in [104]:

Table 12.1. *MATLAB's ODE solvers.*

Solver	Problem type	Type of algorithm
`ode45`	Nonstiff	Explicit Runge–Kutta pair, orders 4 and 5
`ode23`	Nonstiff	Explicit Runge–Kutta pair, orders 2 and 3
`ode113`	Nonstiff	Explicit linear multistep, orders 1 to 13
`ode15s`	Stiff	Implicit linear multistep, orders 1 to 5
`ode15i`	Fully implicit	Implicit linear multistep, orders 1 to 5
`ode23s`	Stiff	Modified Rosenbrock pair (one-step), orders 2 and 3
`ode23t`	Mildly stiff	Trapezoidal rule (implicit), orders 2 and 3
`ode23tb`	Stiff	Implicit Runge–Kutta-type algorithm, orders 2 and 3

> The experiments reported here and others we have made suggest that except in special circumstances, `ode45` should be the code tried first. If there is reason to believe the problem to be stiff, or if the problem turns out to be unexpectedly difficult for `ode45`, the `ode15s` code should be tried.

The stiff solvers in Table 12.1 use information about the Jacobian matrix, $\partial f_i/\partial y_j$, at various points along the solution. By default, they automatically generate approximate Jacobians using finite differences. An option can be set via `odeset` to specify the sparsity pattern of the Jacobian, which aids construction of the finite difference approximation. However, the reliability and efficiency of the solvers is generally improved if a function that evaluates the Jacobian is supplied.

To illustrate how Jacobian information can be encoded, we look at the system of ODEs

$$\frac{d}{dt}y(t) = Ay(t) + y(t).*(1 - y(t)) + v,$$

where A is N-by-N and v is N-by-1 with

$$A = r_1 \begin{bmatrix} 0 & 1 & & & \\ -1 & 0 & 1 & & \\ & \ddots & \ddots & \ddots & \\ & & \ddots & \ddots & 1 \\ & & & -1 & 0 \end{bmatrix} + r_2 \begin{bmatrix} -2 & 1 & & & \\ 1 & -2 & 1 & & \\ & \ddots & \ddots & \ddots & \\ & & \ddots & \ddots & 1 \\ & & & 1 & -2 \end{bmatrix},$$

$v = [r_2 - r_1, 0, \ldots, 0, r_2 + r_1]^T$, $r_1 = -a/(2\Delta x)$, and $r_2 = b/\Delta x^2$. Here, a, b, and Δx are parameters with values $a = 1$, $b = 5 \times 10^{-2}$, and $\Delta x = 1/(N+1)$. This ODE system arises when the method of lines based on central differences is used to semi-discretize the partial differential equation (PDE)

$$\frac{\partial}{\partial t}u(x,t) + a\frac{\partial}{\partial x}u(x,t) = b\frac{\partial^2}{\partial x^2}u(x,t) + u(x,t)(1 - u(x,t)), \quad 0 \le x \le 1,$$

with Dirichlet boundary conditions $u(0,t) = u(1,t) = 1$. This PDE is of reaction-convection-diffusion type (and could be solved directly with `pdepe`, described in Section 12.5). The ODE solution component $y_j(t)$ approximates $u(j\Delta x, t)$. We suppose that the PDE comes with the initial data $u(x,0) = (1+\cos 2\pi x)/2$, for which it can be

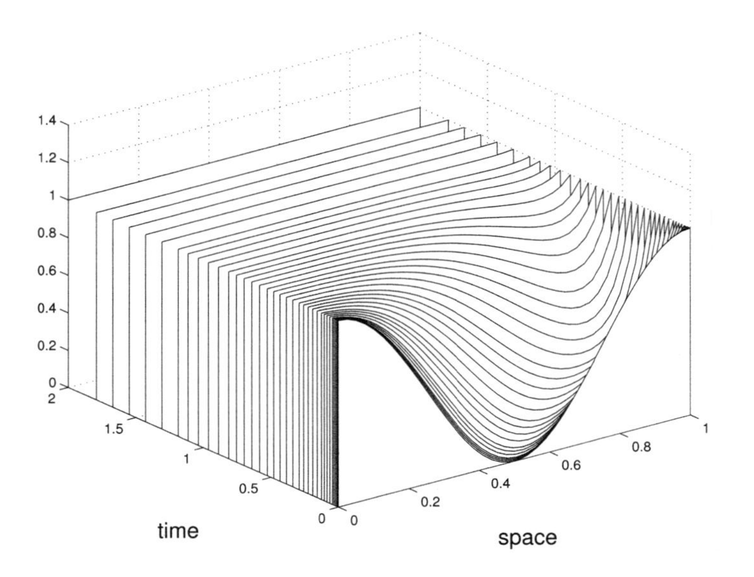

Figure 12.12. *Stiff ODE example, with Jacobian information supplied.*

shown that $u(x,t)$ tends to the steady state $u(x,t) \equiv 1$ as $t \to \infty$. The corresponding ODE initial condition is $(y_0)_j = (1 + \cos(2\pi j/(N+1)))/2$. The Jacobian for this ODE has the form $A + I - 2\,\mathrm{diag}(y(t))$, where I denotes the identity.

Listing 12.6 shows a function `rcd` that implements and solves this system using `ode15s`. We have set $N = 38$ and $0 \le t \le 2$. The `Jacobian` property of `odeset` specifies the nested function `jacobian` that evaluates the Jacobian and that returns it as a `sparse` array. See Chapter 15 for details about sparse matrices and the function `spdiags`. The jth column of the output matrix `y` contains the approximation to $y_j(t)$, and we have created `U` by appending an extra column `ones(size(t))` at each end of `y` to account for the PDE boundary conditions. The plot produced by `rcd` is shown in Figure 12.12.

The ODE solvers can be applied to problems of the form

$$M(t, y(t))\frac{d}{dt}y(t) = f(t, y(t)), \quad y(t_0) = y_0,$$

where the *mass matrix*, $M(t, y(t))$, is square and nonsingular. (The `ode23s` solver applies only when M is independent of t and $y(t)$.) Mass matrices arise naturally when semi-discretization is performed with a finite element method. A mass matrix can be specified in a similar manner to a Jacobian, via `odeset`. The `ode15s` and `ode23t` functions can solve certain problems where M is singular—more precisely, they can be used if the resulting differential-algebraic equation (DAE) is of index 1 and y_0 is sufficiently close to being consistent. DAEs are a class of problems that contain algebraic, as well as differential, constraints on the variables; see [4] or [13] for details.

We now discuss a DAE example: the Chemical Akzo Nobel problem of [112]. This index 1 system has the form

$$M\frac{d}{dt}y(t) = f(y(t)),$$

Listing 12.6. *Function* rcd.

```
function rcd
%RCD Stiff ODE from method of lines on reaction-convection-diffusion problem.

N = 38; a = 1; b = 5e-2;
tspan = [0 2]; space = [1:N]/(N+1);

y0 = 0.5*(1+cos(2*pi*space));
y0 = y0(:);
options = odeset('Jacobian',@jacobian);
options = odeset(options,'RelTol',1e-3,'AbsTol',1e-3);

[t,y] = ode15s(@f,tspan,y0,options);
e = ones(size(t)); U = [e y e];
waterfall([0:1/(N+1):1],t,U)
xlabel('space','FontSize',16), ylabel('time','FontSize',16)

    % -------------------------- Nested functions ---------------------------
    function dydt = f(t,y)
    %F         Differential equation.

    r1 = -a*(N+1)/2;
    r2 = b*(N+1)^2;
    up = [y(2:N);0]; down = [0;y(1:N-1)];
    e1 = [1;zeros(N-1,1)]; eN = [zeros(N-1,1);1];

    dydt = r1*(up-down) + r2*(-2*y+up+down) + (r2-r1)*e1 + ...
           (r2+r1)*eN + y.*(1-y);
    end

    function dfdy = jacobian(t,y)
    %JACOBIAN  Jacobian matrix.

    r1 = -a*(N+1)/2;
    r2 = b*(N+1)^2;
    u = (r2-r1)*ones(N,1);
    v = (-2*r2+1)*ones(N,1) - 2*y;
    w = (r2+r1)*ones(N,1);

    dfdy = spdiags([u v w],[-1 0 1],N,N);
    end

end
```

with $y(t) \in \mathbb{R}^6$, $0 \le t \le 180$,

$$M = \begin{bmatrix} 1 & 0 & 0 & 0 & 0 & 0 \\ 0 & 1 & 0 & 0 & 0 & 0 \\ 0 & 0 & 1 & 0 & 0 & 0 \\ 0 & 0 & 0 & 1 & 0 & 0 \\ 0 & 0 & 0 & 0 & 1 & 0 \\ 0 & 0 & 0 & 0 & 0 & 0 \end{bmatrix} \quad \text{and} \quad f(y) = \begin{bmatrix} -2r_1 + r_2 - r_3 - r_4 \\ -0.5r_1 - r_4 - 0.5r_5 + F_{\text{in}} \\ r_1 - r_2 + r_3 \\ -r_2 + r_3 - 2r_4 \\ r_2 - r_3 + r_5 \\ K_s y_1 y_4 - y_6 \end{bmatrix},$$

where the auxiliary variables are defined as

$$\begin{aligned} r_1 &= k_1 y_1^4 \sqrt{y_2}, & r_2 &= k_2 y_3 y_4, \\ r_3 &= k_2 y_1 y_5 / K, & r_4 &= k_3 y_1 y_4^2, \\ r_5 &= k_4 y_6^2 \sqrt{y_2}, & F_{\text{in}} &= klA\,(p(\text{CO}_2)/H - y_2)\,. \end{aligned}$$

Constants in the system take values $k_1 = 18.7$, $k_2 = 0.58$, $k_3 = 0.09$, $k_4 = 0.42$, $K = 34.4$, $klA = 3.3$, $K_s = 115.83$, $p(\text{CO}_2) = 0.9$, and $H = 737$, and the initial condition is

$$y_0 = [0.444, 0.00123, 0, 0.007, 0, K_s y_1(0) y_4(0)]^T.$$

Details about the mathematical modeling and chemistry issues behind this problem can be found in [112].

The function `chemakzo` in Listing 12.7 solves this DAE using `ode15s`, which can solve DAEs of index 1. We use `odeset` to set up an `options` structure that specifies the mass matrix and reports it as singular. In our case, because the mass matrix M is constant, we are able to specify it as the value of the `Mass` property. More generally, for a mass matrix that depends on t and y, a suitable function, `mymass(t,y)`, say, must be set up and the `Mass` property set to `@mymass`.

After solving the DAE, `chemakzo` plots the six solution components, as shown in Figure 12.13. These agree to visual accuracy with the solution plots in [112]. As a further check, we compare the solution at $t = 180$ with the reference solution supplied in [112], to find that the norm of the difference is `yerr = 2.0626e-6`.

The solver `ode15i` is designed to handle general index 1 DAEs that may be written in the fully implicit form

$$F\left(t, y(t), \frac{dy(t)}{dt}\right) = 0,$$

where F is a given nonlinear function and suitable initial conditions are supplied. A separate function `decic` is available to compute consistent initial conditions for these problems. The reference [99] describes `ode15i` and `decic` and gives examples of their use.

The ODE solvers offer other features that you may find useful. Type `help odeset` to see the full range of properties that can be controlled through the `options` structure. The function `odeget` extracts property values from the `options` structure. The MATLAB ODE solvers are well documented and are supported by a rich variety of example files, some of which we list below. In each case, `help filename` gives an informative description of the file, `type filename` lists the contents of the file, and typing `filename` runs a demonstration. The examples can also be accessed via the `odeexamples` function.

`rigidode`: nonstiff ODE.

`brussode`, `vdpode`: stiff ODEs.

Listing 12.7. *Function* chemakzo.

```
function chemakzo
%CHEMAKZO     Chemical Akzo Nobel problem.
%             Index 1 DAE describing a chemical process.

M = eye(6); M(6,6) = 0;
options = odeset('Mass',M,'MassSingular','yes');

tspan = [0 180];
Ks = 115.83;
y0 = [0.444; 0.00123; 0; 0.007; 0; Ks*0.444*0.007];
[t,y] = ode15s(@chem_rhs,tspan,y0,options);

for i = 1:6
   subplot(2,3,i)
   plot(t,y(:,i),'LineWidth',2), grid on
   title(['y_',int2str(i)]), xlabel('t'), xlim([0 180])
end

% Reference solution at t = 180
yref = [0.1150794920661702;     0.1203831471567715e-2
        0.1611562887407974;     0.3656156421249283e-3
        0.1708010885264404e-1; 0.4873531310307455e-2]';

yerr = norm(y(end,:) - yref)

    % ------------------- Nested function -------------------
    function rhs = chem_rhs(t,y)
    %CHEM_RHS      Right-hand side of DAE

    if y(2) < 0, error('Negative y(2) in DAE function.'), end

    k1 = 18.7; k2 = 0.58; k3 = 0.09; k4 = 0.42;
    K = 34.4; klA = 3.3; pCO2 = 0.9; H = 737;

    r1 = k1*(y(1)^4)*sqrt(y(2));
    r2 = k2*y(3)*y(4);
    r3 = k2*y(1)*y(5)/K;
    r4 = k3*y(1)*(y(4)^2);
    r5 = k4*(y(6)^2)*sqrt(y(2));
    Fin = klA*(pCO2/H - y(2));

    rhs = [-2*r1 + r2 - r3 - r4;
           -0.5*r1 - r4 - 0.5*r5 + Fin;
           r1 - r2 + r3;
           -r2 + r3 - 2*r4;
           r2 - r3 + r5;
           Ks*y(1)*y(4)-y(6)];
    end

end
```

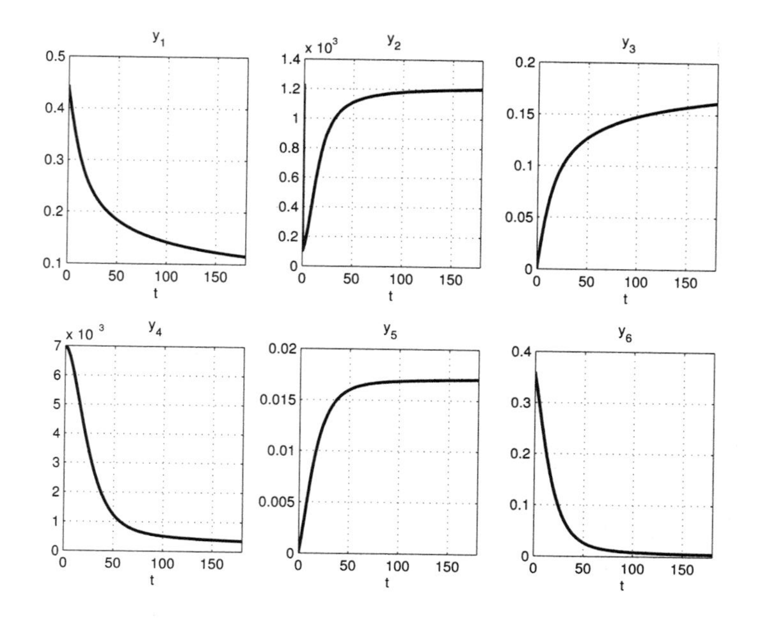

Figure 12.13. *DAE solution components from* `chemakzo` *in Listing* 12.7.

`ballode`: event location problem.

`orbitode`: problem involving event location and the use of an output function (`odephas2`) to process the solution as the integration proceeds.

`fem1ode`, `fem2ode`, `batonode`: ODEs with mass matrices.

`amp1dae`, `hb1dae`, `ihb1dae`, `iburgersode`: DAEs.

12.3. Boundary Value Problems with `bvp4c`

The function `bvp4c` uses a collocation method to solve systems of ODEs in two-point boundary value form. These systems may be written

$$\frac{d}{dx}y(x) = f(x, y(x), p), \quad g(y(a), y(b), p) = 0.$$

Here, as for the initial value problem in the previous section, $y(x)$ is an unknown m-vector and f is a given function of x and y that also produces an m-vector. The vector p, which may be absent, is an unknown vector of parameters to be determined. The solution is required over the range $a \leq x \leq b$ and the given function g specifies the boundary conditions. Note that the independent variable was labeled t in the previous section and is now labeled x. This is consistent with MATLAB's documentation and reflects the fact that two-point boundary value problems (BVPs) usually arise over an interval of space rather than time. Generally, BVPs are more computationally challenging than initial value problems. They may have no solution, and it is common for more than one solution to exist. For these reasons, `bvp4c` requires an initial guess to be supplied for the solution. The initial guess and the final solution are stored in structures. We introduce bvp4c through a simple example before giving more details.

A scalar BVP describing the cross-sectional shape of a water droplet on a flat surface is given by [92]

$$\frac{d^2}{dx^2}h(x) + (1 - h(x))\left(1 + \left(\frac{d}{dx}h(x)\right)^2\right)^{3/2} = 0, \quad h(-1) = 0,\ h(1) = 0.$$

Here, $h(x)$ measures the height of the droplet at point x. We set $y_1(x) = h(x)$ and $y_2(x) = dh(x)/dx$ and rewrite the equation as a system of two first-order equations:

$$\frac{d}{dx}y_1(x) = y_2(x),$$
$$\frac{d}{dx}y_2(x) = (y_1(x) - 1)\left(1 + y_2(x)^2\right)^{3/2}.$$

This system is represented by the function

```
function yprime = drop(x,y)
%DROP    ODE/BVP water droplet example.
%        YPRIME = DROP(X,Y) evaluates derivative.

yprime = [y(2); (y(1)-1)*((1+y(2)^2)^(3/2))];
```

The boundary conditions are specified via a residual function. This function returns zero when evaluated at the boundary values. Our boundary conditions $y_1(-1) = y_1(1) = 0$ can be encoded in the following function:

```
function res = dropbc(ya,yb)
%DROPBC    ODE/BVP water droplet boundary conditions.
%          RES = DROPBC(YA,YB) evaluates residual.

res = [ya(1); yb(1)];
```

As an initial guess for the solution, we use $y_1(x) = \sqrt{1 - x^2}$ and $y_2(x) = -x/(0.1 + \sqrt{1 - x^2})$. This information is set up by the function dropinit:

```
function yinit = dropinit(x)
%DROPINIT    ODE/BVP water droplet initial guess.
%            YINIT = DROPINIT(X) evaluates initial guess at X.

yinit = [sqrt(1-x^2); -x/(0.1+sqrt(1+x^2))];
```

The following code solves the BVP and produces Figure 12.14.

```
solinit = bvpinit(linspace(-1,1,20),@dropinit);
sol = bvp4c(@drop,@dropbc,solinit);
fill(sol.x,sol.y(1,:),[0.7 0.7 0.7])
axis([-1 1 0 1])
xlabel('x','FontSize',16)
ylabel('h','Rotation',0,'FontSize',16)
```

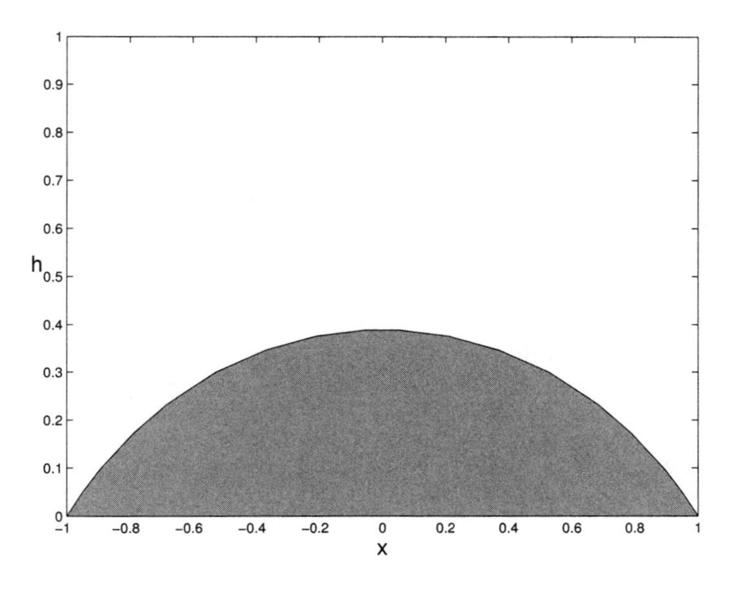

Figure 12.14. *Water droplet BVP solved by* `bvp4c`.

Here, the call to `bvpinit` sets up the structure `solinit`, which contains the data produced by evaluating `dropinit` at 20 equally spaced values between −1 and 1. We then call `bvp4c`, which returns the solution in the structure `sol`. The `fill` command fills the curve that the solution makes in the x-y_1 plane.

In general, `bvp4c` can be called in the form

```
sol = bvp4c(@odefun,@bcfun,solinit,options);
```

Here, `odefun` evaluates the differential equations and `bcfun` gives the residual for the boundary conditions. The function `odefun` has the general form

```
yprime = odefun(x,y)
```

and `bcfun` has the general form

```
res = bcfun(ya,yb)
```

Both functions must return column vectors. The initial guess structure `solinit` has two required fields: `solinit.x` contains the x values at which the initial guess is supplied, ordered from left to right with `solinit.x(1)` and `solinit.x(end)` giving a and b, respectively. Correspondingly, `solinit.y(:,i)` gives the initial guess for the solution at the point `solinit.x(i)`. This structure also allows a guess for a vector of unknown parameters to be specified, as we will see in function `skiprun` below. The helper function `bvpinit` can be used to create the initial guess structure, as in the example above. The remaining arguments for `bvp4c` are optional. The `options` structure allows various properties of the collocation algorithm to be altered from their default values, including the error tolerances and the maximum number of meshpoints allowed. The function `bvpset`, which is similar to `odeset`, can be used to create the required structure; see `doc bvpset` for details.

The output argument `sol` is a structure that contains the numerical solution. The field `sol.x` gives the array of x values at which the solution has been computed. (These points are chosen automatically by `bvp4c`.) The approximate solution at `sol.x(i)` is given by `sol.y(:,i)`. Similarly, an approximate value of the first derivative of the solution at `sol.x(i)` is given by `sol.yp(:,i)`. Unlike for the ODE solvers, the only form in which output from `bvp4c` can be obtained is a structure (and the same is true for the DDE and PDE solvers described in the next two sections).

Note that the structures `solinit` and `sol` above can be given any names, but the field names `x`, `y`, and `yp` must be used.

The function `deval`, described on p. 182, may be used to provide the solution and its derivative at general x values.

Our next example treats a differential equation depending on a parameter and emphasizes that nonlinear BVPs can have nonunique solutions. The equation

$$\frac{d^2}{dx^2}\theta(x) + \lambda \sin\theta(x)\cos\theta(x) = 0, \quad \theta(-1) = 0,\ \theta(1) = 0,$$

arises in liquid crystal theory [66]. Here, $\theta(x)$ quantifies the average local molecular orientation and the constant parameter $\lambda > 0$ is a measure of an applied magnetic field. If λ is small then the only solution to this problem is the trivial one, $\theta(x) \equiv 0$. However, for $\lambda > \pi^2/4 \approx 2.467$ a solution with $\theta(x) > 0$ for $-1 < x < 1$ exists, and $-\theta(x)$ is then also a solution. (Physically, a distorted state of the material may arise if the magnetic field is sufficiently strong.) For the positive solution the midpoint value, $\theta(0)$, increases monotonically with λ and approaches $\pi/2$ as λ tends to infinity. Writing $y_1(x) = \theta(x)$ and $y_2(x) = d\theta(x)/dx$ the ODE becomes

$$\begin{aligned}\frac{d}{dx}y_1(x) &= y_2(x),\\ \frac{d}{dx}y_2(x) &= -\lambda \sin y_1(x)\cos y_1(x).\end{aligned}$$

The function `lcrun` in Listing 12.8 solves the BVP for parameter values $\lambda = 2.4$, 2.5, 3, and 10, producing Figure 12.15. In this example, as for some of the ODE problems in the previous section, we have written a function `lcrun` that has no input or output arguments and created `lc` as a nested function and `lcbc` and `lcinit` as subfunctions of `lcrun`. This allows us to solve the BVP with a single M-file. The nested function `lc` evaluates the ODE right-hand side. The boundary conditions, which are the same as those in the previous example, are coded in `lcbc`. Note that making `lc` a nested function ensures that the parameter `lambda` is known to `lc` when it is called by `bvp4c`. As an initial guess for $\lambda = 10$, we use $y_1(x) = \sin((x+1)\pi/2)$ and $y_2(x) = \pi\cos((x+1)\pi/2)/2$, which is set up by `lcinit`. For the remaining three λ values we use the solution for the previous λ as the starting guess for the next; this is known as continuation in the parameter λ, and it is a valuable (perhaps necessary) technique when solving hard problems. From Figure 12.15 we see that `bvp4c` has found the nontrivial positive solution for each of the three `lambda` values beyond $\pi^2/4$. For continuation to work in this example we need to take the λ values in decreasing order; the increasing order leads to the trivial solution each time.

Our final example involves the equation

$$\frac{d^2}{dx^2}y(x) + \mu y(x) = 0,$$

Listing 12.8. *Function* lcrun.

```
function lcrun
%LCRUN   Liquid crystal BVP.
%        Solves the liquid crystal BVP for four different lambda values.

lambda_vals = [2.4, 2.5, 3, 10];
lambda_vals = lambda_vals(end:-1:1); % Necessary order for continuation.

solinit = bvpinit(linspace(-1,1,20),@lcinit);
lambda = lambda_vals(1); sola = bvp4c(@lc,@lcbc,solinit);
lambda = lambda_vals(2); solb = bvp4c(@lc,@lcbc,sola);
lambda = lambda_vals(3); solc = bvp4c(@lc,@lcbc,solb);
lambda = lambda_vals(4); sold = bvp4c(@lc,@lcbc,solc);
plot(sola.x,sola.y(1,:),'-', 'LineWidth',4), hold on
plot(solb.x,solb.y(1,:),'--','LineWidth',2)
plot(solc.x,solc.y(1,:),'--','LineWidth',4)
plot(sold.x,sold.y(1,:),'--','LineWidth',6), hold off
legend([repmat('\lambda = ',4,1) num2str(lambda_vals')])
xlabel('x','FontSize',16)
ylabel('\theta','Rotation',0,'FontSize',16)
ylim([-0.1 1.5])

    function yprime = lc(x,y)
    %LC      ODE/BVP liquid crystal system.
    yprime = [y(2); -lambda*sin(y(1))*cos(y(1))];
    end

end

function res = lcbc(ya,yb)
%LCBC     ODE/BVP liquid crystal boundary conditions.
res = [ya(1); yb(1)];
end

function yinit = lcinit(x)
%LCINIT   ODE/BVP liquid crystal initial guess.
yinit = [sin(0.5*(x+1)*pi); 0.5*pi*cos(0.5*(x+1)*pi)];
end
```

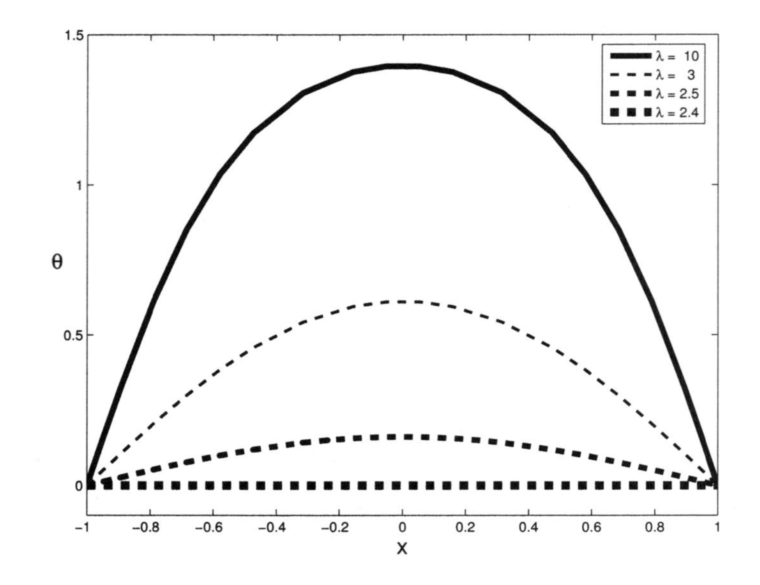

Figure 12.15. *Liquid crystal BVP solved by* `bvp4c`.

with boundary conditions

$$y(0) = 0, \qquad \left(\frac{d}{dx}y(x)\right)_{x=0} = 1, \qquad \left(y(x) + \frac{d}{dx}y(x)\right)_{x=1} = 0.$$

This equation models the displacement of a skipping rope that is fixed at $x = 0$, has elastic support at $x = 1$, and rotates with uniform angular velocity about its equilibrium position along the x-axis [43, Sec. 5.2]. This BVP is an *eigenvalue problem*—we must find a value of the parameter μ for which a solution exists. (We can regard the two conditions at $x = 0$ as defining an initial value problem; we must then find a value of μ for which the solution matches the boundary condition at $x = 1$.) We can use `bvp4c` to solve this eigenvalue problem if we supply a guess for the unknown parameter μ as well as a guess for the corresponding solution $y(x)$. This is done in the function `skiprun` in Listing 12.9. As a first-order system, the differential equation may be written

$$\frac{d}{dx}y_1(x) = y_2(x),$$
$$\frac{d}{dx}y_2(x) = -\mu y_1(x).$$

This system is encoded in the subfunction `skip` and the boundary conditions in `skipbc`. Our initial guess for the solution is $y_1(x) = \sin(x)$, $y_2(x) = \cos(x)$, specified in `skipinit`. Note that the input argument `5` is added in the call to `bvpinit`. This is our guess for μ, and it is stored in the `parameters` field of the structure `solinit` and hence passed to `bvp4c`. Figure 12.16 shows the solution computed by `bvp4c`. The computed value for μ is returned in the `parameters` field of the structure `sol`. We have

```
>> sol = skiprun
```

Listing 12.9. *Function* skiprun.

```
function sol = skiprun
%SKIPRUN  Skipping rope BVP/eigenvalue example.

solinit = bvpinit(linspace(0,1,10),@skipinit,5);
sol = bvp4c(@skip,@skipbc,solinit);
plot(sol.x,sol.y(1,:),'-', sol.x,sol.yp(1,:),'--', 'LineWidth',4)
xlabel('x','FontSize',12)
legend('y_1','y_2')

% ------------------------ Subfunctions -----------------------
function yprime = skip(x,y,mu)
%SKIP     ODE/BVP skipping rope example.
%         YPRIME = SKIP(X,Y,MU) evaluates derivative.
yprime = [y(2); -mu*y(1)];

function res = skipbc(ya,yb,mu)
%SKIPBC   ODE/BVP skipping rope boundary conditions.
%         RES = SKIPBC(YA,YB,MU) evaluates residual.
res = [ya(1); ya(2)-1; yb(1)+yb(2)];

function yinit = skipinit(x)
%SKIPINIT ODE/BVP skipping rope initial guess.
%         YINIT = SKIPINIT(X) evaluates initial guess at X.
yinit = [sin(x); cos(x)];
```

```
sol =

             x: [1x10 double]
             y: [2x10 double]
            yp: [2x10 double]
        solver: 'bvp4c'
    parameters: 4.1159e+000
```

It is known that this BVP has eigenvalues given by $\mu = \gamma^2$, where γ is a solution of $\tan(\gamma) + \gamma = 0$. Using fzero to locate a γ value near 2, we can check the accuracy of the computed μ as follows:

```
>> gam = fzero(@(x)tan(x)+x,2); mu = gam^2;
>> error = abs(sol.parameters - mu)
error =
  2.9343e-005
```

The tutorial [103] gives a range of examples that illustrate the versatility of bvp4c. The examples deal with a number of issues, including

- changing the error tolerances,
- evaluating the solution at any point in the range $[a, b]$,
- choosing appropriate initial guesses by continuation,

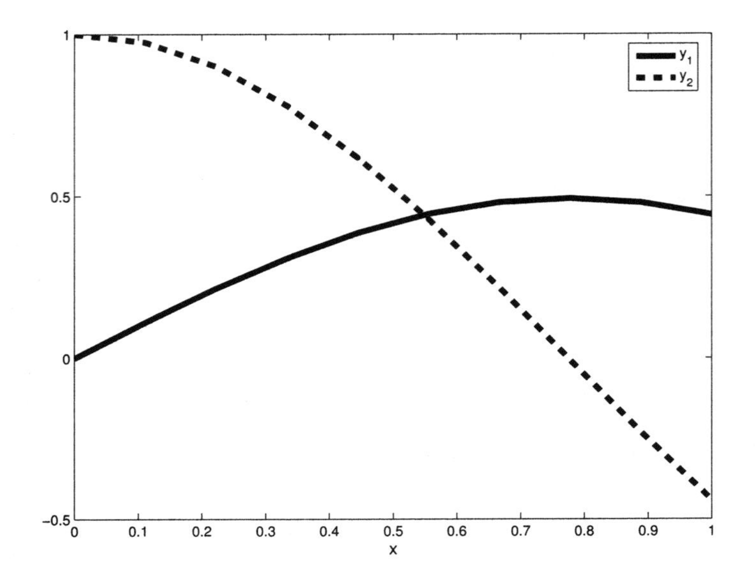

Figure 12.16. *Skipping rope eigenvalue BVP solved by* `bvp4c`*.*

- dealing with singularities,
- solving problems with periodic boundary conditions,
- solving problems over an infinite interval,
- solving multipoint BVPs (where non-endpoint conditions are specified for the solution).

Further information can also be obtained from the help for the functions `bvp4c`, `bvpget`, `bvpinit`, `bvpval`, `bvpset`, from [54], and from the example files

twobvp: solves a BVP that has exactly two solutions;

mat4bvp: finds the fourth eigenvalue of Mathieu's equation;

shockbvp: solves a difficult BVP with a shock layer;

threebvp: solves a three-point BVP.

The function `bvp4c` can also solve a class of boundary value problems over $0 \le x \le b$ that have a singularity at $x = 0$; for details see the online help and [100].

12.4. Delay-Differential Equations with dde23

The function `dde23` solves systems of delay-differential equations (DDEs) of the form

$$\frac{d}{dt}y(t) = f\big(t, y(t), y(t-\tau_1), y(t-\tau_2), \dots, y(t-\tau_k)\big),$$

for $t > t_0$, where $\tau_1, \tau_2, \dots, \tau_k$ are positive constants known as delays or lags. DDEs differ from ODEs in that the right-hand side function, f, depends on the solution value at earlier points $t - \tau_i$ as well as at the current point t. Instead of an initial condition, an *initial function*, $S(t)$, must be specified, such that $y(t) = S(t)$ for $t \leq t_0$. Using dde23 is similar to using one of MATLAB's ODE solvers in the mode where the solution is returned as a structure.

A DDE system for predator-prey populations is [70, (3.11)]

$$\frac{d}{dt}y_1(t) = y_1(t)\left(2\left(1 - \frac{y_1(t)}{50}\right) - \frac{y_2(t)}{y_1(t)+40}\right) - h,$$
$$\frac{d}{dt}y_2(t) = y_2(t)\left(-3 + \frac{6y_1(t-\tau)}{y_1(t-\tau)+40}\right).$$

Here, $y_1(t)$ and $y_2(t)$ denote the densities of the prey and predator populations, respectively, at time t. The delay parameter τ accounts for either (a) a gestation period in the predators or (b) a reaction time in the predators, and the parameter h represents a harvesting rate for the prey. We will take $\tau = 9$, a constant initial function $S(t) = [35, 10]^T$, and try two different harvesting rates, $h = 10$ and $h = 15$. The analysis and computation in [70] show that for $h = 15$ the solution should evolve to the equilibrium state $y_1(t) \equiv 40$, $y_2(t) \equiv 2$, whereas for $h = 10$ a limit cycle is present.

This DDE is solved for $h = 10$ and $h = 15$ by function `harvest` in Listing 12.10, which produces the pictures in Figure 12.17. The first two input arguments for `dde23` are `@f`, the handle of the function that evaluates the right-hand side of the DDE, and `tau`, which specifies the size of the delay. In general, `tau` is a k-vector when there are k lags. Next, `ic = [35;10]` specifies the initial function to be a constant. The vector `tspan` gives the range for the independent variable. In the nested `f`, `Z` is a column vector that represents $y(t - \tau)$.

As for the ODE solvers, the output argument `sol` is a structure, with the field `sol.x` giving an array of t values for which the field `sol.y` is a corresponding array of solution values. A third field, `sol.yp`, provides the first derivative of the solution. Figure 12.17 confirms the expected difference in behavior between $h = 10$ and $h = 15$.

A general call to `dde23` takes the form

```
sol = dde23(@ddefun,delays,history,tspan,options);
```

The function `ddefun` has input arguments `(t,y,Z)` and returns a column vector giving the right-hand side of the DDE. The input argument `Z` is an array whose jth column corresponds to $y(t - \tau_j)$. The second input argument to `dde23`, `delays`, is a row vector that defines the delays; so `delays(j)` is τ_j. Any number of distinct, positive delays may be used. (It is possible to use `dde23` to solve an ODE system—that is, a DDE with no delays. In this case it is best to pass the empty array `[]` as the delay argument.) The function `history` has input argument `t` and returns the value $S(t)$ as a column vector. As we saw in the example above, in the commonly arising case where the initial function is constant it is permissible to supply a vector as the `history` argument, rather than a function name. The input argument `tspan` is used to specify the points at which a solution is required. It has similar functionality to the corresponding argument of the ODE solvers. Similarly, the `options` argument, which may be set by calls to `ddeset`, allows error tolerances and event location requirements to be specified, as in the ODE case. Additionally, points where low-order derivatives of the solution are known to have discontinuities may be specified via `options`. Such information may improve the accuracy and efficiency of `dde23`.

As in the ODE case described on p. 180, the output argument `sol` may be fed into the function `deval` in order to evaluate the solution and its derivative at specified t values.

Listing 12.10. *Function* **harvest**.

```
function harvest
%HARVEST   Predator-prey model with delay and harvesting.

tau = 9;
ic = [35;10];
tspan = [0 250];

h = 10;
sol = dde23(@f,tau,ic,tspan);
subplot(2,1,1)
plot(sol.x,sol.y(1,:),'r-', sol.x,sol.y(2,:),'g--', 'LineWidth',2)
legend('y_1','y_2','Location','East')
title('h = 10','FontSize',12), xlabel t, ylabel('y','Rotation',0)

h = 15;
sol = dde23(@f,tau,ic,tspan);
subplot(2,1,2)
plot(sol.x,sol.y(1,:),'r-', sol.x,sol.y(2,:),'g--', 'LineWidth',2)
legend('y_1','y_2','Location','East')
title('h = 15','FontSize',12), xlabel t, ylabel('y','Rotation',0)

    function v = f(t,y,Z)
    %F            Harvest differential equation.
    v = [y(1)*(2*(1-y(1)/50) - y(2)/(y(1)+40)) - h
         y(2)*(-3 + 6*Z(1)/(Z(1)+40))];
    end

end
```

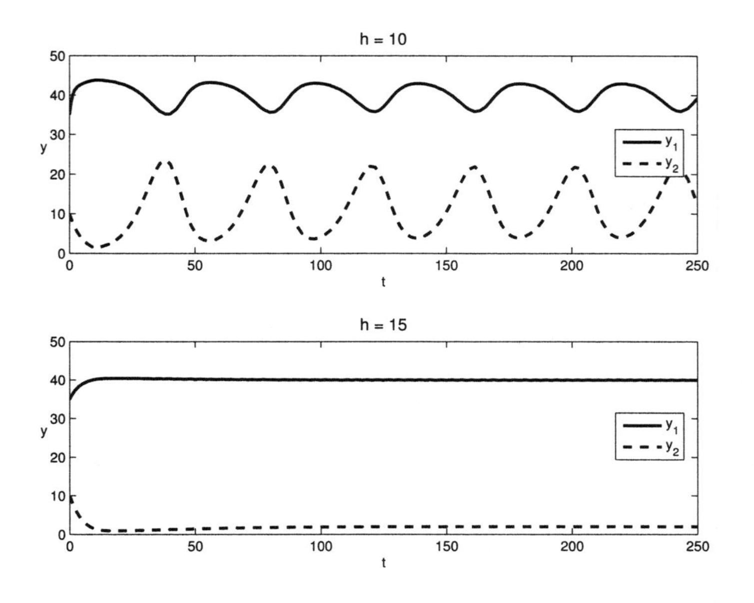

Figure 12.17. *Predator-prey model with delay and harvesting.*

Our second DDE system example is

$$\frac{d}{dt}y_1(t) = -y_1(t) + 2\tanh(y_2(t-\tau_2)),$$
$$\frac{d}{dt}y_2(t) = -y_2(t) - 1.5\tanh(y_1(t-\tau_1)),$$

which models a network of two neurons. Using an initial function of the form $S(t) = 0.1(\sin(t/10), \cos(t/10))^T$, we will examine two pairs of delays: first $\tau_1 = 0.2$ and $\tau_2 = 0.5$ and then $\tau_1 = 0.325$ and $\tau_2 = 0.525$. In the first case, the system should damp down to zero and in the second case a limit cycle should be apparent; see [125] for details. This DDE is solved by function `neural` in Listing 12.11, which produces Figure 12.18.

The tutorial [106] discusses `dde23` more comprehensively and provides downloadable code that illustrates its use, including examples that require event location and discontinuity handling. The theory and algorithmics behind `dde23` are covered in [102] and [105].

12.5. Partial Differential Equations with pdepe

MATLAB's `pdepe` solves a class of parabolic/elliptic PDE systems. These systems involve a vector-valued unknown function u that depends on a scalar space variable, x, and a scalar time variable, t. The general class to which `pdepe` applies has the form

$$c\left(x,t,u,\frac{\partial u}{\partial x}\right)\frac{\partial u}{\partial t} = x^{-m}\frac{\partial}{\partial x}\left(x^m f\left(x,t,u,\frac{\partial u}{\partial x}\right)\right) + s\left(x,t,u,\frac{\partial u}{\partial x}\right),$$

where $a \le x \le b$ and $t_0 \le t \le t_f$. The integer m can be 0, 1, or 2, corresponding to slab, cylindrical, and spherical symmetry, respectively. The function c is a diagonal

Listing 12.11. *Function* neural.

```
function neural
%NEURAL     Neural network model with delays.

tspan = [0 40];
sol = dde23(@f,[0.2,0.5],@history,tspan);
subplot(2,2,1)
plot(sol.x,sol.y(1,:),'r-', sol.x,sol.y(2,:),'g--', 'LineWidth',2)
legend('y_1','y_2')
title('\tau_1 = 0.2, \tau_2 = 0.5','FontSize',12)
xlabel t, ylabel('y','Rotation',0), ylim([-0.2,0.2])

subplot(2,2,3)
plot(sol.y(1,:),sol.y(2,:),'r-')
xlabel y_1, ylabel('y_2','Rotation',0)
xlim([-0.2,0.2]), ylim([-0.1,0.1])

sol = dde23(@f,[0.325,0.525],@history,tspan);
subplot(2,2,2)
plot(sol.x,sol.y(1,:),'r-', sol.x,sol.y(2,:),'g--', 'LineWidth',2)
legend('y_1','y_2')
title('\tau_1 = 0.325, \tau_2 = 0.525','FontSize',12)
xlabel t,  ylabel('y','Rotation',0), ylim([-0.2,0.2])

subplot(2,2,4)
plot(sol.y(1,:),sol.y(2,:),'r-')
xlabel y_1, ylabel('y_2','Rotation',0)
xlim([-0.2,0.2]), ylim([-0.1,0.1])

function v = f(t,y,Z)
%F       Neural network differential equation.
ylag1 = Z(:,1);
ylag2 = Z(:,2);
v = [-y(1) + 2*tanh(ylag2(2))
     -y(2) - 1.5*tanh(ylag1(1))];

function v = history(t)
%HISTORY   Initial function for neural network model
v = 0.1*[sin(t/10);cos(t/10)];
```

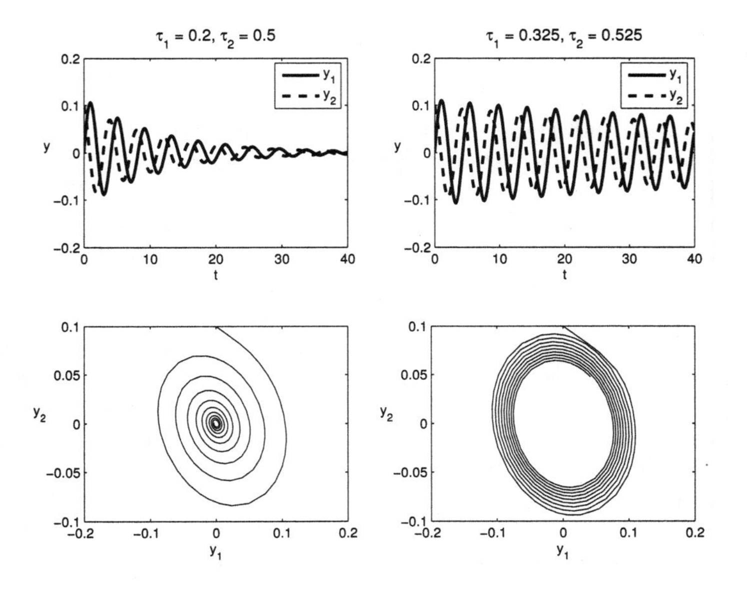

Figure 12.18. *Neural network DDE.*

matrix and the flux and source functions f and s are vector valued. Initial and boundary conditions must be supplied in the following form. For $a \le x \le b$ and $t = t_0$ the solution must satisfy $u(x,t_0) = u_0(x)$ for a specified function u_0. For $x = a$ and $t_0 \le t \le t_f$ the solution must satisfy

$$p_a(x,t,u) + q_a(x,t) f\left(x,t,u,\frac{\partial u}{\partial x}\right) = 0,$$

for specified functions p_a and q_a. Similarly, for $x = b$ and $t_0 \le t \le t_f$,

$$p_b(x,t,u) + q_b(x,t) f\left(x,t,u,\frac{\partial u}{\partial x}\right) = 0$$

must hold for specified functions p_b and q_b. Certain other restrictions are placed on the class of problems that can be solved by pdepe; see doc pdepe for details.

A call to pdepe has the general form

```
sol = pdepe(m,@pdefun,@pdeic,@pdebc,xmesh,tspan,options);
```

which is similar to the syntax for bvp4c. The input argument m can take the values 0, 1, or 2, as described above. The function pdefun has the form

```
function [c,f,s] = pdefun(x,t,u,DuDx)
```

It accepts the space and time variables together with vectors u and DuDx that approximate the solution u and the partial derivative $\partial u/\partial x$, and returns vectors containing the diagonal of the matrix c and the flux and source functions f and s. Initial conditions are encoded in the function pdeic, which takes the form

```
function u0 = pdeic(x)
```

The function pdebc of the form

```
function [pa,qa,pb,qb] = pdebc(xa,ua,xb,ub,t)
```

evaluates p_a, q_a, p_b, and q_b for the boundary conditions at xa $= a$ and xb $= b$. The vector xmesh in the argument list of pdepe is a set of points in $[a, b]$ with xmesh(1) $= a$ and xmesh(end) $= b$, ordered so that xmesh(i) $<$ xmesh(i+1). This defines the x values at which the numerical solution is computed. The algorithm uses a second-order spatial discretization method based on the xmesh values. Hence the choice of xmesh has a strong influence on the accuracy and cost of the numerical solution. Closely spaced xmesh points should be used in regions where the solution is likely to vary rapidly with respect to x. The vector tspan specifies the time points in $[t_0, t_f]$ where the solution is to be returned, with tspan(1) $= t_0$, tspan(end) $= t_f$ and tspan(i) $<$ tspan(i+1). The time integration in pdepe is performed by ode15s and the actual timestep values are chosen dynamically—the tspan points simply determine where the solution is returned and have little impact on the cost or accuracy. The default properties of ode15s can be overridden via the optional input argument options, which can be created with the odeset function (see Section 12.2.1). Altering the defaults is not usually necessary so we do not discuss this further.

The output argument sol is a three-dimensional array such that sol(j,k,i) is the approximation to the ith component of u at the point $t =$ tspan(j), $x =$ xmesh(k). A postprocessing function pdeval is available for computing u and $\partial u/\partial x$ at points that are not in xmesh.

To illustrate the use of pdepe, we begin with the Black–Scholes PDE, famous for modeling derivative prices in financial mathematics. In transformed and dimensionless form [127, Sec. 5.4], using parameter values from [86, Chap. 13], we have

$$\frac{\partial u}{\partial t} = \frac{\partial^2 u}{\partial x^2} + (k-1)\frac{\partial u}{\partial x} - ku, \quad a \le x \le b, \quad t_0 \le t \le t_f,$$

where $k = r/(\sigma^2/2)$, $r = 0.065$, $\sigma = 0.8$, $a = \log(2/5)$, $b = \log(7/5)$, $t_0 = 0$, $t_f = 5$, with initial condition

$$u(x,0) = \max(\exp(x) - 1, 0)$$

and boundary conditions

$$u(a,t) = 0, \quad u(b,t) = \frac{7 - 5\exp(-kt)}{5}.$$

This is of the general form allowed by pdepe with $m = 0$ and

$$c(x,t,u) = 1, \quad f\left(x,t,u,\frac{\partial u}{\partial x}\right) = \frac{\partial u}{\partial x}, \quad s\left(x,t,u,\frac{\partial u}{\partial x}\right) = (k-1)\frac{\partial u}{\partial x} - ku.$$

At $x = a$ the boundary conditions have $p(x,t,u) = u$ and $q(x,t,u) = 0$, and at $x = b$ they have $p(x,t,u) = u - (7 - 5\exp(-kt))/5$ and $q(x,t,u) = 0$. The function bs in Listing 12.12 implements the problem. Here, we have used linspace to generate 40 equally spaced x-values between a and b for the spatial mesh and 20 equally spaced t-values between t_0 and t_f for the output times. The nested function bspde defines the PDE in terms of c, f, and s and bsic specifies the initial condition. Similarly, in nested function bsbc the boundary conditions at $x = a$ and $x = b$ are returned in pa, qa, pb, and qb. We use the 3D plotting function mesh to display the solution. Figure 12.19 shows the resulting picture.

Listing 12.12. *Function* bs.

```
function bs
%BS     Black-Scholes PDE.
%       Solves the transformed Black-Scholes equation.

m = 0;
r = 0.065;
sigma = 0.8;
k = r/(0.5*sigma^2);
a = log(2/5);
b = log(7/5);
t0 = 0;
tf = 5;

xmesh = linspace(a,b,40);
tspan = linspace(t0,tf,20);

sol = pdepe(m,@bspde,@bsic,@bsbc,xmesh,tspan);
u = sol(:,:,1);

mesh(xmesh,tspan,u)
xlabel('x','FontSize',12)
ylabel('t','FontSize',12)
zlabel('u','FontSize',12,'Rotation',0)

    function [c,f,s] = bspde(x,t,u,DuDx)
    %BSPDE  Black-Scholes PDE.
    c = 1;
    f = DuDx;
    s = (k-1)*DuDx-k*u;
    end

    function u0 = bsic(x)
    %BSIC   Initial condition at t = t0.
    u0 = max(exp(x)-1,0);
    end

    function [pa,qa,pb,qb] = bsbc(xa,ua,xb,ub,t)
    %BSBC   Boundary conditions at x = a and x = b.
    pa = ua;
    qa = 0;
    pb = ub - (7 - 5*exp(-k*t))/5;
    qb = 0;
    end

end
```

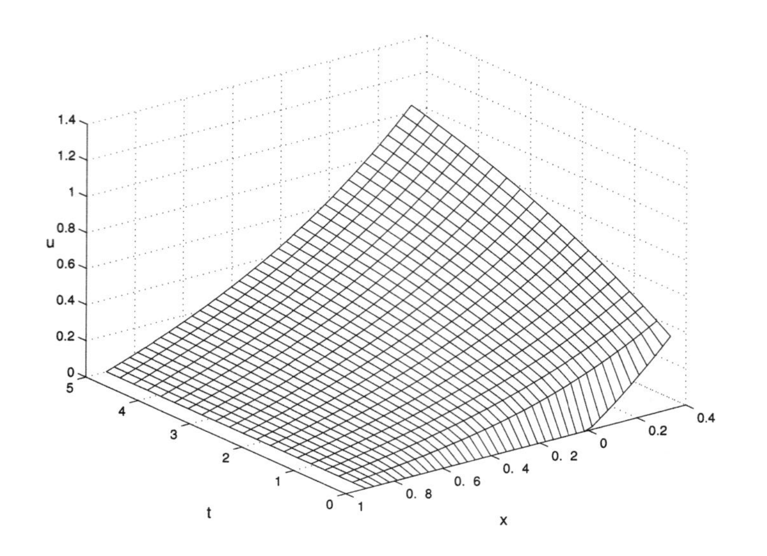

Figure 12.19. *Black–Scholes solution with* `pdepe`.

Next, we look at a system of two reaction-diffusion equations of a type that arises in mathematical biology [45, Chap. 12]:

$$\frac{\partial u}{\partial t} = \frac{1}{2}\frac{\partial^2 u}{\partial x^2} + \frac{1}{1+v^2},$$
$$\frac{\partial v}{\partial t} = \frac{1}{2}\frac{\partial^2 v}{\partial x^2} + \frac{1}{1+u^2},$$

for $0 \le x \le 1$ and $0 \le t \le 0.2$. Our initial conditions are

$$u(x,0) = 1 + \tfrac{1}{2}\cos(2\pi x), \quad v(x,0) = 1 - \tfrac{1}{2}\cos(2\pi x),$$

and our boundary conditions are

$$\frac{\partial u}{\partial x}(0,t) = \frac{\partial u}{\partial x}(1,t) = \frac{\partial v}{\partial x}(0,t) = \frac{\partial v}{\partial x}(1,t) = 0.$$

To put this into the framework of `pdepe` we write (u, v) as (u_1, u_2) and express the PDE as

$$\begin{bmatrix} 1 & 0 \\ 0 & 1 \end{bmatrix} \times \frac{\partial}{\partial t}\begin{bmatrix} u_1 \\ u_2 \end{bmatrix} = \frac{\partial}{\partial x}\begin{bmatrix} \frac{1}{2}\partial u_1/\partial x \\ \frac{1}{2}\partial u_2/\partial x \end{bmatrix} + \begin{bmatrix} 1/(1+u_2^2) \\ 1/(1+u_1^2) \end{bmatrix}.$$

The function `mbiol` in Listing 12.13 solves the PDE system. Note that the output arguments `c`, `f`, `s`, `pa`, `qa`, `pb` and `qb` in the subfunctions `mbpde` and `mbbc` are 2-by-1 arrays, because there are two PDEs in the system. The solutions plotted with `surf` can be seen in the upper part of Figure 12.20. It follows from [45, Ex. 12.5] that the energy

$$E(t) = \frac{1}{2}\int_0^1 \left[\left(\frac{\partial u}{\partial x}\right)^2 + \left(\frac{\partial v}{\partial x}\right)^2\right] dx$$

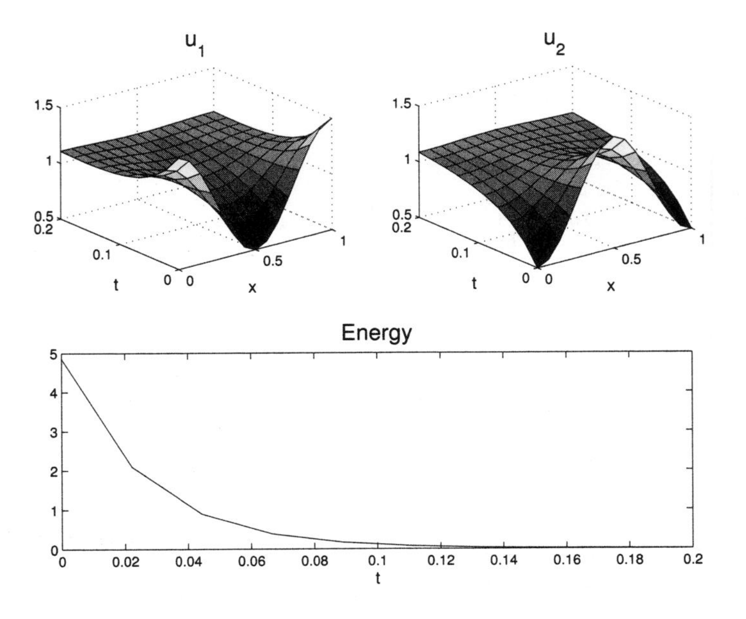

Figure 12.20. *Reaction-diffusion system solution with* `pdepe`.

decays exponentially to zero as $t \to \infty$. To verify this fact numerically, we use simple finite differences and quadrature in `mbiol` to approximate the energy integral. (Alternatively, the function `pdeval` could be used to obtain approximations to $\partial u/\partial x$ and $\partial v/\partial x$.) The resulting plot of $E(t)$ is given in the lower part of Figure 12.20.

Further examples of `pdepe` in use can be found in `doc pdepe`. We note that `pdepe` is designed to solve a subclass of small systems of parabolic and elliptic PDEs to modest accuracy. If your PDE is not suitable for `pdepe` then the Partial Differential Equation Toolbox might be appropriate.

Listing 12.13. *Function* mbiol.

```
function mbiol
%MBIOL   Reaction-diffusion system from mathematical biology.
%        Solves the PDE and tests the energy decay condition.

m = 0;
xmesh = linspace(0,1,15);
tspan = linspace(0,0.2,10);
sol = pdepe(m,@mbpde,@mbic,@mbbc,xmesh,tspan);
u1 = sol(:,:,1);
u2 = sol(:,:,2);

subplot(221)
surf(xmesh,tspan,u1)
xlabel('x','FontSize',12)
ylabel('t','FontSize',12)
title('u_1','FontSize',16)

subplot(222)
surf(xmesh,tspan,u2)
xlabel('x','FontSize',12)
ylabel('t','FontSize',12)
title('u_2','FontSize',16)

% Estimate energy integral.
dx = xmesh(2) - xmesh(1);  % Constant spacing.
energy = 0.5*sum( (diff(u1,1,2)).^2 + (diff(u2,1,2)).^2, 2)/dx;
subplot(212)
plot(tspan',energy)
xlabel('t','FontSize',12)
title('Energy','FontSize',16)

% ----------------------- Subfunctions -----------------------
function [c,f,s] = mbpde(x,t,u,DuDx)
c = [1; 1];
f = DuDx/2;
s = [1/(1+u(2)^2); 1/(1+u(1)^2)];

function u0 = mbic(x);
u0 = [1+0.5*cos(2*pi*x); 1-0.5*cos(2*pi*x)];

function [pa,qa,pb,qb] = mbbc(xa,ua,xb,ub,t)
pa = [0; 0];
qa = [1; 1];
pb = [0; 0];
qb = [1; 1];
```

Multidimensional integrals are another whole multidimensional bag of worms.

— WILLIAM H. PRESS, SAUL A. TEUKOLSKY,
WILLIAM T. VETTERLING, and BRIAN P. FLANNERY,
Numerical Recipes in FORTRAN (1992)

Perhaps the crudest way to evaluate $\int_y^x f(u)du$
is to plot the graph of $f(u)$ *on uniformly squared paper*
and then count the squares that lie inside the desired area.
This method gives numerical integration its other name:
numerical quadrature.
Another way, suitable for chemists,
is to plot the graph on paper of uniform density,
cut out the area in question, and weigh it.

— WILLIAM M. KAHAN, *Handheld Calculator Evaluates Integrals* (1980)

The options vector is optional.

— LAWRENCE F. SHAMPINE and MARK W. REICHELT,
The MATLAB ODE Suite (1997)

Just about any BVP can be formulated for solution with bvp4c.

— LAWRENCE F. SHAMPINE, JACEK KIERZENKA, and MARK W. REICHELT,
Solving Boundary Value Problems for Ordinary
Differential Equations in MATLAB with bvp4c (2000)

Chapter 13
Input and Output

In this chapter we discuss how to obtain input from the user, how to display information on the screen, and how to read and write text files. Note that textual output can be captured into a file (perhaps for subsequent printing) using the **diary** command, as described on p. 31. How to print and save figures is discussed in Section 8.4.

13.1. User Input

User input can be obtained with the **input** function, which displays a prompt and waits for a user response:

```
>> x = input('Starting point: ')
Starting point: 0.5
x =
    0.5000
```

Here, the user has responded by typing "0.5", which is assigned to x. The input is interpreted as a string when an argument 's' is appended:

```
>> mytitle = input('Title for plot: ','s')
Title for plot: Experiment 2
mytitle =
Experiment 2
```

The function **ginput** collects data via mouse clicks. The command

```
[x,y] = ginput(n)
```

returns in the vectors x and y the coordinates of the next n mouse clicks from the current figure window. Input can be terminated before the nth mouse click by pressing the return key. One use of **ginput** is to find the approximate location of points on a graph. For example, with Figure 8.7 in the current figure window, you might type [x,y] = ginput(1) and click on one of the places where the curves intersect. As another example, the first two lines of the Bezier curve example on p. 97 can be replaced by

```
axis([0 1 0 1])
[x,y] = ginput(4);
P = [x';y'];
```

Now the control points are determined by the user's mouse clicks.

The **pause** command suspends execution until a key is pressed, while **pause(n)** waits for n seconds before continuing. Typical use of **pause** is between plots displayed in sequence. In the past it was also used in conjunction with the `echo` command in M-files intended for demonstration, though this usage has been superseded by demonstrations shown in the Help browser.

13.2. Output to the Screen

The results of MATLAB computations are displayed on the screen whenever a semicolon is omitted after an assignment and the format of the output can be varied using the `format` command. But much greater control over the output is available with the use of several functions.

The `disp` function displays the value of a variable, according to the current `format`, without first printing the variable name and "=". If its argument is a string, `disp` displays the string. Example:

```
>> disp('Here is a 3-by-3 magic square'), disp(magic(3))
Here is a 3-by-3 magic square
     8     1     6
     3     5     7
     4     9     2
```

More sophisticated formatting can be done with the `fprintf` function. The syntax is `fprintf(`*format*`,`*list-of-expressions*`)`, where *format* is a string that specifies the precise output format for each expression in the list. In the example

```
>> fprintf('%6.3f\n', pi)
 3.142
```

the `%` character denotes the start of a format specifier requesting a field width of 6 with 3 digits after the decimal point and `\n` denotes a new line (without which subsequent output would continue on the same line). If the specified field width is not large enough MATLAB expands it as necessary:

```
>> fprintf('%6.3f\n', pi^10)
93648.047
```

The fixed point notation produced by `f` is suitable for displaying integers (using `%n.0f`) and when a fixed number of decimal places are required, such as when displaying dollars and cents (using `%n.2f`). If `f` is replaced by `e` then the digit after the period denotes one less than the total number of significant digits to display in exponential notation (there will always be one digit before the decimal point):

```
>> fprintf('%12.3e\n', pi)
  3.142e+000
```

When choosing the field width remember that for a negative number a minus sign occupies one position:

```
>> fprintf('%5.2f\n%5.2f\n',exp(1),-exp(1))
 ?.72
   72
```

'gn just after the % character causes the field to be left-justified. Compare

```
>> fprintf('%5.0f\n%5.0f\n',9,103)
    9
  103

>> fprintf('%-5.0f\n%-5.0f\n',9,103)
9
103
```

The format string can contain characters to be printed literally, as the following example shows:

```
>> iter = 11; m = 5;  U = orth(randn(m)) + 1e-10;

>> fprintf('iter = %2.0f\n', iter)
iter = 11

>> fprintf('norm(U''*U-I) = %11.4e\n', norm(U'*U - eye(m)))
norm(U'*U-I) = 8.4618e-010
```

Note that, within a string, `''` represents a single quote.

To print `%` and `\` use `\%` and `\\` in the format string. Another useful format specifier is g, which uses whichever of `e` and `f` produces the shorter result:

```
>> fprintf('%g %g\n', exp(1), exp(20))
2.71828 4.85165e+008
```

Various other specifiers and special characters are supported by `fprintf`, which behaves similarly to the C function of the same name; see `doc fprintf`.

If more numbers are supplied to be printed than there are format specifiers in the `fprintf` statement then the format specifiers are reused, with elements being taken from a matrix down the first column, then down the second column, and so on. This feature can be used to avoid a loop. Example:

```
>> A = [30 40 60 70];
>> fprintf('%g miles/hour = %g kilometers/hour\n', [A; 8*A/5])
30 miles/hour = 48 kilometers/hour
40 miles/hour = 64 kilometers/hour
60 miles/hour = 96 kilometers/hour
70 miles/hour = 112 kilometers/hour
```

To print a string variable use the `s` format specifier. This example makes use of a cell array (see Section 18.3):

```
>> data = {'Alan Turing', 1912, 1954};
>> fprintf('%s (%4.0f-%4.0f)\n', data{1:3})
Alan Turing (1912-1954)
```

The function `sprintf` is analogous to `fprintf` but returns its output as a string. It is useful for producing labels for plots. A simpler to use but less versatile alternati is `num2str`: `num2str(x,n)` converts `x` to a string with `n` significant digits, defaulting to 4. For converting integers to strings, `int2str` can be used. three examples, the second and third of which make use of string concat Section 18.1).

```
>> n = 16;
>> err_msg = sprintf('Must supply a %d-by-%d matrix', n, n)
err_msg =
Must supply a 16-by-16 matrix

>> disp(['Pi is given to 6 significant figures by ' num2str(pi,6)])
Pi is given to 6 significant figures by 3.14159

>> i = 3;
>> title_str = ['Result of experiment ' int2str(i)]
title_str =
Result of experiment 3
```

13.3. File Input and Output

A number of functions are provided for reading and writing binary and formatted text files; type **help iofun** to see the complete list.

We show by example how to write data to a formatted text file and then read it back in. Before operating on a file it must be opened with the **fopen** function, whose first argument is the filename and whose second argument is a file permission, which has several possible values including **'r'** for read and **'w'** for write. A file identifier is returned by **fopen**; it is used in subsequent read and write statements to specify the file. Data is written using the **fprintf** function, which takes as its first argument the file identifier. Thus the code

```
A = [30 40 60 70];
fid = fopen('myoutput','w');
fprintf(fid,'%g miles/hour = %g kilometers/hour\n', [A; 8*A/5]);
fclose(fid);
```

creates a file **myoutput** containing

```
30 miles/hour = 48 kilometers/hour
40 miles/hour = 64 kilometers/hour
60 miles/hour = 96 kilometers/hour
70 miles/hour = 112 kilometers/hour
```

The file can be read in as follows.

```
>> fid = fopen('myoutput','r');
>> X = fscanf(fid,'%g miles/hour = %g kilometers/hour')
```

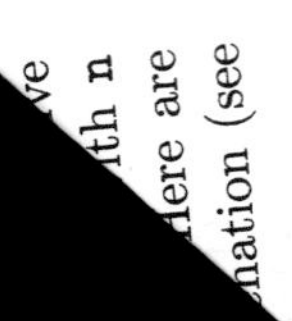

The `fscanf` function reads data formatted according to the specified format string, which in this example says, "read a general floating point number (%g), skip over the string `' miles/hour = '`, read another general floating point number and skip over the string `' kilometers/hour'`. The format string is recycled until the entire file has been read and the output is returned in a vector. We can convert the vector to the original matrix format using

```
>> X = reshape(X,2,4)'
X =
    30    48
    40    64
    60    96
    70   112
```

Alternatively, a matrix of the required shape can be obtained directly:

```
>> X = fscanf(fid,'%g miles/hour = %g kilometers/hour',[2 inf])'
X =
    30    48
    40    64
    60    96
    70   112
```

The third argument to `fscanf` specifies the dimensions of the output matrix, which is filled column by column. We specify `inf` for the number of columns, to allow for any number of lines in the file, and transpose to recover the original format.

Another way to read a text file is with the powerful `textscan` function, which uses a different syntax than `fscanf` for the format string and returns the output in a cell array:

```
>> fid = fopen('myoutput','r');
>> C = textscan(fid,'%f miles/hour = %f kilometers/hour')
C =
    [4x1 double]    [4x1 double]
>> fclose(fid);

>> C{1}'
ans =
    30    40    60    70
>> C{2}'
ans =
    48    64    96   112
```

`textscan` has a number of configurable parameters and is recommended for reading large files.

Binary files are created and read using the functions `fread` and `fwrite`. See the online help for details of their usage.

Make input easy to prepare and output self-explanatory.

— BRIAN W. KERNIGHAN and P. J. PLAUGER, *The Elements of Programming Style* (1978)

Output is almost *like input but it's not input, it's output.*
To correlate the output with the input and
to verify that the input was put in correctly,
it's a good idea to output the input along with the output.

— ROGER EMANUEL KAUFMAN, *A FORTRAN Coloring Book* (1978)

On two occasions I have been asked [by members of Parliament],
"Pray, Mr. Babbage,
if you put into the machine wrong figures,
will the right answers come out?"
I am not able rightly to apprehend
the kind of confusion of ideas
that could provoke such a question.

— CHARLES BABBAGE

Chapter 14 Troubleshooting

14.1. Errors

Errors in MATLAB are of two types: syntax errors and runtime errors. A syntax error is illustrated by

```
>> for i=1#10, x(i) = 1/i; end
??? for i=1#10, x(i) = 1/i; end
             |
Error: Missing variable or function.
```

Here a # has been typed instead of a colon and the error message pinpoints where the problem occurs. If an error occurs in an M-file then the name of the M-file and the line on which the error occurred are shown.

A runtime error occurs with the script `fib` in Listing 14.1. The loop should begin at `i = 3` to avoid referencing `x(0)`. When we run the script, MATLAB produces an informative error message:

```
>> fib
??? Attempted to access x(0); index must be a positive integer
    or logical.

Error in ==> fib at 4
    x(i) = x(i-1) + x(i-2);
```

In the Command Window, the phrase `fib at 4` is an underlined hyperlink to line 4 of the script `fib`. If you click on the hyperlink then `fib.m` is opened in the MATLAB Editor/Debugger (see Section 7.2) and the cursor is placed on line 4.

When an error occurs in a nested sequence of M-file calls, the history of the calls is shown in the error message. The first "`Error in`" line is the one describing the M-file in which the error is located.

MATLAB's error messages are sometimes rather unhelpful and occasionally misleading. It is perhaps inevitable with such a powerful language that an error message does not always make it immediately clear what has gone wrong. We give a few examples illustrating error messages generated for reasons that are perhaps not obvious.

• `This statement is incomplete.` The message is produced by the following code, which is one way of implementing the sign function (MATLAB's `sign`):

```
if x > 0
   f = 1;
else if x == 0
   f = 0;
```

Listing 14.1. *Script* `fib` *that generates a runtime error.*

```
%FIB        Fibonacci numbers.
x = ones(50,1);
for i = 2:50
    x(i) = x(i-1) + x(i-2);
end
```

```
else
   f = -1;
end
```

The problem is an unwanted space between `else` and `if`. MATLAB (correctly) interprets the `if` after the `else` as starting a new `if` statement and then complains when it runs out of `end`s to match the `if`s.

• `Undefined function or variable.` Several commands, such as `clear`, `load`, and `global`, take a list of arguments separated by spaces. If a comma is used in the list it is interpreted as separating statements, not arguments. For example, the command `clear a,b` clears `a` and prints `b`, so if `b` is undefined the above error message is produced.

• `Matrix must be square.` This message is produced when an attempt is made to exponentiate a nonsquare matrix, and can be puzzling. For example, it is generated by the expression `(1:5)^3`, which was presumably meant to be an elementwise cubing operation and thus should be expressed as `(1:5).^3`.

• `At least one operand must be scalar.` This message is generated when elementwise exponentiation is intended but `^` is typed instead of `.^`, as in `(1:5)^(1:5)`.

Many functions check for error conditions, issuing an error message and terminating when one occurs. For example:

```
>> mod(3,sqrt(-2))
??? Error using ==> mod
Arguments must be real.
```

In an M-file this behavior can be achieved with the `error` command:

```
if ~isreal(arg2), error('Arguments must be real.'), end
```

produces the result just shown when `arg2` is not real. An invocation of `error` can also give arguments and use a format conversion string, just as with `fprintf` (see Section 13.2). For example, assuming that `A = zeros(3)`:

```
>> error('Matrix dimension is %g, but should be even.', length(A))
??? Matrix dimension is 3, but should be even.
```

The most recent error message can be recalled with the `lasterr` function.

Errors support message identifiers that identify the error. Identifiers are more commonly used with warnings, and they are described in the next section.

14.2. Warnings

The function `warning`, like `error`, displays its string argument, but execution continues instead of stopping. The reason for using `warning` rather than displaying a string with `disp` (for example) is that the display of warning messages can be controlled via certain special string arguments to `warning`. In particular, `warning('off')` or `warning off` turns off the display of warning messages and `warning('on')` or `warning on` turns them back on again.

MATLAB allows message identifiers to be attached to warnings in order to identify the source of the warning and to allow warnings to be turned off and on individually. For example:

```
>> 1/0
Warning: Divide by zero.
ans =
   Inf
```

To suppress this warning, we need to find its identifier, which can be obtained as the second output argument of `lastwarn`:

```
>> [warnmsg,msg_id] = lastwarn
warnmsg =
Divide by zero.
msg_id =
MATLAB:divideByZero
```

We can turn this warning off, without affecting the status of any other warnings, with

```
>> warning off MATLAB:divideByZero % Or warning('off',msg_id)
>> 1/0
ans =
   Inf
```

In MATLAB 6.5, the identifiers of warnings were automatically displayed. This behavior can be obtained in MATLAB 7 with `warning on verbose`:

```
>> warning on verbose
>> inv([1 1; 1 1])
Warning: Matrix is singular to working precision.
(Type "warning off MATLAB:singularMatrix" to suppress this warning.)
ans =
   Inf   Inf
   Inf   Inf
```

The status of all warnings can be viewed:

```
>> warning query all
The default warning state is 'on'. Warnings not set to the
default are

  State  Warning Identifier

    off  MATLAB:UsingLongNames
```

```
    off  MATLAB:divideByZero
    off  MATLAB:intConvertNaN
    off  MATLAB:intConvertNonIntVal
    ...
```

Here, we have truncated the list. Apart from the second warning, all the other warnings are in the state they were in when our MATLAB session started.

You can define your own warnings with identifiers using the syntax

```
warning('msg_id','warnmsg')
```

(or by making `'warnmsg'` a format string and following it by a list of arguments). Here, `'msg_id'` is a string comprising a component field, which might identify a product or a toolbox, followed by a mnemonic field relating to the message. The actual warning message is in the string `'warnmsg'`. For example, the function `fd_deriv` in Listing 10.1 is likely to return inaccurate results if `h` is close to `eps`, so we could append to the function the lines

```
if h <= 1e-14
   warning('MATLABGuide:fd_deriv:RoundingErrorMayDominate',...
           'Difference H of order EPS may produce inaccurate result.')
end
```

If we do so, then the following behavior is observed:

```
>> fd_deriv(@exp,2,1e-15);
Warning: Difference H of order EPS may produce inaccurate result.
> In fd_deriv at 11
```

A quick way to turn off a warning that has just been invoked is to type `warning off last`.

If you change the `warning` state in an M-file it is good practice to save the old state and restore it before the end of the M-file, as in the following example:

```
warns = warning;   % or "warns = warning('query','all')"
warning('off')     % or "warning off all"
...
warning(warns)
```

14.3. Debugging

Debugging MATLAB M-files is in principle no different to debugging any other type of computer program, but several facilities are available to ease the task. When an M-file runs but does not perform as expected it is often helpful to print out the values of key variables, which can be done by removing semicolons from assignment statements or adding statements consisting of the relevant variable names.

When it is necessary to inspect several variables and the relations between them the `keyboard` statement is invaluable. When a `keyboard` statement is encountered in an M-file execution halts and a command line with the special prompt `K>>` appears. Any MATLAB command can be executed and variables in the workspace can be inspected or changed. When keyboard mode is invoked from within a function the visible workspace is that of the function. The command `dbup` changes the workspace to that of the calling function or the main workspace; `dbdown` reverses the effect of `dbup`. Typing `return` followed by the return key causes execution of the M-file to be resumed. The `dbcont` command has the same effect. Alternatively, the `dbquit` command quits keyboard mode and terminates the M-file.

Another way to invoke keyboard mode is via the debugger. Typing

```
dbstop in foo at 5
```

sets a breakpoint at line 5 of `foo.m`; this causes subsequent execution of `foo.m` to stop just before line 5 and keyboard mode to be entered. The command

```
dbstop in foo at 3 if i==5
```

sets a conditional breakpoint that causes execution to stop only if the given expression evaluates to true, that is, if $i = 5$. A listing of `foo.m` with line numbers is obtained with `dbtype foo`. Breakpoints are cleared using the `dbclear` command. Breakpoints can also be set from the MATLAB Editor/Debugger.

We illustrate the use of the debugger on the script `fib` discussed in the last section (Listing 14.1). Here, we set a breakpoint on a runtime error and then inspect the value of the loop index when the error occurs:

```
>> dbstop if error
>> fib
??? Attempted to access x(0); index must be a positive integer
    or logical.

Error in ==> fib at 4
    x(i) = x(i-1) + x(i-2);

K>> i
i =
     2
K>> dbquit
```

MATLAB's debugger is a powerful tool with several other features that are described in the online documentation. In addition to the command line interface to the debugger illustrated above, an Editor/Debugger window is available that provides a visual interface (see Section 7.2).

A useful tip for debugging is to execute

```
>> clear all
```

and one of

```
>> clf
>> close all
```

before executing the code with which you are having trouble. The first command clears variables and functions from memory. This is useful when, for example, you are working with scripts because it is possible for existing variables to cause unexpected behavior or to mask the fact that a variable is accessed before being initialized in the script. The other commands are useful for clearing the effects of previous graphics operations.

14.4. Pitfalls

Here are some suggestions to help avoid pitfalls particular to MATLAB.

- If you use functions `i` or `j` for the imaginary unit, make sure that they have not previously been overridden by variables of the same name (`clear i` or `clear j` clears the variable and reverts to the functional form). In general it is not advisable to choose variable names that are the names of MATLAB functions. For example, if you assign

```
>> rand = 1;
```

then subsequent attempts to use the `rand` function generate an error:

```
>> A = rand(3)
??? Index exceeds matrix dimensions.
```

In fact, MATLAB is still aware of the function `rand`, but the variable takes precedence, as can be seen from

```
>> which -all rand
rand is a variable.
C:\MATLAB7\toolbox\matlab\elmat\rand.bi
C:\MATLAB7\toolbox\matlab\elmat\rand.m   % Shadowed
```

The function can be reinstated by clearing the variable:

```
>> clear rand
>> rand
ans =
    0.6068
```

- Confusing behavior can sometimes result from the fact that `max`, `min`, and `sort` behave differently for real and for complex data—in the complex case they work with the absolute values of the data. For example, suppose we compute the following 4-vector, which should be real but has a tiny nonzero imaginary part due to rounding errors:

```
e =
   4.0076e+000 -2.7756e-016i
  -6.2906e+000 +3.8858e-016i
  -2.9444e+000 +4.9061e-017i
   9.3624e-001 +1.6575e-016i
```

To find the most negative element we need to use `min(real(e))` rather than `min(e)`:

```
>> min(e)
ans =
   9.3624e-001 +1.6575e-016i

>> min(real(e))
ans =
  -6.2906e+000
```

- Mathematical formulae and descriptions of algorithms often index vectors and matrices so that their subscripts start at 0. Since subscripts of MATLAB arrays start at 1, translation of subscripts is necessary when implementing such formulae and algorithms in MATLAB.

The road to wisdom?
Well, it's plain and simple to express:
Err
and err
and err again
but less
and less
and less.
— PIET HEIN, *Grooks* (1966)

Beware of bugs in the above code;
I have only proved it correct, not tried it.
— DONALD E. KNUTH[8] (1977)

Test programs at their boundary values.
— BRIAN W. KERNIGHAN and P. J. PLAUGER,
The Elements of Programming Style (1978)

By June 1949 people had begun to realize that
it was not so easy to get a program right as had at one time appeared...
The realization came over me with full force that
a good part of the remainder of my life was going to be spent in
finding errors in my own programs.
— MAURICE WILKES, *Memoirs of a Computer Pioneer* (1985)

[8]See `http://www-cs-faculty.stanford.edu/~knuth/faq.html`

Chapter 15
Sparse Matrices

A sparse matrix is one with a large percentage of zero elements. When dealing with large, sparse matrices, it is desirable to take advantage of the sparsity by storing and operating only on the nonzeros. MATLAB arrays of dimension up to 2 can have a `sparse` attribute, in which case just the nonzero entries of the array together with their row and column indices are stored. Currently, sparse arrays are supported only for the `double` data type. In this chapter we will use the term "sparse matrix" for a two-dimensional `double` array having the `sparse` attribute and "full matrix" for such an array having the (default) `full` attribute.

15.1. Sparse Matrix Generation

Sparse matrices can be created in various ways, several of which involve the `sparse` function. Given a `t`-vector `s` of matrix entries and `t`-vectors `i` and `j` of indices, the command `A = sparse(i,j,s)` defines a sparse matrix `A` of dimension `max(i)`-by-`max(j)` with `A(i(k),j(k)) = s(k)`, for `k=1:t` and all other elements zero. Example:

```
>> A = sparse([1 2 2 4 4],[3 1 4 2 4],1:5)
A =
   (2,1)        2
   (4,2)        4
   (1,3)        1
   (2,4)        3
   (4,4)        5
```

MATLAB displays a sparse matrix by listing the nonzero entries preceded by their indices, sorted by columns. If an index `i(k),j(k)` is supplied more than once then the corresponding entries are added:

```
>> sparse([1 2 2 4 1],[3 1 4 2 3],1:5)
ans =
   (2,1)        2
   (4,2)        4
   (1,3)        6
   (2,4)        3
```

A sparse matrix can be converted to a full one using the `full` function:

```
>> B = full(A)
B =
     0     0     1     0
     2     0     0     3
```

```
     0     0     0     0
     0     4     0     5
```

Conversely, a full matrix `B` is converted to the sparse storage format by `A = sparse(B)`. The number of nonzeros in a sparse (or full) matrix is returned by `nnz`:

```
>> nnz(A)
ans =
     5
```

After defining `A` and `B`, we can use the `whos` command to check the amount of storage used:

```
whos
  Name          Size                    Bytes  Class

  A             4x4                        80  double array (sparse)
  B             4x4                       128  double array
  ans           1x1                         8  double array

Grand total is 22 elements using 216 bytes
```

The matrix `B` comprises 16 double precision numbers of 8 bytes each, making a total of 128 bytes. The storage required for a sparse `n`-by-`n` matrix with `nnz` nonzeros is `8*nnz + 4*(nnz+n+1)` bytes, which includes the `nnz` double precision numbers plus some 4-byte integers. The same formula applies to `m`-by-`n` matrices, since the number of rows does not affect the required storage.

The `sparse` function accepts three extra arguments. The command

```
A = sparse(i,j,s,m,n)
```

constructs an `m`-by-`n` sparse matrix; the last two arguments are necessary when the last row or column of `A` is all zero. The command

```
A = sparse(i,j,s,m,n,nzmax)
```

allocates space for `nzmax` nonzeros, which is useful if extra nonzeros, not in `s`, are to be introduced later, for example when `A` is generated column by column.

A sparse matrix of zeros is produced by `sparse(m,n)` (both arguments must be specified), which is an abbreviation for `sparse([],[],[],m,n,0)`.

The sparse identity matrix is produced by `speye(n)` or `speye(m,n)`, while the command `spones(A)` produces a matrix with the same sparsity pattern as `A` and with ones in the nonzero positions.

The arguments that `sparse` would need to reconstruct an existing matrix `A` via `sparse(i,j,s,m,n)` can be obtained using

```
[i,j,s] = find(A);
[m,n] = size(A);
```

If just `s` is required, then `s = nonzeros(A)` can be used. The number of storage locations allocated for nonzeros in `A` can be obtained with `nzmax(A)`. The inequality `nnz(A) <= nzmax(A)` always holds.

The function `spdiags` is an analogue of `diag` for sparse matrices. The command `A = spdiags(B,d,m,n)` creates an `m`-by-`n` matrix `A` whose diagonals indexed by `d` are taken from the columns of `B`. This function is best understood by looking at examples. Given

```
B =
        1     2     0
        1     2     3
        0     2     3
        0     2     3
d =
       -2     0     1
```

we can define

```
>> A = spdiags(B,d,4,4)
A =
   (1,1)        2
   (3,1)        1
   (1,2)        3
   (2,2)        2
   (4,2)        1
   (2,3)        3
   (3,3)        2
   (3,4)        3
   (4,4)        2

>> full(A)
ans =
     2     3     0     0
     0     2     3     0
     1     0     2     3
     0     1     0     2
```

Note that the subdiagonals are taken from the leading parts of the columns of `B` and the superdiagonals from the trailing parts. Diagonals can be extracted with `spdiags`: `[B,d] = spdiags(A)` recovers `B` and `d` above. The next example sets up a particular tridiagonal matrix:

```
>> n = 5; e = ones(n,1);
>> A = spdiags([-e 4*e -e],[-1 0 1],n,n);
>> full(A)
ans =
     4    -1     0     0     0
    -1     4    -1     0     0
     0    -1     4    -1     0
     0     0    -1     4    -1
     0     0     0    -1     4
```

Random sparse matrices are generated with `sprand` and `sprandn`. The command `A = sprand(S)` generates a matrix with the same sparsity pattern as `S` and with nonzero entries uniformly distributed on $[0,1]$. Alternatively, `A = sprand(m,n,density)` generates an `m`-by-`n` matrix of a random sparsity pattern containing approximately `density*m*n` nonzero entries uniformly distributed on $[0,1]$. With four input arguments, `A = sprand(m,n,density,rc)` produces a matrix for which the reciprocal of the condition number is about `rc`. The syntax for `sprandn` is the same, but random numbers from the normal (0,1) distribution are produced.

An invaluable command for visualizing sparse matrices is `spy`, which plots the sparsity pattern with a dot representing a nonzero; see the plots in the next section.

A sparse array can be distinguished from a full one using the logical function `issparse` (there is no "`isfull`" function); see Table 6.1.

15.2. Linear Algebra

MATLAB is able to solve sparse linear equation, eigenvalue, and singular value problems, taking advantage of sparsity.

As for full matrices (see Section 9.2.1), the backslash operator \ can be used to solve linear systems. The effect of `x = A\b` when `A` is sparse is roughly as follows; for full details, type `doc 'arithmetic operators'`. If `A` is square and banded then a banded solver is used. If `A` is a permutation of a triangular matrix, substitution is used. If `A` is a Hermitian positive definite matrix a Cholesky factorization with a minimum degree reordering is used. For a square matrix with no special properties, a sparse LU factorization with reordering is computed using UMFPACK [19] with a reordering produced by a modified version of `colamd` (see below). If `A` is rectangular then QR factorization is used; a rank-deficiency test is performed based on the diagonal elements of the triangular factor.

The importance of avoiding matrix inversion where possible was explained in Section 9.3 in the context of full matrices. Matrix inversion is deprecated even more strongly for sparse matrices, because the inverse of a matrix containing many zeros usually has far fewer zeros—often none. Hence although the `inv` function will invert a sparse matrix (and return the result in the sparse format), it should almost never be used to do so.

To compute or estimate the condition number of a sparse matrix `condest` should be used (see Section 9.1), as `cond` and `rcond` are designed only for full matrices.

The `chol` function for Cholesky factorization behaves in a similar way for sparse matrices as for full matrices, but the computations are done using sparse data structures.

The `lu` function for LU factorization has some extra options not present in the full case. With up to three output arguments, the same mathematical factorization is produced as in the full case, and `lu` works in the same way it has since MATLAB 4. A second input argument `thresh`, as in `lu(A,thresh)`, sets a pivoting threshold, which must lie between 0 and 1. The pivoting strategy requires that the pivot element have magnitude at least `thresh` times the magnitude of the largest element below the diagonal in the pivot column. The default is 1, corresponding to partial pivoting, and a threshold of 0 forces no pivoting.

With a fourth output argument `lu` uses UMFPACK. The general syntax is

```
[L,U,P,Q] = lu(A,thresh)
```

Here, a factorization $PAQ = LU$ is produced where L is unit lower triangular, U is upper triangular, and P and Q are permutation matrices produced by a modified version of `colamd`. Rectangular A are supported. The row permutations in P are used for numerical stability and the column permutations in Q are used to reduce fill-in. The threshold `thresh` has the same meaning as when there are less than four output arguments, but its default is now 0.1.

Since `lu` (with two output arguments) and `chol` do not pivot for sparsity (that is, they do not use row or column interchanges in order to try to reduce the cost of the factorizations), it is advisable to consider reordering the matrix before factorizing it. A full discussion of reordering algorithms is beyond the scope of this book, but we give some examples.

We illustrate reorderings with the Wathen matrix:

```
A = gallery('wathen',8,8);
subplot(121), spy(A), subplot(122), spy(chol(A))
```

The spy plots of A and its Cholesky factor are shown in Figure 15.1. Now we reorder the matrix using the symmetric reverse Cuthill–McKee permutation and refactorize:

```
r = symrcm(A);
subplot(121), spy(A(r,r)), subplot(122), spy(chol(A(r,r)))
```

Note that all the reordering functions return an integer permutation vector rather than a permutation matrix (see Section 21.3 for more on permutation vectors and matrices). The spy plots are shown in Figure 15.2. Finally, we try the symmetric approximate minimum degree ordering:

```
m = symamd(A);
subplot(121), spy(A(m,m)), subplot(122), spy(chol(A(m,m)))
```

The spy plots are shown in Figure 15.3. For this matrix the minimum degree ordering leads to the sparsest Cholesky factor—the one with the least nonzeros.

For LU factorization, possible reorderings include

```
p = colamd(A); p = colperm(A);
```

after which `A(:,p)` is factorized.

In the QR factorization `[Q,R] = qr(A)` of a sparse rectangular matrix `A` the orthogonal factor `Q` can be much less sparse than `A`, so it is usual to try to avoid explicitly forming `Q`. When given a sparse matrix and one output argument, the `qr` function returns just the upper triangular factor `R`: `R = qr(A)`. When called as `[C,R] = qr(A,B)`, the matrix `C = Q'*B` is returned along with `R`. This enables an overdetermined system `Ax = b` to be solved in the least squares sense by

```
[c,R] = qr(A,b);
x = R\c;
```

The backslash operator (`A\b`) uses this method for rectangular `A`.

The iterative linear system solvers in Table 9.2 are also designed to handle large sparse systems. See Section 9.8 for details of how to use them. Sparse eigenvalue and singular value problems can be solved using `eigs` and `svds`, which are also described in Section 9.8. The eigenvalues (only) of a real, sparse symmetric matrix can be computed by `eig`, but for any other type of matrix (including complex Hermitian) it is necessary to use `eigs`.

The function `spparms` helps determine or change the internal workings of some of the sparse factorization functions. Typing `spparms` by itself prints the current settings:

```
>> spparms
No SParse MONItor output.
mmd: threshold = 1.1 * mindegree + 1,
     using approximate degrees in A'*A,
```

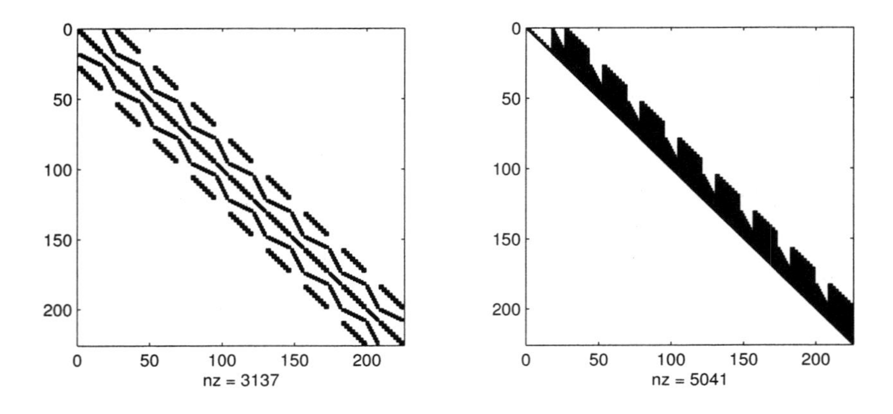

Figure 15.1. *Wathen matrix (left) and its Cholesky factor (right).*

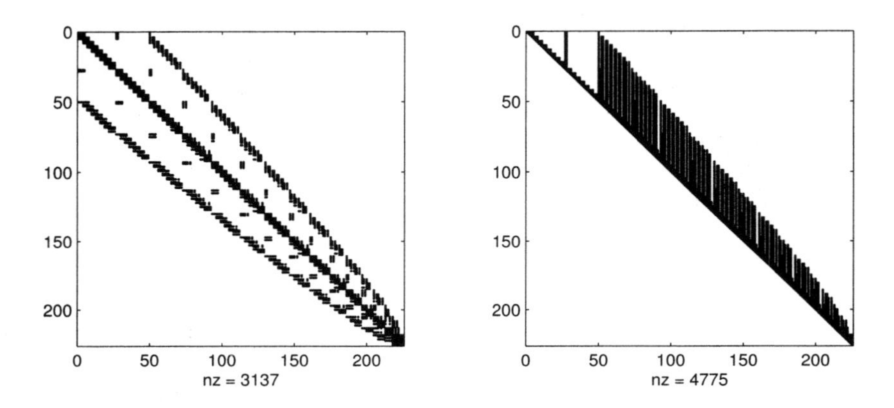

Figure 15.2. *Wathen matrix (left) and its Cholesky factor (right) with symmetric reverse Cuthill–McKee ordering* (`symrcm`).

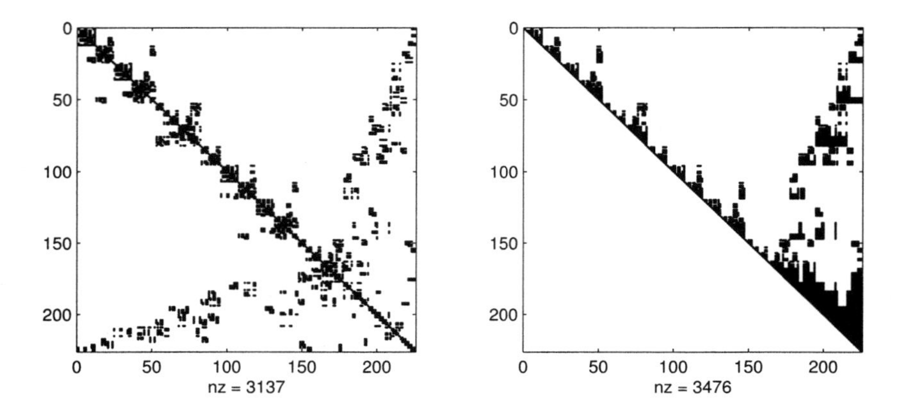

Figure 15.3. *Wathen matrix (left) and its Cholesky factor (right) with symmetric minimum degree ordering* (`symamd`).

```
     supernode amalgamation every 3 stages,
     row reduction every 3 stages,
     withhold rows at least 50% dense in colmmd.
Minimum degree orderings used with chol, v4 lu and qr in \ and /.
Approximate minimum degree orderings used with UMFPACK in \ and /.
Pivot tolerance of 0.1 used by UMFPACK in \ and /.
Backslash uses band solver if band density is > 0.5
UMFPACK used for lu in \ and /.
```

How much of the matrix must be zero for it to be considered sparse
depends on the computation to be performed,
the pattern of the nonzeros,
and even the architecture of the computer.
Generally, we say that a matrix is sparse
if there is an advantage in exploiting its zeros.

— I. S. DUFF, A. M. ERISMAN and J. K. REID,
Direct Methods for Sparse Matrices (1986)

Sparse matrices are created explicitly rather than automatically.
If you don't need them, you won't see them mysteriously appear.

— *The MATLAB EXPO: An Introduction to MATLAB,*
SIMULINK and the MATLAB Application Toolboxes (1993)

An objective of a good sparse matrix algorithm should be:
The time required for a sparse matrix operation should be
proportional to the number of arithmetic operations on nonzero quantities.
We call this the "time is proportional to flops" rule;
it is a fundamental tenet of our design.

— JOHN R. GILBERT, CLEVE B. MOLER, and ROBERT S. SCHREIBER,
Sparse Matrices in MATLAB: Design and Implementation (1992)

Chapter 16
Further M-Files

16.1. Elements of M-File Style

As you use MATLAB you will build up your own collection of M-files. Some may be short scripts that are intended to be used only once, but others will be of potential use in future work. Based on our experience with MATLAB we offer some guidelines on making M-files easy to use, understand, and maintain.

In Chapter 7 we explained the structure of the leading comment lines of a function, including the H1 line. Adhering to this format and fully documenting the function in the leading comment lines is vital if you are to be able to reuse and perhaps modify the function some time after writing it. A further benefit is that writing the comment lines forces you to think carefully about the design of the function, including the number and ordering of the input and output arguments.

It is helpful to include in the leading comment lines an example of how the function is used, in a form that can be cut and pasted into the command line (hence function names should not be given in capitals). MATLAB functions that provide such examples include `fzero`, `meshgrid`, `null`, and `texlabel`.

In formatting the code, it is advisable to follow the example of the M-files provided with MATLAB, and to use

- spaces around logical operators and = in assignment statements,
- one statement per line (with exceptions such as a short `if`),
- indentation to emphasize `if`, `for`, `switch`, and `while` structures (as provided automatically by MATLAB's Editor/Debugger—see Section 7.2),
- variable names beginning with capital letters for matrices.

Compare the code segment

```
if stopit(4)==1
% Right-angled simplex based on coordinate axes.
alpha=norm(x0,inf)*ones(n+1,1);
for j=2:n+1, V(:,j)=x0+alpha(j)*V(:,j); end
end
```

with the more readable

```
if stopit(4) == 1
   % Right-angled simplex based on coordinate axes.
   alpha = norm(x0,inf)*ones(n+1,1);
   for j = 2:n+1
```

```
        V(:,j) = x0 + alpha(j)*V(:,j);
    end
end
```

In this book we usually follow these rules, occasionally breaking them to save space.

A rough guide to choosing variable names is that the length and complexity of a name should be proportional to the variable's scope (the region in which it is used). Loop index variables are typically one character long because they have local scope and are easily recognized. Constants used throughout an M-file merit longer, more descriptive names.

A MATLAB function that helps in choosing variable names, especially in an automated way, is `genvarname`. See `doc genvarname` for some interesting examples of its use.

16.2. Checking and Comparing M-Files

MATLAB provides some useful tools for automatically checking and comparing M-files. The `mlint` function reads an M-file and produces a report of potential errors and problems, and makes suggestions for improving the efficiency and maintainability of the code. The M-file `badfun` in Listing 16.1 is perfectly legal MATLAB:

```
>> badfun(1,2);
x =
    2.2361
```

However, it contains several weaknesses that `mlint` detects:

```
>> mlint badfun
L 1 (C 13): "y"  is an output argument that is never set
L 1 (C 18-22): "badfu" : function name should agree with filename
L 4 (C 15): | is deprecated in conditionals -- use ||
L 4 (C 29): "c"  assignment value is never used
L 6 (C 3): = value is implicitly printed
```

The output refers to lines (L) and columns (C). The third line refers to the fact that `||` is preferred for tests between scalars; see Section 6.1. As this example shows, `mlint` is good at detecting variables that are never assigned or used, which is useful because these problems can be hard to spot in longer M-files.

The companion function `mlintrpt` runs `mlint` on all the M-files in the current directory and reports the results in the MATLAB Web browser, with hypertext links to the M-file names and line numbers. It can also be called on a single M-file.

Two M-files can be compared with the `visdiff` function, which produces a report in the MATLAB Web browser containing listings of the two M-files with differences marked.

If you want to give someone else an M-file `myfun` that you have written, you also need to give them all the M-files that it calls that are not provided with MATLAB. This list can be determined by typing

```
[Mfiles,builtins] = depfun('myfun')
```

which returns a list of the M-files that are called by `myfun` or by a function called by `myfun`, and so on; the optional second argument contains the built-in functions that

Listing 16.1. *Script* badfun.

```
function [x,y] = badfu(a,b,c)
%BADFUN   Function on which to illustrate MLINT.

if nargin < 3 | isempty(c), c = 1; end

x = sqrt(a^2+b^2)
```

are called. Note that the input to **depfun** must be a string, not a function handle. Another way to obtain this information is with the **inmem** command, which lists all M-files that have been parsed into memory. If you begin by clearing all functions (`clear functions`), run the M-file in question, and then invoke **inmem**, you can deduce which M-files have been called. Also useful in this context is **depdir**:

```
list = depdir('myfun')
```

lists the directories in which M-files dependent on `myfun` reside. Finally, **deprpt** lists dependencies for all the M-files in the current directory, showing the results in the MATLAB Web browser.

All the tools mentioned in this section can also be invoked from the Directory Reports menu in the Current Directory browser.

16.3. Profiling

MATLAB has a profiler that reports, for a given sequence of computations, how much time is spent in each line of each M-file and how many times each line is executed, how many times each M-file is called, and the results of running `mlint` (see Section 16.2). Profiling has several uses.

- Identifying "hot spots"—those parts of a computation that dominate the execution time. If you wish to optimize the code then you should concentrate on the hot spots.
- Spotting inefficiencies, such as code that can be taken outside a loop.
- Revealing lines in an M-file that are never executed. This enables you to spot unnecessary code and to check whether your test data fully exercises the code.

To illustrate the use of the profiler, we apply it to MATLAB's **membrane** function (used on p. 104):

```
profile on
A = membrane(1,50);
profile viewer
profile off
```

The **profile viewer** command generates an html report that is displayed in the MATLAB Web browser. Figure 16.1 shows the result. Clicking on the **membrane** link

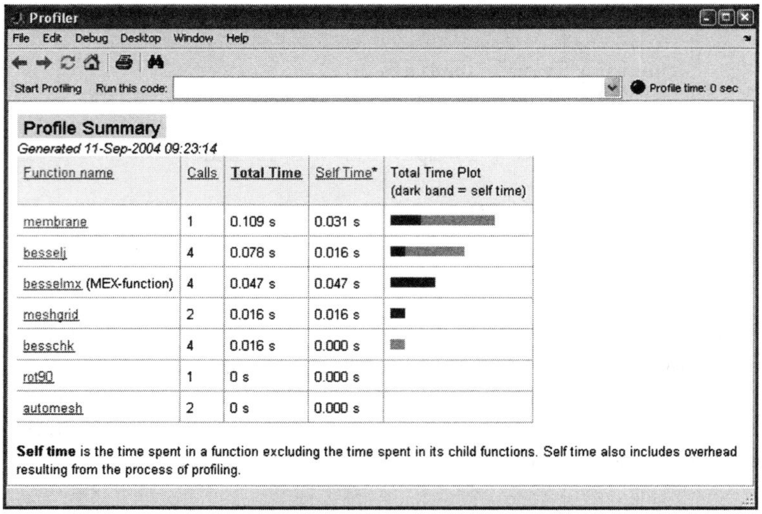

Profiler

File Edit Debug Desktop Window Help

Start Profiling Run this code: Profile time: 0 sec

Profile Summary

Generated 11-Sep-2004 09:23:14

Function name	Calls	Total Time	Self Time*	Total Time Plot (dark band = self time)
membrane	1	0.109 s	0.031 s	
besselj	4	0.078 s	0.016 s	
besselmx (MEX-function)	4	0.047 s	0.047 s	
meshgrid	2	0.016 s	0.016 s	
besschk	4	0.016 s	0.000 s	
rot90	1	0 s	0.000 s	
automesh	2	0 s	0.000 s	

Self time is the time spent in a function excluding the time spent in its child functions. Self time also includes overhead resulting from the process of profiling.

Figure 16.1. `profile viewer` *report for* `membrane` *example.*

Profiler

File Edit Debug Desktop Window Help

Start Profiling Run this code: Profile time: 0 sec

membrane (1 call, 0.156 sec)

Generated 11-Sep-2004 09:37:21

M-function in file E:\MATLAB7final\toolbox\matlab\demos\membrane.m

[Copy to new window for comparing multiple runs]

Refresh

Show parent files Show busy lines Show child files

Show M-Lint results Show file coverage Show file listing

Parents (calling functions)

No parent

Lines where the most time was spent

Line Number	Code	Calls	Total Time	% Time	Time Plot
70	`bl = besselj(alfl,t);`	1	0.063 s	40.0%	
98	`S = S + c(j) * besselj(alfa(j)...`	2	0.047 s	30.0%	
28	`if nargin < 1, k = 1; end`	2	0.016 s	10.0%	
106	`L = rot90(L/max(max(abs(L))),-...`	1	0.016 s	10.0%	
73	`[ignore,k] = sort([kl k2]);`	1	0.016 s	10.0%	
Other lines & overhead			0 s	0%	
Totals			0.156 s	100%	

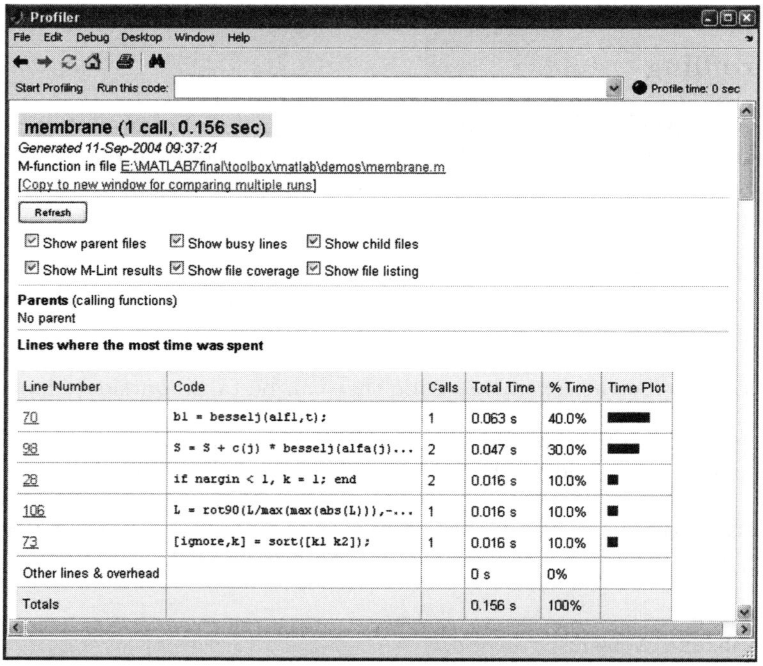

Figure 16.2. *More from* `profile viewer` *report for* `membrane` *example.*

Listing 16.2. *Script* ops.

```
%OPS   Profile this file to check costs of various elementary ops and funs.

rand('state',1), randn('state',1)
n = 500;
a = 100*rand(n);
b = randn(n);

for i = 1:100
    a+b;
    a-b;
    a.*b;
    a./b;
    sqrt(a);
    exp(a);
    sin(a);
    tan(a);
end
```

produces Figure 16.2, which shows only the first screen of a long document. The profile reveals that `membrane` spends most of its time evaluating Bessel functions.

Next, consider the script `ops` in Listing 16.2. We profiled the script in order to compare the relative costs of the elementary operations `+`, `-`, `*`, `/` and the elementary functions `sqrt`, `exp`, `sin`, `tan`.

```
profile on
ops
profile viewer
profile off
```

We selected just the file listing to be shown in the report, giving the report shown in Figure 16.3. The precise results will vary with the computer. As expected, the exponential and trigonometric functions are much more costly than the four elementary operations.

16.4. Creating a Toolbox

If you develop a number of functions with a common theme that you wish to group together, and possibly distribute to others, you should consider creating a toolbox. A toolbox is simply a collection of M-files living in the same directory, along with the special files `Contents.m` and, possibly, `readme.m`. The file `Contents.m` is an M-file of comments containing the names of the M-files in the toolbox with a short description of each, and its leading lines give the version and copyright information for the toolbox.

For example, we could create a toolbox in a directory `mytoolbox` on the MATLAB path, with `Contents.m` file beginning as follows:

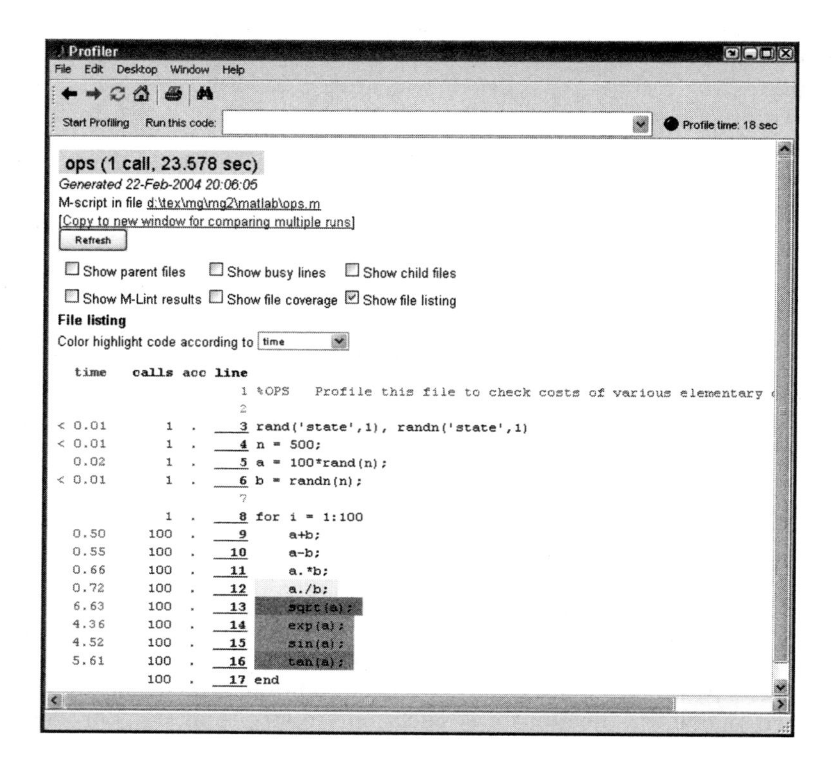

Figure 16.3. `profile viewer` *report for* `ops` *example.*

```
% Mytoolbox.
% Version 1.1             5-May-2004
% Copyright (c) 2004 by A. N. Other
%
% myfun1 - My first useful function.
% myfun2 - My second useful function.
```

Provided the precise format of the first two lines is followed, typing `ver mytoolbox` lists some version information about MATLAB followed by

```
Mytoolbox                                     Version 1.1
```

Typing `help mytoolbox` lists `Contents.m` to the screen, while `helpwin mytoolbox` lists it in the Help browser. Typing `whatsnew mytoolbox` lists `readme.m` to the screen.

A `Contents.m` file can be generated automatically—and an existing file checked against the H1 lines of the M-files in the directory—with the `contentsrpt` function. An alternative invocation is by selecting Directory Reports-Contents Report in the Current Directory browser.

I've become convinced that all compilers written from now on should be designed to provide all programmers with feedback indicating what parts of their programs are costing the most.

— DONALD E. KNUTH, *Structured Programming with* `go to` *Statements* (1974)

Instrument your programs.
Measure before making "efficiency" changes.

— BRIAN W. KERNIGHAN and P. J. PLAUGER, *The Elements of Programming Style* (1978)

Arnold was unhappily aware that the complete Jurassic Park program contained more than half a million lines of code, most of it undocumented, without explanation.

— MICHAEL CRICHTON, *Jurassic Park* (1990)

If we wish to count lines of code, we should not regard them as lines produced but as lines spent.

— Edsger W. Dijkstra

Chapter 17
Handle Graphics

The graphics functions described in Chapter 8 can produce a wide range of output and are sufficient to satisfy the needs of many MATLAB users. These functions are part of an object-oriented graphics system known as Handle Graphics that provides full control over the way MATLAB displays data. A knowledge of Handle Graphics is useful if you want to fine-tune the appearance of your plots, and it enables you to produce displays that are not possible with the existing functions. This chapter provides a brief introduction to Handle Graphics. More information can be found in [76].

17.1. Objects and Properties

Handle Graphics builds graphs out of objects organized in a hierarchy, as shown in Figure 17.1. The Root object corresponds to the whole screen and a Figure object to a figure window. Of the two classes of objects on the third level of the tree we will be concerned only with the Axes object, which is a region of the figure window in which objects from the bottom level of the tree are displayed. A figure window may contain more than one Axes object, as we will see in an example below. From the bottom level of the tree we will be concerned only with the Line, Surface, and Text objects. In fact, Figure 17.1 is greatly simplified: the objects on the third level belong to the class of "core objects", and there are further classes of "plot objects", "group objects", and "annotation objects" that we have not shown. For the full tree, see "Handle Graphics Property Browser" in the Help browser.

Each object has a unique identifier called a handle, which is a floating point number (sometimes an integer). The handle of the Root object is always 0. The handle of a Figure object is, by default, the figure number displayed on the title bar (but this can be changed). To use Handle Graphics you create objects and manipulate their properties by reference to their handles, making use of the `get` and `set` functions.

We begin with a simple example:

```
>> plot(1:10,'o-')
```

This produces the left-hand plot in Figure 17.2. Now we interactively investigate the objects comprising the plot, beginning by using the `findobj` function to obtain the handles of all the objects:

```
>> h = findobj
h =
         0
    1.0000
   73.0011
```

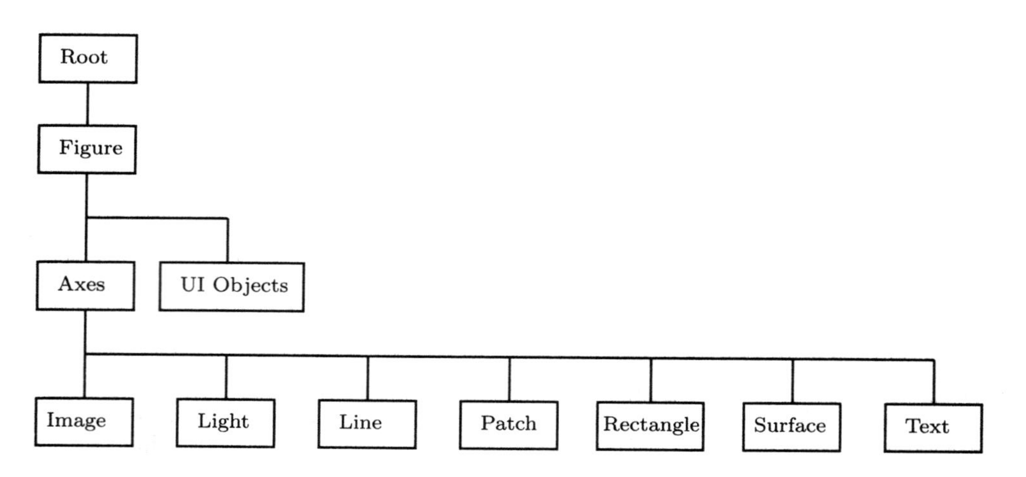

Figure 17.1. *Hierarchical structure of Handle Graphics objects (simplified).*

```
    1.0050
```

We know from the conventions that the first handle, 0, is that of the root, and the second, 1, is that of the figure. We can determine the types of all the objects that these handles represent using the **get** function:

```
>> get(h,'type')
ans =
    'root'
    'figure'
    'axes'
    'line'
```

Thus `h(3)` is the handle to an Axes object and `h(4)` that to a Line object.

A handle provides access to the various properties of an object that govern its appearance. A list of properties can be obtained by calling the **set** function with the appropriate handle. For the Axes object the properties are illustrated by

```
>> set(h(3))
ActivePositionProperty: [ position | {outerposition} ]
ALim
ALimMode: [ {auto} | manual ]
AmbientLightColor
Box: [ on | {off} ]
CameraPosition
        ...
Visible: [ {on} | off ]
```

Here we have replaced about 100 lines of output with "...". The property names are listed one per line. For those properties that take string values the possible values are listed in square brackets; the default is enclosed in curly braces. For the Line object the properties are listed by

```
>> set(h(4))
```

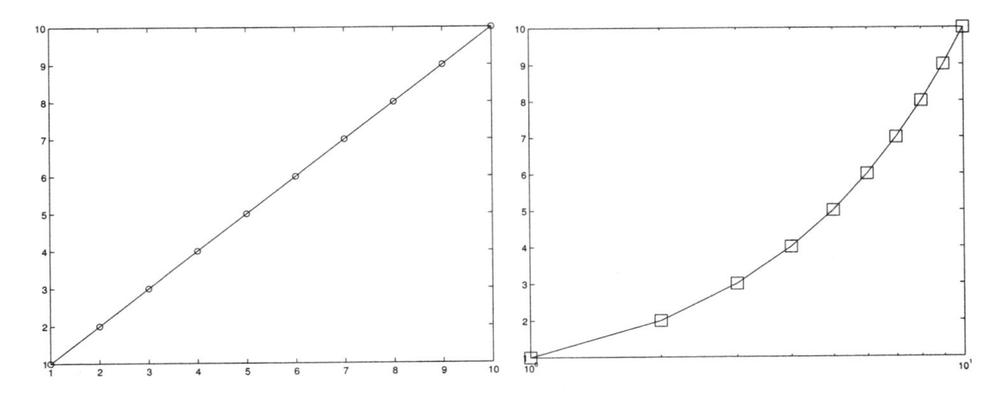

Figure 17.2. *Left: original. Right: modified by* `set` *commands.*

```
ans =

              Color: {}
          EraseMode: {4x1 cell}
          LineStyle: {5x1 cell}
          LineWidth: {}
          Marker: {14x1 cell}
          ...
        ZDataSource: {}
```

The possible values, which this time are not shown, can be obtained by providing the property name as a second argument to `set`:

```
>> set(h(4),'Marker')
[ + | o | * | . | x | square | diamond | v | ^ | > | < | ...
pentagram | hexagram | {none} ]
```

We see that the `Marker` property of the Line object has fourteen possible values (those listed in Table 8.1 together with `none`, for use when only a line is required) and the default is `none`. Full documentation of all object properties is available under "Handle Graphics Property Browser" in the Help browser. For direct access to the relevant help pages you can type `doc name_props`, where `name` is replaced by `rootobject` or by any of the names of objects below the root in Figure 17.1. Thus `doc line_props` displays information about Line object properties.

Property values are assigned by providing `set` with the handle and pairs of property names and property values. Thus the command

```
>> set(h(4),'Marker','s','MarkerSize',16)
```

replaces the original `'o'` marker by a square of size 16 in the line object in our example. It is only necessary to provide enough characters of the property name or value to uniquely identify it, so `'s'` above is equivalent to `'square'`, and any mixture of upper and lower case letters can be used. Next, we check the possible values of the `XScale` property for the Axes object:

```
>> set(h(3),'XScale')
[ {linear} | log ]
```

We set this property to `log` to make the x-axis scale logarithmic (as for `semilogx`):

```
set(h(3),'XScale','log')
```

The modified plot is shown on the right-hand side of Figure 17.2.

Tab completion works for property names and property values and can be a great time-saver. For example, in the `set(h(4),...)` command above, `MarkerSize` can be entered as `MarkerS` followed by tab, and either `Marker` or `MarkerSize` can be entered by typing `M`, tab, and then selecting the appropriate entry from the resulting menu of four possible completions.

For a further example, we consider the following code, which produces Figure 17.3:

```
x = linspace(0,2*pi,35);

a1 = subplot(2,1,1);                        % Axes object.
l1 = plot(x,sin(x),'x');                    % Line object.

a2 = subplot(2,1,2);                        % Axes object.
l2 = plot(x,cos(x).*sin(x));                % Line object.
tx2 = xlabel('x'); ty2 = ylabel('y');  % Text objects.
```

When the following code is executed it modifies properties of objects to produce Figure 17.4.

```
set(a1,'Box','off')                                % box off.
set(a1,'XTick',[])
set(a1,'YAxisLocation','right')
set(a1,'TickDir','out')
set(l1,'Marker','<')

set(a2,'Position',[0.2 0.15 0.65 0.35])
set(a2,'XLim',[0 2*pi])                            % xlim([0 2*pi]).
set(a2,'FontSize',14)
set(a2,'XTick',[0 pi/2 pi 2*pi])
set(a2,'XTickLabel','0|pi/2|pi|2pi')
set(a2,'XGrid','on')
set(a2,'XScale','log')
set(l2,'LineWidth',6)
set(tx2,'FontAngle','italic','FontSize',20)
set(ty2,'Rotation',0,'FontAngle','italic','FontSize',20)
```

Some of the effects of these `set` commands can be produced using commands discussed in Chapter 8, as indicated in the comments, or by appending property name-value pairs to argument lists of `plot` and `text`. For example, `box off` can be used in place of `set(a1,'Box','off')`, provided that the first Axes is current. However, certain effects can be conveniently achieved only by using `set`.

The properties altered here are mostly self-explanatory. An exception is the `Position` property of Axes, which is specified by a vector of the form `[left bottom width height]`, where `left` and `bottom` are the distances from the left edge and bottom edge, respectively, of the Figure window to the bottom left corner of the Axes rectangle, and `width` and `height` define the dimensions of the rectangle. The units of measurement are defined by the `Units` property, whose default is `normalized`, which

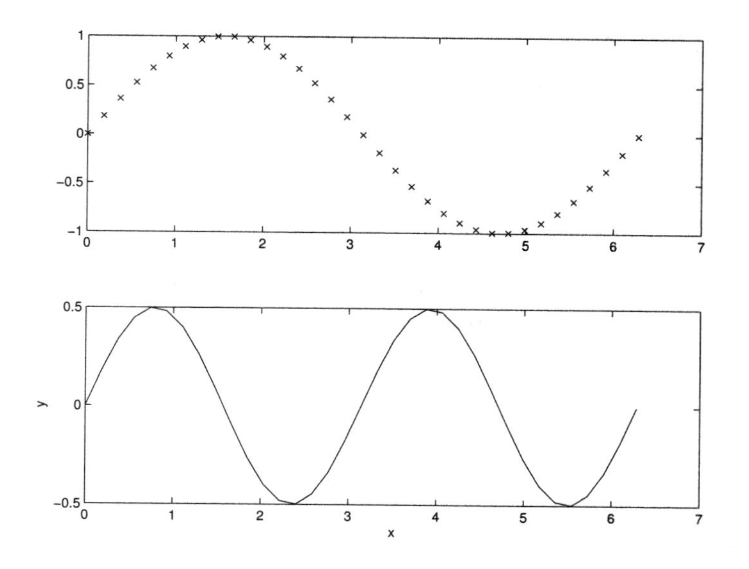

Figure 17.3. *Straightforward use of* **subplot**.

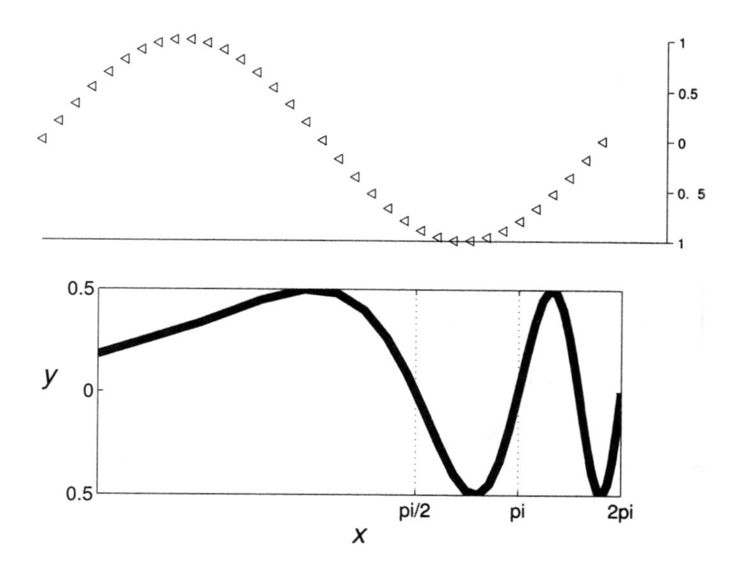

Figure 17.4. *Modified version of Figure* 17.3 *postprocessed using Handle Graphics.*

maps the lower left corner of the figure window to $(0,0)$ and the upper right corner to $(1.0, 1.0)$. Note that tick labels do not support TeX notation, so we could not produce the symbol π in the x-axis labels specified by the `XTickLabel` property.

A counterpart to the `set` function is `get`, which queries the current values of properties. With just a handle as argument, `get` lists all the properties:

```
>> get(l1)
                  Color: [0 0 1]
              EraseMode: 'normal'
              LineStyle: 'none'
              LineWidth: 0.5000
                 Marker: '<'
             MarkerSize: 6
              ...
            ZDataSource: ''
```

When invoked with a second argument specifying a property name, `get` lists just that value:

```
>> get(a2,'XTick')
ans =
         0    1.5708    3.1416    6.2832
```

The `delete` function deletes an object with the specified handle. Thus `delete(l1)` removes the sine curve from the top plot in Figure 17.4 and `delete(tx2)` removes the x-axis label from the bottom plot.

Generally, if you plan to change the properties of an object after creating it then you should save the handle when you create it, as in the example above. However, handles of existing objects can be retrieved using `gca`, `gcf`, and `gco`, which return the handles of the current Axes, the current Figure, and the current Object, respectively. In the following example we check the current and possible values of the `FontWeight` property for the current Axes and then change the property to `bold`:

```
>> get(gca,'FontWeight')
ans =
normal

>> set(gca,'FontWeight')
[ light | {normal} | demi | bold ]

>> set(gca,'FontWeight','bold')
```

The "current Object" whose handle is returned by `gco` is the object last clicked on with the mouse. Thus if we want to change the marker to `'*'` for the curve in the upper plot of Figure 17.4 we can click on the curve and then type

```
>> set(gco,'Marker','*')
```

In addition to setting graphics properties from the command line or in M-files it is possible to set them interactively using the Property Editor. The Property Editor is invoked on a particular graphic object by first enabling plot editing, by clicking on the plot editing icon in the figure window toolbar, and then double-clicking on the object. Experimenting with the Property Editor is an excellent way to learn about Handle Graphics.

The importance of the hierarchical nature of the Handle Graphics structure is not completely apparent in the simple examples described above. A particular object, say the Root, contains the handles of all its children, which makes it possible to traverse the tree structure, using `get(h,'Children')`, `get(h,'Parent')`, and the `findobj` and `findall` functions. Furthermore, it is possible to set default values for properties, and if these are set on a particular Axes, for example, they are inherited by all the children of that Axes. Some of these aspects are described in the following sections; see also [76] for details.

Also beyond the scope of this book are MATLAB's Graphical User Interface (GUI) tools, described in [71] (type `help uitools` for a list of the relevant functions). However, we mention one GUI function that is of broad interest: `waitbar` displays a graphical bar in a window that can be used to show the progress of a computation. Its usage is illustrated by (see also Listing 1.4)

```
h = waitbar(0,'Computing...')
for j = 1:n
  % Some computation ...
  waitbar(j/n) % Set bar to show fraction j/n complete.
end
close(h)
```

17.2. Root and Default Properties

As mentioned in the previous section, the handle of the Root object is 0. The assignable root properties can be determined using `get(0)`. The list includes not only graphics properties, but also properties such as the status of `diary` and `format`, the screen size, and the recursion limit. Thus, for example, we can change the format as follows:

```
>> set(0,'Format','bank'), x = 1.23456
x =
          1.23
>> set(0,'Format','short'), x
x =
    1.2346
```

We can determine the current format using

```
>> get(0,'Format')
ans =
short
```

Accessing the format via a handle can be useful inside a function, where we might want to save the current format before changing it and then restore the saved format before returning control from the function.

Typing `get(0,'Factory')` returns in a structure all the factory-defined values of all user-settable properties. The default value for the object property *ObjectTypePropertyName* can be obtained with `get(0,'Default`*ObjectTypePropertyName*`')`. For example:

```
>> get(0,'DefaultLineMarkerSize')
ans =
     6
```

The command `set(0,'Default`*`ObjectTypePropertyName`*`')` sets the default value for *ObjectTypePropertyName*, and since the root handle is specified this default applies to all levels in the hierarchy. This command is useful for setting default property values at the start of a session. For example, before giving a presentation with a data projector we might type

```
set(0,'defaulttextfontsize',12)
set(0,'defaultaxesfontsize',12)
set(0,'defaultlinemarkersize',10)
set(0,'defaultlinelinewidth',2)
set(0,'defaulttextfontweight','bold')
set(0,'defaultaxesfontweight','bold')
```

in order to make the MATLAB graphics more readable to the audience. (It obviously makes sense to create an M-file containing these commands.) On the other hand, before generating encapsulated PostScript figures for inclusion in a paper we might type

```
set(0,'defaulttextfontsize',12)
set(0,'defaultaxesfontsize',12)
set(0,'defaultlinemarkersize',8)
set(0,'defaultlinelinewidth',1)
```

The advantage of these commands is that they make it unnecessary to append modifiers such as `'FontSize',12` to every `title`, `xlabel`, and so on.

The factory settings can be restored with commands of the form

```
set(0,'defaultlinelinewidth','factory')
```

17.3. Animation

Two types of animation are possible in MATLAB. A sequence of figures can be saved and then replayed as a movie, and an animated plot can be produced by manipulating the `XData`, `YData`, and `ZData` properties of objects. We give one example of each type. For further details see [76].

To create a movie, you draw the figures one at a time, use the `getframe` function to save each one as a pixel snapshot in a structure, and then invoke the `movie` function to replay the figures. Here is an example:[9]

```
clear  % Remove existing variables.
Z = peaks; surf(Z)
axis tight
set(gca,'nextplot','replacechildren')
disp('Creating the movie...')
```

[9]The fact that `F` in this example is not preallocated (cf. Section 20.2) does not cause any loss of efficiency. Since `getframe` returns a structure, `F` is a vector of `structs` and it is only pointers that need to be deallocated and reallocated.

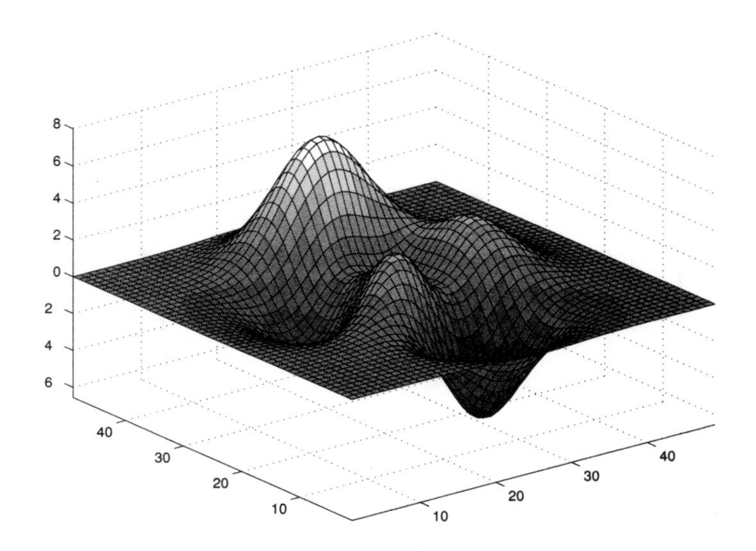

Figure 17.5. *One frame from a movie.*

```
for j = 1:11
    surf(cos(2*pi*(j-1)/10).*Z,Z)
    F(j) = getframe;
end
disp('Playing the movie...')
movie(F)
```

Figure 17.5 shows one intermediate frame from the movie. The **set** command causes all **surf** plots after the first to leave unaltered the Axes properties, such as **axis tight** and the grid lines. The movie is replayed **n** times with **movie(F,n)**. The amount of storage required by the movie depends on the window size but not on the contents of the window.

The second type of animation is most easily obtained using the functions **comet** and **comet3**. They behave like limited versions of **plot** and **plot3**, differing in that the plot is traced out by a "comet" consisting of a head (a circle), a body (in one color), and a tail (in another color). For example, try

```
x = linspace(-2,2,500);
y = exp(x).*sin(1./x);
comet(x,y)
```

We give a simple example to illustrate the principle used by **comet**. This example can be adapted for use in situations in which the data must be plotted as it is generated, as when solving a differential equation, for example (see the MATLAB demonstration function `lorenz`, mentioned on p. 11).

```
x = linspace(-pi,pi,2000);
y = cos(tan(x))-tan(sin(x));
p = plot(x(1),y(1),'.','EraseMode','none','MarkerSize',5);
```

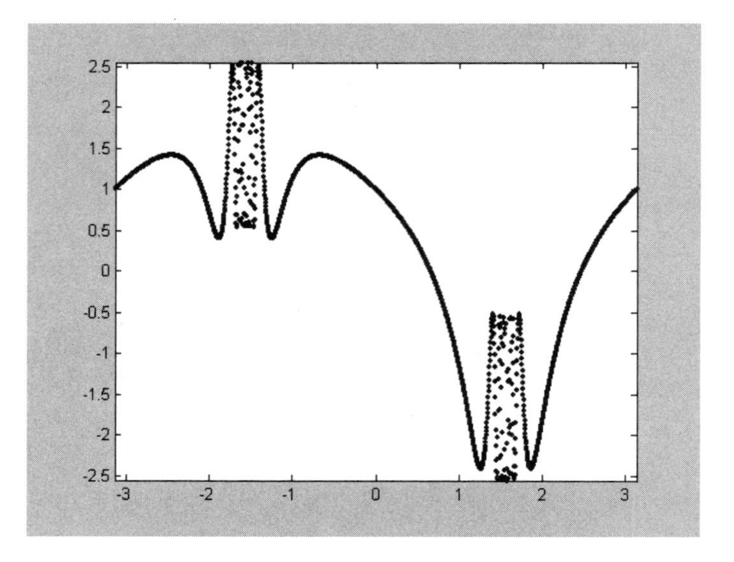

Figure 17.6. *Animated figure upon completion.*

```
axis([min(x) max(x) min(y) max(y)])
hold on
for i = 2:length(x)
    set(p,'XData',x(i),'YData',y(i))
    drawnow
end
hold off
```

This code creates a plot of just one point and then keeps redrawing the point by changing the XData and YData properties of the corresponding Line object. The key is to set the EraseMode property to none so that MATLAB does not erase existing objects when the plot is redrawn by the drawnow command. If EraseMode is set to background then the old point is erased as the new one is plotted, so a moving dot is seen. Figure 17.6 shows the final result. This figure is lower resolution than the others in the book because it was produced by using getframe to save the original figure, redisplaying it with image and then saving in the usual way. The reason we could not save the original figure directly is that it contains only one dot, the others being from unerased earlier plots.

17.4. Examples

In this section we give some practical examples in which Handle Graphics is used to create customized graphics.

MATLAB's choices of tick marks and axis limits are not always the most appropriate. The upper plot in Figure 17.7 shows the relative distance from IEEE single precision numbers $x \in [1, 16]$ to the next larger floating point number. The tick marks on the x-axis do not emphasize the important fact that interesting changes happen at a power of 2. The lower plot in Figure 17.7 (which is [42, Fig. 2.1]) differs from the upper one in that the following Handle Graphics commands were appended:

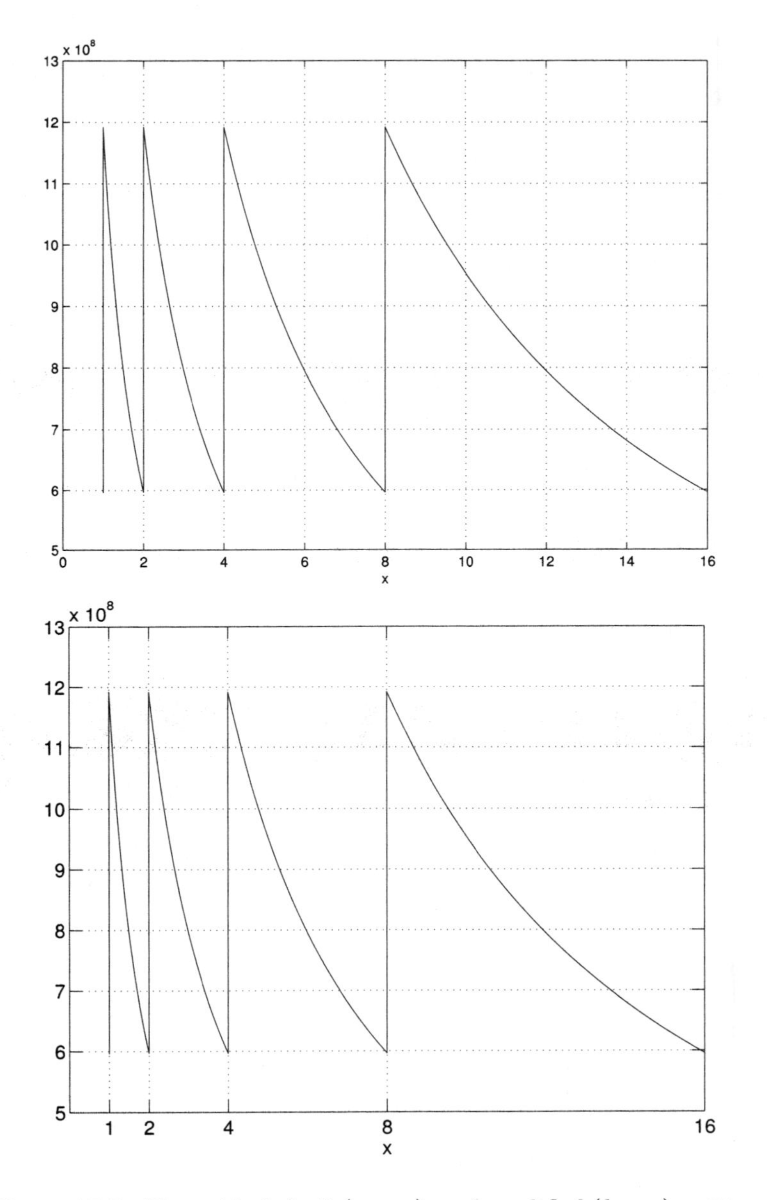

Figure 17.7. *Plot with default (upper) and modified (lower) settings.*

```
set(gca,'XTick',[1 2 4 8 16])
set(gca,'TickLength',[0.02 0.025])
set(gca,'FontSize',14);
set(get(gca,'xlabel'),'FontSize',14);
```

The first `set` command specifies the location of the ticks on the x-axis and the second increases the length of the ticks (to 0.02 for 2D plots and 0.025 for 3D plots, in units normalized relative to the longest of the visible x-, y-, or z-axis lines). The third and fourth commands set a 14-point font size for the tick labels and x-axis label.

Suppose that you wish to use a nonstandard font size (say, 16) throughout a Figure object. Explicitly setting the `FontSize` property for each Text object and each Axes is tedious. Instead, after creating the figure, you can type

```
h = findall(gcf,'type','text'); set(h,'FontSize',16)
h = findall(gcf,'type','axes'); set(h,'FontSize',16)
```

Note that using `findobj` in the first line would not produce any change to the `xlabel`, `ylabel`, or `title`. The reason is that these text objects are created with the `HandleVisibility` property set to `off`, which makes them invisible to `findobj`, but not to `findall`. (Look at the code with `type findall` to gain some insight.) For this reason, however, `findall` should be used with caution as it may expose to view handles that have intentionally been hidden by an application, so manipulating the corresponding objects could produce strange results.

The next example illustrates the use of a cell array (see Section 18.3) to specify the `YTickLabel` data and the `YDir` property to reverse the order of the y-axis values. The script file in Listing 17.1 produces Figure 17.8, which shows the most frequently used words of four letters or more, and their frequencies of occurrence, in a draft of this book.

The script `hist_ex` in Listing 17.2 produces the histogram of samples from the normal (0,1) distribution shown in Figure 17.9. It illustrates how a little fine-tuning can greatly improve the appearance of a plot. By default, `hist` produces histograms that print very dark and have little separation between the bars (see Figure 8.24). We therefore set the faces of the bars to mid gray and the edges to black. We also change the x-axis tick marks to point outwards from the graph; the default of pointing inwards causes the tick marks to be invisible for a histogram. Finally, we increase the font size for the x- and y-axis labels.

The script `cheb3plot` in Listing 17.3 plots seven Chebyshev polynomials in three dimensions, producing Figure 17.10 (this plot reproduces part of one in [26, Fig. A-1]). The script uses the `cheby` function from Listing 7.4. Note the use of the `DataAspectRatio` Axes property to set the aspect ratios of the axes, and the adjustment of the x- and y-axis labels, which by default are placed rather non-centrally.

Handle Graphics can be used to superimpose two different Axes, using the left y-axis for one set of data and the right y-axis for another. This is done by the script `garden` in Listing 17.4, which produces Figure 17.11. The comments in the code explain how it works. Note that the function `plotyy` automates this process of producing different left and right y-axes in the case of simple plots.

The final example illustrates how diagrams, as opposed to plots of data or functions, can be generated. The script `sqrt_ex` in Listing 17.5 produces Figure 17.12. It uses the `line` function, which is a low-level routine that creates a line object in the current Axes. Several of MATLAB's higher level graphics routines make use of `line`. Somewhat oxymoronically, the script also uses the `rectangle` function to draw

Listing 17.1. *Script* `wfreq`.

```
%WFREQ

% Cell array z stores the data:
z = {492, 'matrix'
     475, 'that'
     456, 'function'
     420, 'with'
     280, 'this'
     273, 'figure'
     261, 'example'
     226, 'which'
     201, 'functions'
     169, 'plot'
     158, 'using'
     154, 'file'
     150, 'command'
     140, 'from'
     135, 'vector'};
% Draw bar graph of first column of z.  CAT converts to column vector.
barh(cat(1,z{:,1}))
n = length(z);
set(gca,'YTick',1:n,'YTickLabel',z(:,2))
set(gca,'YDir','reverse')  % Reverse order of y-values.
ylim([0 n+1])
grid
```

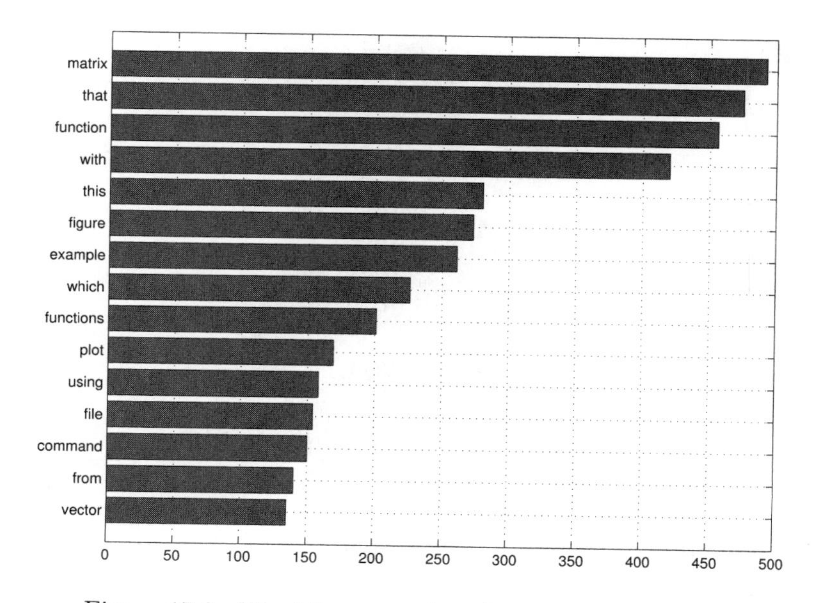

Figure 17.8. *Word frequency bar chart created by* `wfreq`.

Listing 17.2. *Script* hist_ex.

```
%HIST_EX

randn('state',1)    % To make histogram reproducible.
y = randn(10000,1);

hist(y,min(y):0.1:max(y))

h = findobj(gca,'Type','patch');
set(h,'EdgeColor','k','Facecolor',[1 1 1]/2)
set(gca,'TickDir','out')
xlim([min(y) max(y)])

set(gca,'FontSize',14)
```

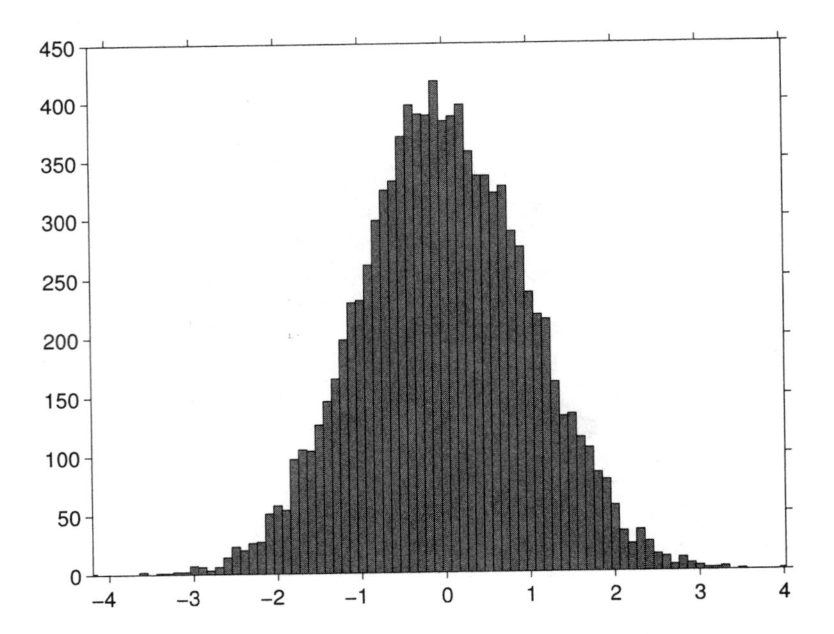

Figure 17.9. *Histogram of samples from normal* $(0, 1)$ *distribution.*

Listing 17.3. *Script* cheb3plot.

```
%CHEB3PLOT

y = linspace(-1,1,1500)';
Z = cheby(y,61);

k = [0 2 4 10 20 40 60];
z = Z(:,k+1);

for j = 1:length(k)
    plot3(j*ones(size(y)),y,z(:,j),'LineWidth',1.5);
    hold on
end
hold off
box on
set(gca,'DataAspectRatio',[1 0.75 4])  % Change shape of box.
view(-72,28)
set(gca,'XTickLabel',k)

% Labels, with adjustment of position.
xlabel('k','FontSize',14)
h = get(gca,'Xlabel'); set(h,'Position',get(h,'Position') + [1.5 0.1 0])
ylabel('x','FontSize',14)
h = get(gca,'Ylabel'); set(h,'Position',get(h,'Position') + [0 0.25 0])
set(gca,'FontSize',14)
```

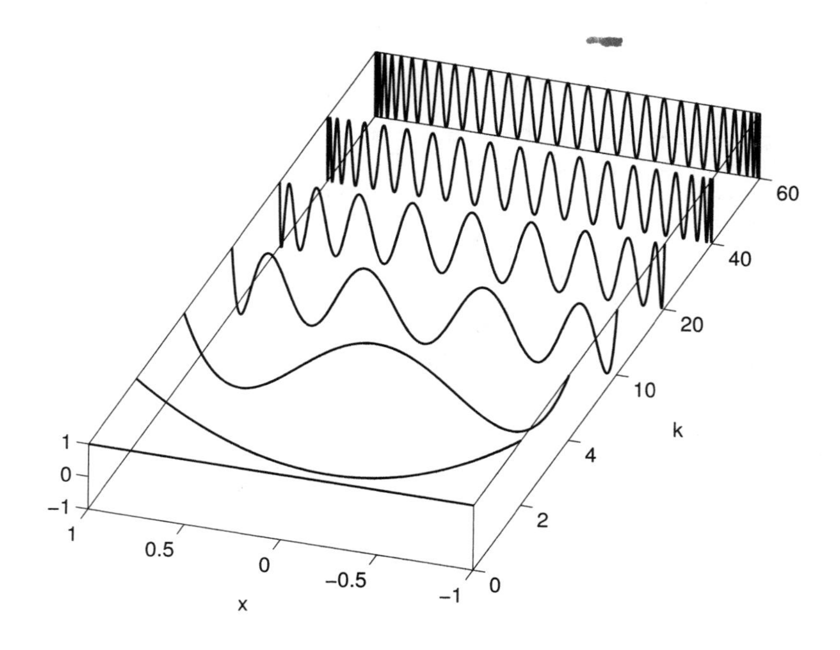

Figure 17.10. *Selected Chebyshev polynomials* $T_k(x)$ *on* $[-1,1]$.

Listing 17.4. *Script* garden.

```
%GARDEN

% Cols: Carrots|Broccoli|Green Beans|Cucumbers|Chard.  Rows are months.
Y = [0.4 0.3 0.0 0.0 0.9
     0.6 0.4 0.0 0.0 1.0
     0.7 0.8 0.3 0.2 1.2
     0.6 0.5 0.9 0.4 1.1
     0.4 0.4 0.7 0.6 0.9];

t = [13 15 22 24 18]; % Temperature.

b = bar(Y,'stacked');
ylabel('Yield (kg)'), ylim([0 4])

h1 = gca; % Handle of first axis.
set(h1,'XTickLabel','May|June|July|August|September')

% Create a second axis at same location as first and plot to it.
h2 = axes('Position',get(h1,'Position'));
p = plot(t,'Marker','square','MarkerSize',12,'LineStyle','-',...
          'LineWidth',2,'MarkerFaceColor',[.6 .6 .6]);
ylabel('Degrees (Celsius)')
title('Fran''s vegetable garden','FontSize',14)

% Align second x-axis with first and remove tick marks and tick labels.
set(h2,'Xlim',get(h1,'XLim'),'XTick',[],'XTickLabel',[])
% Locate second y-axis on right, make background transparent.
set(h2,'YAxisLocation','right','Color','none')

% Make second y-axis tick marks line up with those of first.
ylimits = get(h2,'YLim');
yinc = (ylimits(2)-ylimits(1))/4;
set(h2,'Ytick',[ylimits(1):yinc:ylimits(2)])

% Give legend the Axes handles and place top left.
legend([b,p],'Carrots','Broccoli','Green Beans','Cucumbers',...
       'Swiss Chard','Temperature','Location','NW')
```

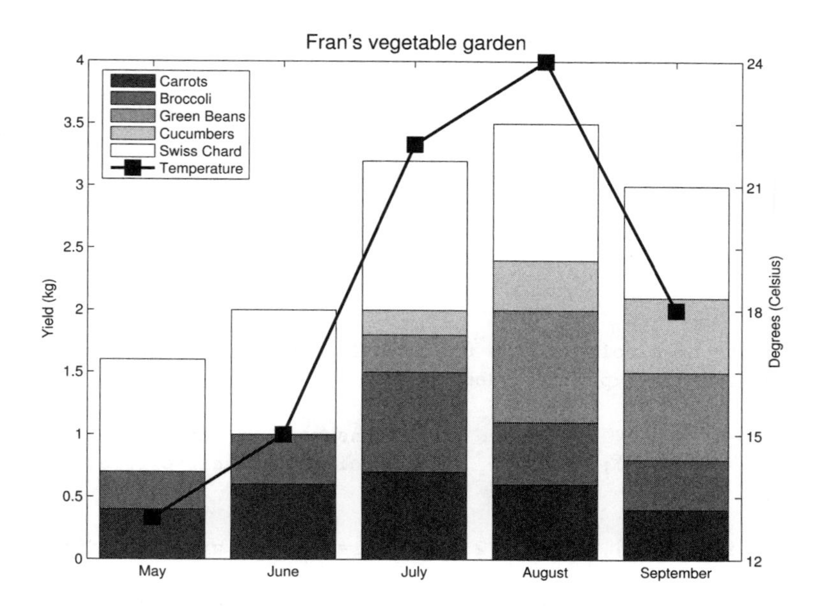

Figure 17.11. *Example with superimposed Axes created by script* `garden`.

a circle. The `Position` property of `rectangle` is a vector `[x y w h]` that specifies a rectangle of width `w` and height `h` with bottom left corner at the point `x`, `y`, all in Axes data units. The `Curvature` property determines the curvature of the sides of the rectangle, with extremes `[0 0]` for square sides and `[1 1]` for an ellipse. The `HorizontalAlignment` and `VerticalAlignment` text properties have been used to help position the text.

Listing 17.5. *Script* sqrt_ex.

```
%SQRT_EX
% Script plotting a point on the unit circle and its two square roots,
% with the right half-plane shaded.

clear i                                  % Ensure i is function, not variable.
z = -1+i; z = z/abs(z);                  % Point z on unit circle.
s = sqrt(z);

h = axes('XLim',[-2 2],'YLim',[-2 2]); % Create Axes with specified range.

fill([0 2 2 0],[-2 -2 2 2],[.8 .8 .8]) % Shade right half-plane.
hold on

plot(z,'s','MarkerSize',8), line([0 real(z)],[0 imag(z)])
plot(s,'d','MarkerSize',8), line([0 real(s)],[0 imag(s)])
plot(-s,'d','MarkerSize',8), line([0 -real(s)],[0 -imag(s)],'LineStyle',':')

% Unit circle.
rectangle('Position',[-1,-1,2,2],'Curvature',[1,1],'LineStyle','--')
axis square

% Draw x- and y-axes through origin.
plot([-2 2], [0 0], '-'), plot([0 0], [-2 2], '-')
set(h,'XTick',[],'YTick',[])

xlabel('Re \lambda')
ylabel('Im \lambda','Rotation',0,'HorizontalAlignment','right')

text(real(z),imag(z)+0.2,'\lambda','HorizontalAlignment','center')
text(0,0,'0','HorizontalAlignment','right','VerticalAlignment','top')
text(real(s),imag(s)+0.2,'\lambda^{1/2}')
text(-real(s),-imag(s)-0.2,'-\lambda^{1/2}','HorizontalAlignment','right')
hold off

% Reset FontSize for all text.
g = findall(gcf,'type','text'); set(g,'Fontsize',16)
```

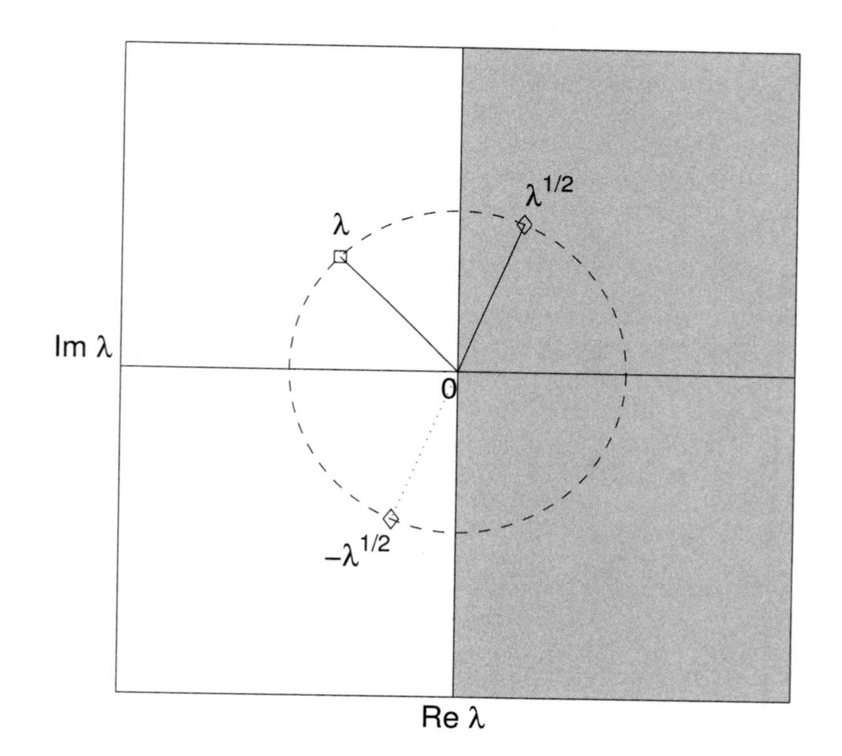

Figure 17.12. *Diagram created by* `sqrt_ex`.

Words with most meanings in the Oxford English Dictionary*:*
1. set
⋮
6. get

— RUSSELL ASH, The Top 10 of Everything (1994)

Handle Graphics . . .
allows you to display your data and then
"reach in" and manipulate any part of the image you've created,
whether that means changing a color, a line style, or a font.

— *The MATLAB EXPO: An Introduction to MATLAB, SIMULINK and the MATLAB Application Toolboxes* (1993)

The best designs . . .
are intriguing and curiosity-provoking,
drawing the viewer into the wonder of the data,
sometimes by narrative power,
sometimes by immense detail,
and sometimes by elegant presentation of simple but interesting data.

— EDWARD R. TUFTE, *The Visual Display of Quantitative Information* (1983)

Did we really want to clutter the text with
endless formatting and Handle Graphics commands such as
`fontsize`, `markersize`, `subplot`, *and* `pbaspect`,
which have nothing to do with the mathematics?
In the end I decided that yes, we did.
I want you to be able to download these programs
and get beautiful results immediately.

— LLOYD N. TREFETHEN, *Spectral Methods in MATLAB* (2000)

Chapter 18
Other Data Types and Multidimensional Arrays

So far in this book we have used several of MATLAB's fundamental data types (or classes): **double**, **single**, **int***, **uint***, **logical**, and **function handle**. In this chapter we describe three further data types: **char**, **struct**, and **cell**. Most of the fundamental types are, in general, multidimensional arrays; we describe multidimensional arrays in the second section. Figure 18.1 shows the complete picture of MATLAB data types. Inline objects, which we met in Section 10.3, fit into the diagram under "user classes".

If you want to determine the data type of an object you can use the class function, which provides essentially the same information as the last column of the output from whos. For example,

```
>> class(pi)
ans =
double

>> class(@sin)
ans =
function_handle

>> class(true)
ans =
logical
```

You can also use the isa function to test whether a variable is of a particular class:

```
>> isa(rand(2),'double')
ans =
     1

>> isa(eye(2),'logical')
ans =
     0
```

18.1. Strings

A string, or character array (char array), is an array of characters represented internally in MATLAB by the corresponding ASCII values. Consider the following example:

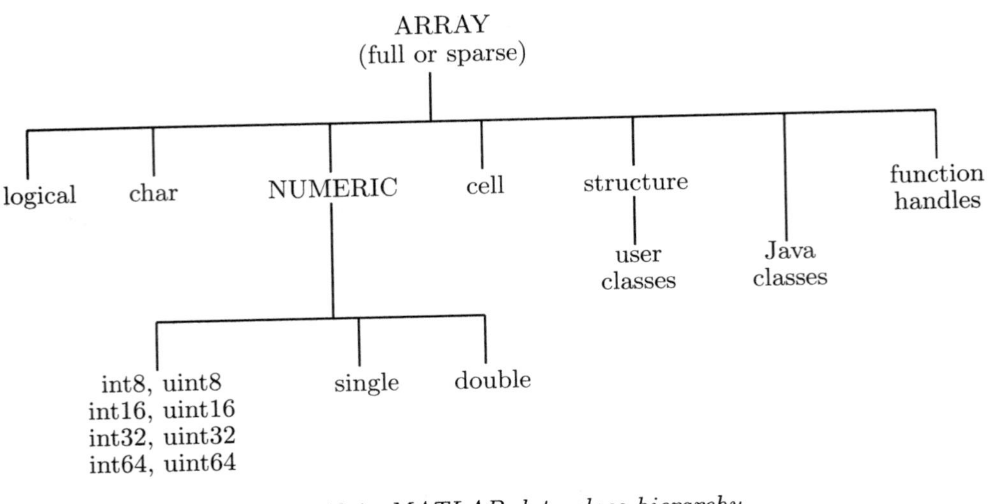

Figure 18.1. *MATLAB data class hierarchy.*

```
>> s = 'ABCabc'
s =
ABCabc

>> sd = double(s)
sd =
    65    66    67    97    98    99

>> s2 = char(sd)
s2 =
ABCabc

>> whos
  Name      Size           Bytes  Class

  s         1x6               12  char array
  s2        1x6               12  char array
  sd        1x6               48  double array

Grand total is 18 elements using 72 bytes
```

We see that a string can be specified by placing characters between single quotes or by applying the `char` function to an array of positive integers. Each character in a string occupies 2 bytes. Converting a string to a `double` array produces an array of ASCII values occupying 8 bytes per element; for example, `double('A')` is 65.

Strings are indexed just like any other array:

```
>> s(6:-1:4)
ans =
cba
```

Strings can also be created by formatting the values of numeric variables, using `int2str`, `num2str`, or `sprintf`, as described in Section 13.2.

MATLAB has several functions for working with strings. Function `strcat` concatenates two strings into one longer string. It removes trailing spaces but leaves leading spaces:

```
>> strcat('Hello',' world')
ans =
Hello world
```

A similar effect can be achieved using the square bracket notation:

```
>> ['Hello ' 'world']
ans =
Hello world
```

Two strings can be compared using `strcmp`: `strcmp(s,t)` returns 1 (true) if `s` and `t` are identical and 0 (false) otherwise. Function `strcmpi` does likewise but treats upper and lower case letters as equivalent. Note the difference between using `strcmp` and the relational operator `==`:

```
>> strcmp('Matlab6','Matlab7')
ans =
     0

>> 'Matlab6' == 'Matlab7'
ans =
     1     1     1     1     1     1     0
```

The relational operator can be used only to compare strings of equal length and it returns a logical vector showing which characters match. To test whether one string is contained in another use `findstr`: `findstr(s,t)` returns a vector of indices of locations where the shorter string appears in the longer:

```
>> findstr('bc','abcd')
ans =
     2

>> findstr('abacad','a')
ans =
     1     3     5
```

A string can be tested for with logical function `ischar`.

Function `eval` executes a string containing any MATLAB expression. Suppose we want to set up matrices `A1`, `A2`, `A3`, `A4`, the pth of which is `A - p*eye(n)`. Instead of writing four assignment statements this can be done in a loop using `eval`:

```
for p = 1:4
    eval(['A', int2str(p), ' = A - p*eye(n)'])
end
```

When `p = 2`, for example, the argument to `eval` is the string `'A2 = A - p*eye(n)'` and `eval` executes the assignment.

Two-dimensional character arrays must be rectangular. This can be achieved by padding with spaces or by the use of `char`:

```
>> subjects = ['Chemistry';'Physics']
??? Error using ==> vertcat
All rows in the bracketed expression must have the same
number of columns.

>> subjects = ['Chemistry';'Physics  ']
subjects =
Chemistry
Physics

>> subjects = char('Chemistry','Physics')
subjects =
Chemistry
Physics
```

An alternative is to make `subjects` a cell array (see Section 18.3):

```
>> subjects = {'Chemistry';'Physics'}
subjects =
    'Chemistry'
    'Physics'
```

For more functions relating to strings see `help strfun`.

18.2. Multidimensional Arrays

Full (but not sparse) arrays of type `double`, `single`, `int*`, `uint*`, `char`, `logical`, `cell`, and `struct` can have more than two dimensions. Multidimensional arrays are defined and manipulated using natural generalizations of the techniques for matrices. For example we can set up a 3-by-2-by-2 array of random normal numbers as follows:

```
>> A = randn(3,2,2)
A(:,:,1) =
    0.8644    0.8735
    0.0942   -0.4380
   -0.8519   -0.4297
A(:,:,2) =
   -1.1027    0.1684
    0.3962   -1.9654
   -0.9649   -0.7443

>> whos
  Name      Size           Bytes  Class

  A         3x2x2             96  double array

Grand total is 12 elements using 96 bytes
```

Notice that MATLAB displays this three-dimensional array a two-dimensional slice at a time. Functions `rand`, `randn`, `zeros`, and `ones` all accept an argument list of the form `(n_1,n_2,...,n_p)` or `([n_1,n_2,...,n_p])` in order to set up an array of dimension n_1-by-n_2-...-by-n_p. An existing two-dimensional array can have its dimensionality extended by assigning to elements in a higher dimension; MATLAB automatically increases the dimensions:

```
>> B = [1 2 3; 4 5 6];
>> B(:,:,2) = ones(2,3)
B(:,:,1) =
     1     2     3
     4     5     6
B(:,:,2) =
     1     1     1
     1     1     1
```

The number of dimensions can be queried using **ndims**, and the **size** function returns the number of elements in each dimension:

```
>> ndims(B)
ans =
     3

>> size(B)
ans =
     2     3     2
```

To build a multidimensional array by listing elements in one statement use the **cat** function, whose first argument specifies the dimension along which to concatenate the arrays comprising its remaining arguments:

```
>> C = cat(3,[1 2 3; 0 -1 -2],[-5 -3 -1; 10 5 0])
C(:,:,1) =
     1     2     3
     0    -1    -2
C(:,:,2) =
    -5    -3    -1
    10     5     0
```

Functions that operate in an elementwise sense can be applied to multidimensional arrays, as can arithmetic, logical, and relational operators. Thus, for example, B-ones(size(B)), B.*B, exp(B), 2.^B, and B > 0 all return the expected results. The data analysis functions in Table 5.7 all operate along the first nonsingleton dimension by default and accept an extra argument **dim** that specifies the dimension over which they are to operate. For B as above, compare

```
>> sum(B)
ans(:,:,1) =
     5     7     9
ans(:,:,2) =
     2     2     2

>> sum(B,3)
ans =
     2     3     4
     5     6     7
```

Table 18.1. *Multidimensional array functions.*

`cat`	Concatenate arrays
`ndims`	Number of dimensions
`ndgrid`	Generate arrays for multidimensional functions and interpolation
`permute`	Permute array dimensions
`ipermute`	Inverse permute array dimensions
`shiftdim`	Shift dimensions
`squeeze`	Remove singleton dimensions

The transpose operator and the linear algebra operations such as **diag**, **inv**, **eig**, and \ are undefined for arrays of dimension greater than 2; they can be applied to two-dimensional sections only.

Table 18.1 lists some functions designed specifically for manipulating multidimensional arrays.

18.3. Structures and Cell Arrays

Structures and cell arrays both provide a way to collect arrays of different types and sizes into a single array. They are MATLAB features of growing importance, used in many places within MATLAB. For example, structures are used by **spline** (p. 162), by **solve** in the next chapter (p. 279), and to set options for the nonlinear equation and optimization solvers (Section 11.2) and the differential equation solvers (Sections 12.2–12.5). Structures also play an important role in object-oriented programming in MATLAB (which is not discussed in this book). Cell arrays are used by the **varargin** and **varargout** functions (Section 10.6), to specify text in graphics commands (p. 113), and in the **switch-case** construct (Section 6.2).

We give only a brief introduction to structures and cell arrays here. See **help datatypes** for a list of functions associated with structures and cell arrays, and see [75] for a tutorial.

Suppose we want to build a collection of 4×4 test matrices, recording for each matrix its name, the matrix elements, and the eigenvalues. We can build an array structure **testmat** having three fields, **name**, **mat**, and **eig**:

```
n = 4;
testmat(1).name = 'Hilbert';
testmat(1).mat = hilb(n);
testmat(1).eig = eig(hilb(n));
testmat(2).name = 'Pascal';
testmat(2).mat = pascal(n);
testmat(2).eig = eig(pascal(n));
```

Displaying the structure gives the field names but not the contents:

```
>> testmat
testmat =
1x2 struct array with fields:
    name
```

```
     mat
     eig
```

We can access individual fields using a period:

```
>> testmat(2).name
ans =
Pascal

>> testmat(1).mat
ans =
    1.0000    0.5000    0.3333    0.2500
    0.5000    0.3333    0.2500    0.2000
    0.3333    0.2500    0.2000    0.1667
    0.2500    0.2000    0.1667    0.1429

>> testmat(2).eig
ans =
    0.0380
    0.4538
    2.2034
   26.3047
```

For array fields, array subscripts can be appended to the field specifier:

```
>> testmat(1).mat(1:2,1:2)
ans =
    1.0000    0.5000
    0.5000    0.3333
```

Another way to set up the testmat structure is using the struct command:

```
testmat = struct('name',{'Hilbert','Pascal'},...
                 'mat',{hilb(n),pascal(n)}, ...
                 'eig',{eig(hilb(n)),eig(pascal(n))})
```

The arguments to the struct function are the field names, with each field name followed by the field contents listed within curly braces (that is, the field contents are cell arrays, which are described next). If the entire structure cannot be assigned with one struct statement then it can be created with fields initialized to a particular value using repmat. For example, we can set up a test matrix structure for five matrices initialized with empty names and zero matrix entries and eigenvalues with

```
>> testmat = repmat(struct('name',{''}, 'mat',{zeros(n)}, ...
                           'eig',{zeros(n,1)}),5,1)
testmat =
5x1 struct array with fields:
    name
    mat
    eig

>> testmat(5) % Check last element of structure.
ans =
```

```
    name: ''
     mat: [4x4 double]
     eig: [4x1 double]
```

For the benefits of such preallocation see Section 20.2.

Cell arrays differ from structures in that they are accessed using array indexing rather than named fields. One way to set up a cell array is by using curly braces as cell array constructors. In this example we set up a 2-by-2 cell array:

```
>> C = {1:3, pi; magic(2), 'A string'}
C =
    [1x3 double]    [  3.1416]
    [2x2 double]    'A string'
```

Cell array contents are indexed using curly braces, and the colon notation can be used in the same way as for other arrays:

```
>> C{1,1}
ans =
     1     2     3

>> C{2,:}
ans =
     1     3
     4     2
ans =
A string
```

The test matrix example can be recast as a cell array as follows:

```
clear testmat
testmat{1,1} = 'Hilbert';
testmat{2,1} = hilb(n);
testmat{3,1} = eig(hilb(n));
testmat{1,2} = 'Pascal';
testmat{2,2} = pascal(n);
testmat{3,2} = eig(pascal(n));
```

The `clear` statement is necessary to remove the previous structure of the same name. Here each collection of test matrix information occupies a column of the cell array, as can be seen from

```
>> testmat
testmat =
    'Hilbert'       'Pascal'
    [4x4 double]    [4x4 double]
    [4x1 double]    [4x1 double]
```

The `celldisp` function can be used to display the contents of a cell array:

```
>> celldisp(testmat)
testmat{1,1} =
Hilbert
```

```
testmat{2,1} =
    1.0000    0.5000    0.3333    0.2500
    0.5000    0.3333    0.2500    0.2000
    0.3333    0.2500    0.2000    0.1667
    0.2500    0.2000    0.1667    0.1429
testmat{3,1} =
    0.0001
    0.0067
    0.1691
    1.5002
testmat{1,2} =
Pascal
testmat{2,2} =
     1     1     1     1
     1     2     3     4
     1     3     6    10
     1     4    10    20
testmat{3,2} =
    0.0380
    0.4538
    2.2034
   26.3047
```

Another way to express the assignments to `testmat` above is by using standard array subscripting, as illustrated by

```
testmat(1,1) = {'Hilbert'};
```

Curly braces must appear on either the left or the right side of the assignment statement in order for the assignment to be valid.

When a component of a cell array is itself an array, its elements can be accessed using parentheses:

```
>> testmat{2,1}(4,4)
ans =
    0.1429
```

Although it was not necessary in our example, we could have preallocated the `testmat` cell array with the `cell` command:

```
testmat = cell(3,2);
```

After this assignment `testmat` is a 3-by-2 cell array of empty matrices.

Useful for visualizing the structure of a cell array is `cellplot`. Figure 18.2 was produced by `cellplot(testmat)`.

Cell arrays can replace comma-separated lists of variables. The `varargin` and `varargout` functions (see Section 10.6) provide good examples of this usage. To illustrate, consider

```
>> testmat{1,:}
ans =
Hilbert
ans =
Pascal
```

Two separate outputs are produced, and by feeding these into `char` we obtain a character array:

```
>> names = char(testmat{1,:})
names =
Hilbert
Pascal

>> whos names
  Name         Size                          Bytes  Class

  names        2x7                              28  char array

Grand total is 14 elements using 28 bytes
```

The functions `cell2struct` and `struct2cell` convert between cell arrays and structures, while `num2cell` creates a cell array of the same size as the given numeric array. The `cat` function, discussed in Section 18.2, provides an elegant way to produce a numeric vector from a structure or cell array. In our test matrix example, if we want to produce a matrix having as its columns the vectors of eigenvalues, we can type

```
cat(2,testmat.eig)
```

for the structure `testmat`, or

```
cat(2,testmat{3,:})
```

for the cell array `testmat`, in both cases obtaining the result

```
ans =
     0.0001     0.0380
     0.0067     0.4538
     0.1691     2.2034
     1.5002    26.3047
```

Here, the first argument of `cat` causes concatenation in the second dimension, that is, columnwise. If this argument is replaced by `1` then the concatenation is row-wise and a long vector is produced. An example of this use of `cat` is in Listing 17.1, where it extracts from a cell array a vector that can then be plotted.

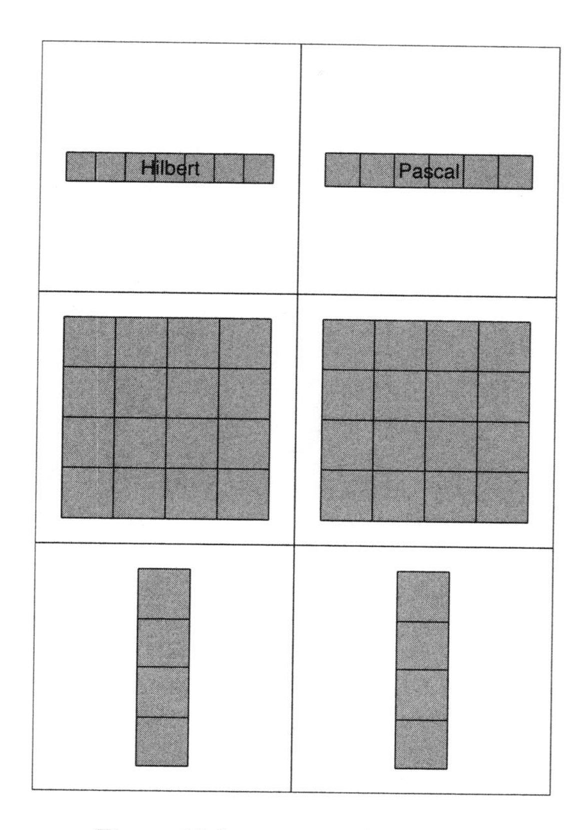

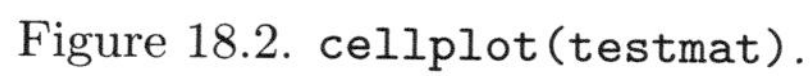
Figure 18.2. `cellplot(testmat)`.

For many applications,
the choice of the proper data structure is really
the only major decision involved in the implementation;
once the choice has been made,
only very simple algorithms are needed.

— ROBERT SEDGEWICK, *Algorithms* (1988)

Chapter 19
The Symbolic Math Toolbox

The Symbolic Math Toolbox is one of the many toolboxes that extend the functionality of MATLAB, and perhaps the one that does so in the most fundamental way. The toolbox is provided with the MATLAB Student Version, but must be purchased as an extra with other versions of MATLAB. You can tell if your MATLAB installation contains the toolbox by issuing the `ver` command and seeing if the toolbox is listed.

The toolbox is based upon the Maple®[10] kernel, which performs all the symbolic and variable precision computations. Maple is a symbolic manipulation package produced by Waterloo Maple, Inc.

To obtain an overview of the functions in the toolbox type `help symbolic`.

19.1. Equation Solving

The Symbolic Math Toolbox defines a new datatype: a symbolic object, denoted by `sym`. Symbolic objects can be created with the `sym` and `syms` commands. Suppose we wish to solve the quadratic equation $ax^2 + bx + c = 0$. We define symbolic variables:

```
>> syms a b c x

>> whos
  Name        Size             Bytes  Class

  a           1x1                126  sym object
  b           1x1                126  sym object
  c           1x1                126  sym object
  x           1x1                126  sym object

Grand total is 8 elements using 504 bytes
```

The same effect can be achieved using

```
>> a = sym('a'); b = sym('b'); c = sym('c'); x = sym('x');
```

We recommend using the shorter `syms` form. Now we can solve the quadratic using the powerful `solve` command:

```
>> y = solve(a*x^2+b*x+c)
y =
[ 1/2/a*(-b+(b^2-4*a*c)^(1/2))]
[ 1/2/a*(-b-(b^2-4*a*c)^(1/2))]
```

[10]Maple is a registered trademark of Waterloo Maple, Inc.

MATLAB creates a 2-by-1 symbolic object y to hold the two solutions. We have used the shortest way to invoke solve. We could also have typed

```
>> y = solve('a*x^2+b*x+c=0');
>> y = solve(a*x^2+b*x+c,x);
```

Since we did not specify an equals sign, MATLAB assumed the expression we provided was to be equated to zero; if an equals sign is explicitly given then the whole expression must be placed in quotes. Less obvious is how MATLAB knew to solve for x and not one of the other symbolic variables. MATLAB applied its findsym function to the expression a*x^2+b*x+c to determine the variable closest alphabetically to x, and solved for that variable. The same procedure is used by other functions in the toolbox. In each case, MATLAB's choice can be overridden by specifying the variable or variables as extra arguments. Thus we can solve the same equation for a as follows:

```
>> solve(a*x^2+b*x+c,a)
ans =
-(b*x+c)/x^2
```

Suppose we now wish to check that the components of y really do satisfy the quadratic equation. We evaluate the quadratic at y, using elementwise squaring since y is a vector:

```
>> a*y.^2+b*y+c
ans =
[ 1/4/a*(-b+(b^2-4*a*c)^(1/2))^2+1/2*b/a*(-b+(b^2-4*a*c)^(1/2))+c]
[ 1/4/a*(-b-(b^2-4*a*c)^(1/2))^2+1/2*b/a*(-b-(b^2-4*a*c)^(1/2))+c]
```

The result is not displayed as zero, but we can apply the simplify function to try to reduce it to zero:

```
>> simplify(ans)
ans =
[ 0]
[ 0]
```

It is characteristic of symbolic manipulation packages that postprocessing is often required to put the results in the most useful form.

Having computed a symbolic solution, a common requirement is to evaluate it for numerical values of the parameters. This can be done using the subs function, which replaces all occurrences of symbolic variables by specified expressions. To find the roots of the quadratic $x^2 - x - 1$ (cf. p. 160) we can type

```
>> a = 1; b = -1; c = -1;
>> subs(y)
ans =
    1.6180
   -0.6180
```

When given one symbolic argument the subs command returns that argument with all variables replaced by their values (if any) from the workspace. Alternatively, subs can be called with three arguments in order to assign values to variables without changing those variables in the workspace:

```
>> subs(y, {a, b, c}, {1, -1, -1})
ans =
    1.6180
   -0.6180
```

Note that the second and third arguments are cell arrays (see Section 18.3).

Simultaneous equations can be specified one at a time to the **solve** function. In general, the number of solutions cannot be predicted. There are two ways to collect the output. As in the next example, if the same number of output arguments as unknowns is supplied then the results are assigned to the outputs (alphabetically):

```
>> syms x y
>> [x,y] = solve('x^2+y^2 = 1','x^3-y^3 = 1')
x =
[                 0]
[                 1]
[ -1+1/2*i*2^(1/2)]
[ -1-1/2*i*2^(1/2)]
y =
[               -1]
[                0]
[ 1+1/2*i*2^(1/2)]
[ 1-1/2*i*2^(1/2)]
```

Alternatively, a single output argument can be provided, in which case a structure (see Section 18.3) containing the solutions is returned:

```
>> S = solve('y = 1/(1+x^2)','y = 1.001 - 0.5*x')
S =
    x: [3x1 sym]
    y: [3x1 sym]

>> [S.x(1), S.y(1)]
ans =
[ 1.0633051173985148109357033343229, .46934744130074259453214833283854]
```

The fields of the structure have the names of the variables, and in this example we looked at the first of the three solutions. This example illustrates that if **solve** cannot find a symbolic solution it will try to find a numeric one. The number of digits computed is controlled by the **digits** function described in Section 19.5; the default is 32 digits.

When interpreting the results of symbolic computations the precedence rules for arithmetic operators need to be kept in mind (see Table 4.1). For example:

```
>> syms a b
>> b=a/2
b =
1/2*a
```

Parentheses are not needed around the `1/2`, since `/` and `*` have the same precedence, but we are used to seeing them included for clarity.

The **sym** and **syms** commands have optional arguments for specifying that a variable is real or positive:

```
syms x real, syms a positive
```

Both statuses can be cleared with

```
syms x a unreal
```

The information that a variable is real or positive can be vital in symbolic computations. For example, consider

```
>> syms p x y
>> y = ((x^p)^(p+1))/x^(p-1);
>> simplify(y)
ans =
(x^p)^p*x
```

The Symbolic Math Toolbox assumes that the variables `x` and `p` are complex and is unable to simplify `y` further. With the additional information that `x` and `p` are positive, further simplification is obtained:

```
>> syms p x positive
>> simplify(y)
ans =
x^(p^2+1)
```

The function `complex` does not accept symbolic arguments, so to set up complex symbolic expressions you must use `sym(sqrt(-1))` or `sym(i)`. For example:

```
>> syms x y
>> z = x + sym(sqrt(-1))*y;
>> expand(z^2)
ans =
x^2+2*i*x*y-y^2
```

19.2. Calculus

The Symbolic Math Toolbox provides symbolic integration and differentiation through the `int` and `diff` functions.

Here is a quick test that the MATLAB developers use to make sure that the Symbolic Math Toolbox is "online":

```
>> int('x')
ans =
1/2*x^2
```

Note that the constant of integration is always omitted. A more complicated example is

```
>> int('sqrt(tan(x))')
ans =
1/2*tan(x)^(1/2)/(cos(x)*sin(x))^(1/2)*cos(x)*2^(1/2)*(pi-...
acos(sin(x)-cos(x)))-1/2*2^(1/2)*log(cos(x)+2^(1/2)*...
tan(x)^(1/2)*cos(x)+sin(x))
```

This answer is easier to read if we "prettyprint" it:

```
>> pretty(ans)

             1/2         1/2
       tan(x)    cos(x) 2    (pi - acos(sin(x) - cos(x)))
   1/2 ----------------------------------------------------
                                    1/2
                        (cos(x) sin(x))

                  1/2                1/2         1/2
         - 1/2 2     log(cos(x) + 2     tan(x)      cos(x) + sin(x))
```

Note that we have not defined `x` to be a symbolic variable, so the argument to `int` must be enclosed in quotes. Alternatively we can define `syms x` and omit the quotes.

Definite integrals $\int_a^b f(x)\,dx$ can be evaluated by appending the limits of integration a and b. Here is an integral that has a singularity at the left endpoint, but which nevertheless has a finite value:

```
>> int('arctan(x)/x^(3/2)',0,1)
ans =
-1/2*pi+1/2*2^(1/2)*log(2+2^(1/2))-1/2*2^(1/2)*log(2-2^(1/2))+...
1/2*2^(1/2)*pi
```

The answer is exact and is rather complicated. We can convert it to numeric form:

```
>> double(ans)
ans =
    1.8971
```

It is important to realize that symbolic manipulation packages cannot "do" all integrals. This may be because the integral does not have a closed form solution in terms of elementary functions, or because it has a closed form solution that the package cannot find. Here is an example of the first kind:

```
>> int('sqrt(1+cos(x)^2)')
ans =
-(sin(x)^2)^(1/2)/sin(x)*EllipticE(cos(x),i)
```

The integral is expressed in terms of an elliptic integral of the second kind, which itself is not expressible in terms of elementary functions. If we evaluate the same integral in definite form we obtain

```
>> int('sqrt(1+cos(x)^2)',0,48)
ans =
30*2^(1/2)*EllipticE(1/2*2^(1/2))+...
2^(1/2)*EllipticE(-sin(48),1/2*2^(1/2))
```

and MATLAB can evaluate the elliptic integrals therein:

```
>> double(ans)
ans =
   58.4705
```

Next we give some examples of symbolic differentiation. We first set up the appropriate symbolic variables and so can omit the quotes from the argument to `diff`:

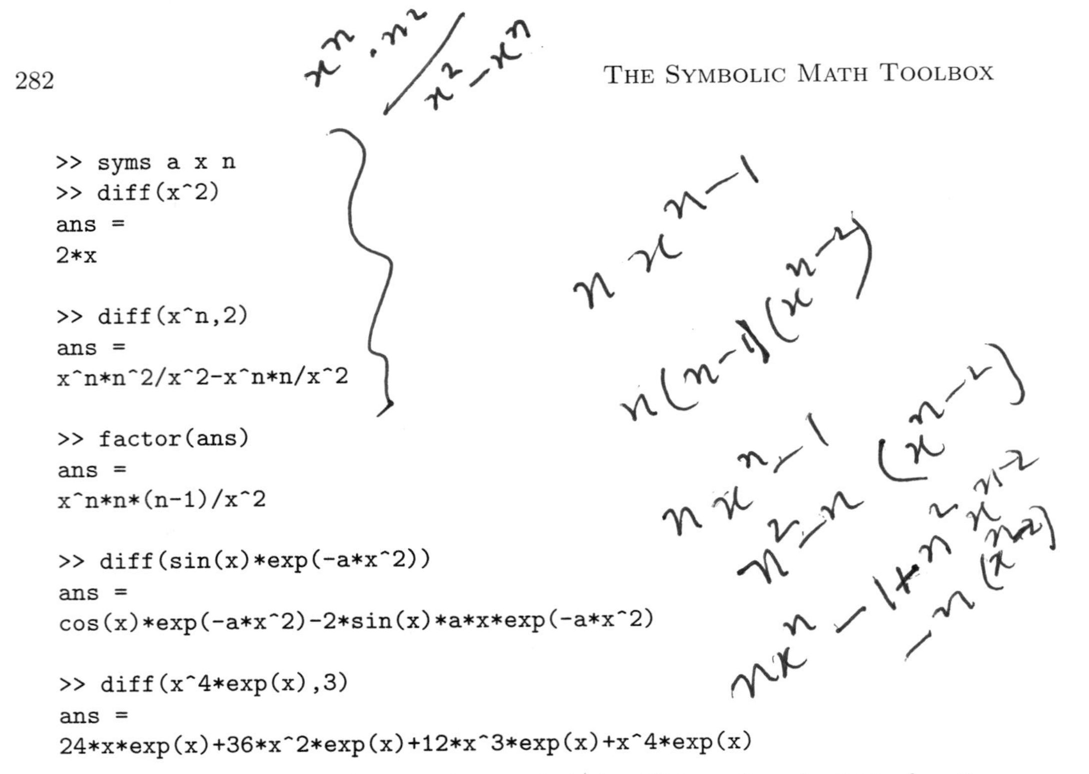

```
>> syms a x n
>> diff(x^2)
ans =
2*x

>> diff(x^n,2)
ans =
x^n*n^2/x^2-x^n*n/x^2

>> factor(ans)
ans =
x^n*n*(n-1)/x^2

>> diff(sin(x)*exp(-a*x^2))
ans =
cos(x)*exp(-a*x^2)-2*sin(x)*a*x*exp(-a*x^2)

>> diff(x^4*exp(x),3)
ans =
24*x*exp(x)+36*x^2*exp(x)+12*x^3*exp(x)+x^4*exp(x)
```

The result of the second differentiation needed simplifying; the `simplify` function does not help in this case so we used `factor`. In the second and last examples a second argument to `diff` specifies the order of the required derivative; the default is the first derivative.

We can obtain mixed partial derivatives of functions of more than one variable by explicitly specifying the variable with respect to which each differentiation is done:

```
>> syms x y
>> f = x^2*exp(-y^2)-y/x
f =
x^2*exp(-y^2)-y/x

>> f_xy = diff(diff(f,x),y)
f_xy =
-4*x*y*exp(-y^2)+1/x^2

>> f_yx = diff(diff(f,y),x)
f_yx =
-4*x*y*exp(-y^2)+1/x^2
```

Functions `int` and `diff` can both be applied to matrices, in which case they operate elementwise.

Differential equations can be solved symbolically with `dsolve`. The equations are specified by expressions in which the letter `D` denotes differentiation, with `D2` denoting a second derivative, `D3` a third derivative, and so on. The default independent variable is `t`. Initial conditions can optionally be specified after the equations, using the syntax `y(a) = b`, `Dy(a) = c`, etc.; if none are specified then the solutions contain arbitrary constants of integration, denoted `C1`, `C2`, etc. For our first example we take the logistic differential equation

$$\frac{d}{dt}y(t) = cy - by^2,$$

solving it first with arbitrary c and b and then with particular values of these parameters as an initial value problem:

```
>> syms b c y t
>> y = dsolve('Dy=c*y-b*y^2')
y =
c/(b+exp(-c*t)*C1*c)

>> y = dsolve('Dy=10*y-y^2','y(0)=0.01')
y =
10/(1+999*exp(-10*t))
```

We now check that the latter solution satisfies the initial condition and the differential equation:

```
>> subs(y,t,0)
ans =
    0.0100

>> res = diff(y,t)-(10*y-y^2)
res =
99900/(1+999*exp(-10*t))^2*exp(-10*t)-100/(1+999*exp(-10*t))+...
100/(1+999*exp(-10*t))^2

>> simplify(res)
ans =
0
```

Next we try to find the general solution to the pendulum equation, which we solved numerically on p. 178:

```
>> y = dsolve('D2theta + sin(theta) = 0')
Warning: Explicit solution could not be found; implicit solution
returned.
> In dsolve at 310
y =
     Int(1/(2*cos(_a)+C1)^(1/2),_a =  .. theta)-t-C2 = 0
    Int(-1/(2*cos(_a)+C1)^(1/2),_a =  .. theta)-t-C2 = 0
```

No explicit solution could be found. If θ is small we can approximate $\sin\theta$ by θ, and in this case `dsolve` is able to find both general and particular solutions:

```
>> y = dsolve('D2theta + theta = 0')
y =
C1*cos(t)+C2*sin(t)

>> y = dsolve('D2theta + theta = 0','theta(0) = 1','Dtheta(0) = 1')
y =
cos(t)+sin(t)
```

If the independent variable is other than the default, `t`, it is important to specify the variable as the last input argument—otherwise, it will be treated as a constant:

Table 19.1. *Calculus functions.*

`diff`	Differentiate
`int`	Integrate
`limit`	Limit
`taylor`	Taylor series
`jacobian`	Jacobian matrix
`symsum`	Summation of series

```
>> y = dsolve('Dy-y*cos(x)=0','y(0)=1')  % Incorrect if Dy = dy/dx.
y =
exp(cos(x)*t)

>> y = dsolve('Dy-y*cos(x)=0','y(0)=1','x')
y =
exp(sin(x))
```

Finally, we emphasize that the results from functions such as `solve` and `dsolve` need to be interpreted with care. For example, when we attempt to solve the differential equation $\frac{d}{dt}y = y^{2/3}$ we obtain

```
>> y = dsolve('Dy = y^(2/3)')
y =
1/27*t^3+1/3*t^2*C1+t*C1^2+C1^3
```

This is a solution for any value of the constant `C1`, but it does not represent all solutions: $y(t) = 0$ is another solution.

Taylor series can be computed using the function `taylor`:

```
>> syms x
>> taylor(log(1+x))
ans =
x-1/2*x^2+1/3*x^3-1/4*x^4+1/5*x^5
```

By default the Taylor series about 0 up to terms of order 5 is produced. A second argument specifies the required number of terms (which is one more than the order) and a third argument the point about which to expand:

```
>> pretty(taylor(exp(-sin(x)),3,1))

  exp(-sin(1)) - exp(-sin(1)) cos(1) (x - 1)

                                                        2            2
             + exp(-sin(1)) (1/2 sin(1) + 1/2 cos(1) ) (x - 1)
```

A function `taylortool` provides a graphical interface to `taylor`, plotting both the function and the Taylor series. See Figure 19.1, which shows the interesting function $\sin(\tan x) - \tan(\sin x)$.

The Symbolic Math Toolbox contains some other calculus functions; see Table 19.1.

To finish this section we describe three further examples of Taylor series, integration, and differentiation, taken from [122]. The input

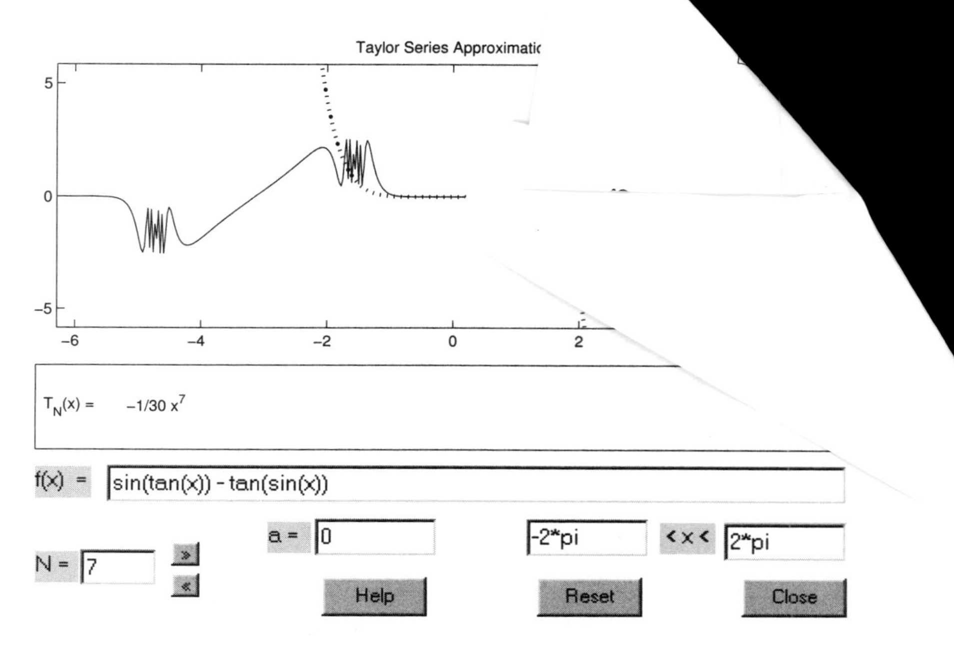

Figure 19.1. `taylortool` *window.*

```
syms x, ezplot(sin(x)+asin(x),-0.8,0.8)
```

produces the plot in Figure 19.2. The curve looks straight, yet sin and arcsin have curvature. What is the explanation? Consider the following three Taylor series up to terms x^5:

```
>> taylor(sin(x)), taylor(asin(x)), taylor(sin(x)+asin(x))
ans =
x-1/6*x^3+1/120*x^5
ans =
x+1/6*x^3+3/40*x^5
ans =
2*x+1/12*x^5
```

The x^3 terms in the Taylor series for sin and arcsin cancel. Hence $\sin(x) + \arcsin(x)$ agrees with $2x$ up to terms of order x^5, and x^5 is small on $[-0.8, 0.8]$.

Next, consider the integral

$$\int_0^1 \frac{x^4(1-x)^4}{1+x^2}\,dx, \tag{19.1}$$

which is clearly positive. Evaluation of the integral produces a surprising result:

```
>> int( x^4*(1-x)^4/(1+x^2),0,1 )
ans =
22/7-pi
```

Hence we have an unexpected demonstration that the well-known approximation 22/7 to π is a strict overestimate.

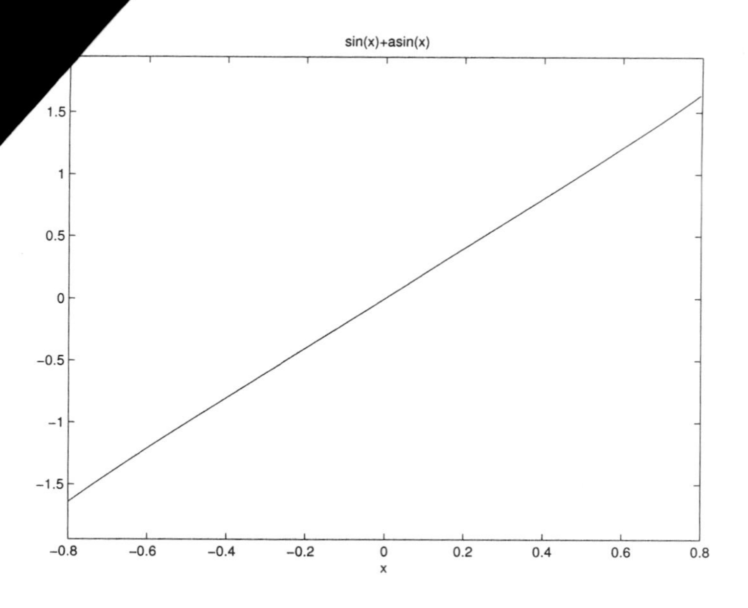

Figure 19.2. $\sin(x) + \arcsin(x)$

Finally, Figure 19.3 plots the integrand in (19.1). The plot appears symmetric and so the maximum might be thought to be at $x = 0.5$. To check, we can find the maximum symbolically:

```
>> g = diff( x^4*(1-x)^4/(1+x^2) ); s = double(solve(g))
s =
        0
        0
        0
   1.0000
   1.0000
   1.0000
   0.4758
  -0.0712 + 1.1815i
  -0.0712 - 1.1815i
```

Three real stationary points have been found and the one that is the desired maximum is at 0.4758, not 0.5.

19.3. Linear Algebra

Several of MATLAB's linear algebra functions have counterparts in the Symbolic Math Toolbox that take symbolic arguments. To illustrate we take the numeric and symbolic representations of the 5-by-5 Frank matrix:

```
>> A_num = gallery('frank',5); A_sym = sym(A_num);
```

This illustrates a different usage of the `sym` function: to convert from a numeric datatype to symbolic form. Since the Frank matrix has small integer entries the

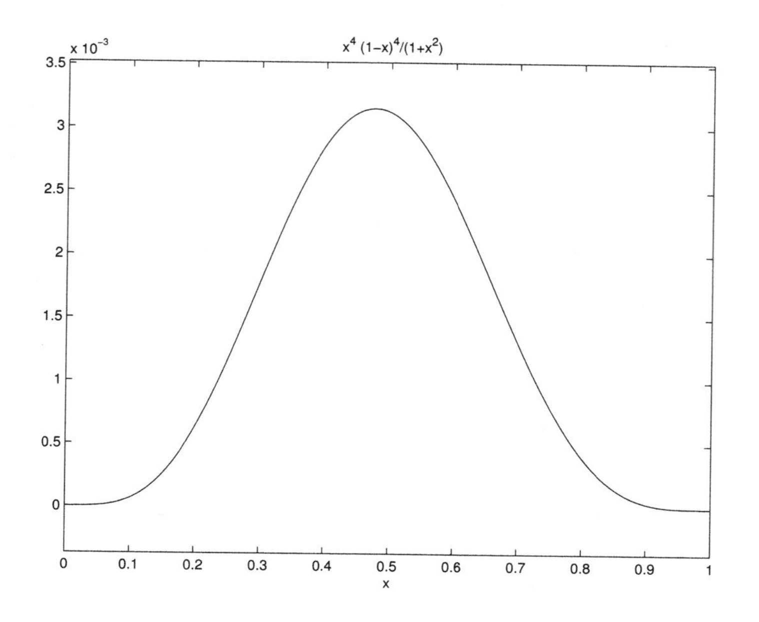

Figure 19.3. *The integrand in* (19.1).

conversion is done exactly. In general, when given a floating point number as argument `sym` tries to express it as a nearby rational number. For example:

```
>> t = 1/3; t, sym(t)
t =
    0.3333
ans =
1/3
```

Here, `t` is a floating point approximation to 1/3, whereas `sym(t)` exactly represents 1/3. For the precise rules used by `sym`, and details of arguments that allow control of the conversion, see `help sym` and also Section 19.5.

Continuing our Frank matrix example we can invert the `double` array `A_num` in the usual way:

```
inv(A_num)
ans =
    1.0000   -1.0000   -0.0000    0.0000         0
   -4.0000    5.0000   -1.0000   -0.0000         0
   12.0000  -15.0000    4.0000   -1.0000         0
  -24.0000   30.0000   -8.0000    3.0000   -1.0000
   24.0000  -30.0000    8.0000   -3.0000    2.0000
```

The trailing zeros show that the computed elements are not exactly integers. We can obtain the exact inverse by applying `inv` to `A_sym`:

```
inv(A_sym)
ans =
[   1,  -1,    0,    0,    0]
```

```
[  -4,   5,  -1,   0,   0]
[  12, -15,   4,  -1,   0]
[ -24,  30,  -8,   3,  -1]
[  24, -30,   8,  -3,   2]
```

Here, MATLAB has recognized that `inv` is being called with a symbolic argument and has invoked a version of `inv` that is part of the Symbolic Math Toolbox. The mechanism that allows different versions of a function to handle different types of arguments is called overloading. You can tell whether a given function is overloaded from its help entry. Assuming the Symbolic Math Toolbox is present, `help inv` produces

```
>> help inv
 INV    Matrix inverse.
    INV(X) is the inverse of the square matrix X.
    A warning message is printed if X is badly scaled or
    nearly singular.

    See also slash, pinv, cond, condest, lsqnonneg, lscov.

    Overloaded functions or methods (ones with the same name in other
    directories)
       help sym/inv.m

    Reference page in Help browser
       doc inv
```

As indicated, to obtain help for the version of `inv` called for a symbolic argument we type `help sym/inv`. We have already used an overloaded function in this chapter: `diff` in the previous section.

Just as for numeric matrices, the backslash operator can be used to solve linear systems with a symbolic coefficient matrix. For example, we can compute the (5,1) element of the inverse of the Frank matrix with

```
>> [0 0 0 0 1]*(A_sym\[1 0 0 0 0]')
ans =
24
```

For a symbolic argument the `eig` function tries to compute the exact eigensystem. We know from Galois theory that this is not always possible in a finite number of operations for matrices of order 5 or more. For the 5-by-5 Frank matrix `eig` succeeds:

```
>> e = eig(A_sym)
e =
[                                             1]
[ 7/2+1/2*10^(1/2)+1/2*(55+14*10^(1/2))^(1/2)]
[ 7/2+1/2*10^(1/2)-1/2*(55+14*10^(1/2))^(1/2)]
[ 7/2-1/2*10^(1/2)+1/2*(55-14*10^(1/2))^(1/2)]
[ 7/2-1/2*10^(1/2)-1/2*(55-14*10^(1/2))^(1/2)]

>> double(e)
ans =
```

```
     1.0000
    10.0629
     0.0994
     3.5566
     0.2812
```

As we noted in the example in Section 9.7, the eigenvalues come in reciprocal pairs. To check we can type

```
>> [e(2)*e(3); e(4)*e(5)]
ans =
[ (7/2+1/2*10^(1/2)+1/2*(55+14*10^(1/2))^(1/2))*...
[ (7/2-1/2*10^(1/2)+1/2*(55-14*10^(1/2))^(1/2))*...
```

Note that we have had to truncate the output. Attempting to simplify these expressions using **simplify** fails. Instead we use the function **simple**, which tries several different simplification methods and reports the shortest answer:

```
>> s = simple(ans)
s =
[ 1]
[ 1]
```

(If **simple(ans)** is typed without an output argument then all intermediate attempted simplifications are displayed.) Finally, while we computed the characteristic polynomial numerically in Section 9.7, we can now obtain it exactly:

```
>> poly(A_sym)
ans =
x^5-15*x^4+55*x^3-55*x^2+15*x-1
```

A complete list of linear algebra functions in the toolbox is given in Table 19.2.

19.4. Polynomials and Rationals

Symbolic polynomial and rational expressions are easily formed using symbolic variables and the usual MATLAB notation. The function **coeffs** returns the (symbolic) coefficients and corresponding terms of a polynomial (in no particular order). The functions **sym2poly** and **poly2sym** convert between a symbolic polynomial and a numeric vector of coefficients of the polynomial, ordering the coefficients in the standard MATLAB way from "highest power down to lowest power". Examples:

```
>> syms x
>> p = (2/3)*x^3-x^2-3*x+1
p =
2/3*x^3-x^2-3*x+1

>> [c,terms] = coeffs(p,x)
c =
[    1,   -3, 2/3,   -1]
terms =
[    1,    x, x^3, x^2]
```

Table 19.2. *Linear algebra functions.*

diag	Diagonal matrices and diagonals of matrix
tril	Extract lower triangular part
triu	Extract upper triangular part
inv	Matrix inverse
det	Determinant
rank	Rank
rref	Reduced row echelon form
null	Basis for null space (not orthonormal)
eig	Eigenvalues and eigenvectors
svd	Singular values and singular vectors
poly	Characteristic polynomial
expm	Matrix exponential
colspace*	Basis for column space
jordan*	Jordan canonical (normal) form

* Functions existing in Symbolic Math Toolbox only.

```
>> a = sym2poly(p)
a =
    0.6667   -1.0000   -3.0000    1.0000

>> q = poly2sym(a)
q =
2/3*x^3-x^2-3*x+1
```

As the coefficient of x^3 in this example illustrates, poly2sym (which calls sym) attempts to convert floating point numbers to nearby rationals.

Division of one polynomial by another is done by quorem, which returns the quotient and remainder (cf. deconv on p. 160):

```
>> [q,r] = quorem(p,x^2)
q =
2/3*x-1
r =
1-3*x
```

The function numden converts a rational into a normal form where the numerator and denominator are polynomials with integer coefficients and then returns the numerator and denominator. For a numeric argument, but not necessarily for symbolic ones, the returned numerator and denominator will be relatively prime:

```
>> [n,d] = numden(sym(2*4)/sym(10))
n =
4
d =
5
```

```
>> r = 1 + x^2/(3+x^2/5);
>> [p,q] = numden(r)
p =
15+6*x^2
q =
15+x^2
```

The `sort` function is overloaded for symbolic arguments. It sorts a polynomial in decreasing order of the powers:

```
>> p = x^2-3-3*x^3+x/2;
>> p = sort(p)
p =
-3*x^3+x^2+1/2*x-3
```

Apart from the functions just described, the Symbolic Math Toolbox offers limited support for manipulating polynomials and rationals. To carry out other tasks we can either call Maple directly or manipulate the symbolic expressions explicitly. Maple has many functions connected with polynomials and they can be viewed by typing `mhelp polynomial`. These functions can be invoked with the toolbox's `maple` function. For example, we can find the degree and leading coefficient of a polynomial as follows:

```
>> maple('degree',p)
ans =
3

>> maple('lcoeff',p)
ans =
-3
```

Suppose we wish to replace the coefficients of a polynomial by their absolute values. Rather than invoke in order `sym2poly`, `abs`, and `poly2sym`, which could lead to small changes in the coefficients, we can convert to and from a string and simply replace minuses in the string by pluses:

```
>> p_abs = sym(strrep(char(p),'-','+'))
p_abs =
+3*x^3+x^2+1/2*x+3
```

(This technique assumes that the polynomial does not have any coefficients displayed in exponential format, such as `2e-4`.)

Finally, suppose we wish to evaluate a symbolic polynomial at a matrix argument. Take for example the polynomial $p(x) = x^2 + x - 1$ and the matrix `diag(1,2,3)`. Here are two different attempts:

```
>> syms x; p = x^2 + x - 1;
>> A = sym(diag([1,2,3]));

>> P1 = polyvalm(sym2poly(p),A)
P1 =
[  1,  0,  0]
[  0,  5,  0]
[  0,  0, 11]
```

```
>> P2 = subs(p,x,A)
P2 =
[  1, -1, -1]
[ -1,  5, -1]
[ -1, -1, 11]
```

The first result, P1, is the correct evaluation $p(A) = A^2 + A - I$ in the matrix sense. The second evaluation gives a different answer because the subs function expands scalars into matrices: it converts the 1 into a matrix of 1s, whereas the first evaluation converts 1 into an identity matrix.

19.5. Variable Precision Arithmetic

In addition to MATLAB's double precision floating point arithmetic and symbolic arithmetic, the Symbolic Math Toolbox supports variable precision floating point arithmetic, which is carried out within the Maple kernel. This is useful for problems where an accurate solution is required and an exact solution is impossible or too time-consuming to obtain. It can also be used to experiment with the effect of varying the precision of a computation.

The function **digits** returns the number of significant decimal digits to which variable precision computations are carried out:

```
>> digits

Digits = 32
```

The default of 32 digits can be changed to **n** by the command **digits(n)**. Variable precision computations are based on the **vpa** command. The simplest usage is to evaluate constants to variable accuracy:

```
>> pi_1 = vpa(pi)
pi_1 =
3.1415926535897932384626433832795
```

It is important to note the distinction between pi_1, a 32-digit approximation to π, and the exact representation

```
>> pi_2 = sym(pi)
pi_2 =
pi
```

The difference is apparent from

```
>> sin(pi_1)
ans =
.28841971693993751058209749445920e-32
>> sin(pi_2)
ans =
0
```

Note, however that both pi_1 and pi_2 are syms:

```
>> whos pi*
  Name        Size                         Bytes  Class

  pi_1        1x1                            190  sym object
  pi_2        1x1                            128  sym object

Grand total is 37 elements using 318 bytes
```

The vpa function takes a second argument that overrides the current number of digits specified by digits:

```
>> vpa(pi,50)
ans =
3.1415926535897932384626433832795028841971693993751
```

In the next example we compute e to 40 digits and then check that taking the logarithm gives back 1 (to within 40 digits):

```
>> digits(40)
>> x = vpa('exp(1)')
x =
2.718281828459045235360287471352662497757
>> vpa(log(x))
ans =
.9999999999999999999999999999999999999999
```

Two minor modifications of this example illustrate pitfalls. First, consider

```
% Incorrect code.
>> digits(40)
>> y = vpa(exp(1))
y =
2.718281828459045534884808148490265011787
>> vpa(log(y))
ans =
1.000000000000000110188913283849495821818
```

We omitted the quotes around exp(1), so MATLAB evaluated exp(1) in double precision floating point arithmetic, converted that 16 digit result to 40 digits—thereby adding 24 meaningless digits—and then evaluated the exponential. In the original version the quotes enable exp(1) to pass through the MATLAB interpreter to be evaluated by Maple. The second modification is:

```
% Incorrect code.
>> digits(32)  % Restore default value.
>> d = 40;
>> x = vpa('exp(1)',d)
x =
2.718281828459045235360287471352662497757
>> vpa(log(x),d)
ans =
1.0000000000000000000000000000000
```

Here, instead of setting the precision via `digits` we specify it within the `vpa` calls. The computation of `x` is done correctly to 40 digits, but the statement `vpa(log(x),d)` computes to 32 digits only (albeit producing the "exact answer", fortuitously). The problem is that MATLAB evaluates `log(x)` in the current 32-digit precision before the `vpa` function has the chance to parse its second argument and force 40-digit precision. It is generally best, therefore, to set the appropriate precision using the `digits` function rather than attempt to do so within `vpa` invocations.

Variable precision linear algebra computations are performed by calling functions with variable precision arguments. For example, we can compute the eigensystem of `pascal(4)` to 32 digits by

```
>> [V,E] = eig(vpa(pascal(4))); diag(E)
ans =
[ .38016015229139947237513500399910e-1]
[     26.304703267097871286055226455525]
[     .45383455002566546509718436703856]
[     2.2034461676473233016100756770374]
```

Another pitfall concerns the use of `vpa` arithmetic to compute the "exact" solution to a floating point problem. Suppose we wish to know the error in the eigenvalues computed by `eig` in the following example:

```
>> A = gallery('chebspec',4,1)
A =
   -0.7071   -1.4142    0.7071   -0.2929
    1.4142   -0.0000   -1.4142    0.5000
   -0.7071    1.4142    0.7071   -1.7071
    1.1716   -2.0000    6.8284   -5.5000

>> e = eig(A)
e =
  -0.9588 + 2.3308i
  -0.9588 -2.3308i
  -1.7912 + 0.7551i
  -1.7912 -0.7551i
```

We compute the eigenvalues in 50-digit precision, convert the result back to double precision (hence obtaining the rounded version of the exact answer), and then compute the error:

```
>> digits(50)
>> ex = eig(vpa(sym(A,'f')));
>> format short e
>> e-double(ex)
ans =
 -1.1102e-015 +1.3323e-015i
 -1.1102e-015 -1.3323e-015i
 -8.8818e-016 -3.3307e-016i
 -8.8818e-016 +3.3307e-016i
```

The use of `vpa(sym(A,'f'))` ensures that the floating point matrix `A` is converted exactly to symbolic form. If we use `vpa(A)`, which is the same as `vpa(sym(A,'r'))`, then, as noted in Section 19.3, MATLAB tries to express the elements of `A` as nearby rational numbers, and this causes a small difference in the matrix and hence in the computed eigenvalues:

```
>> ex1 = eig(vpa(A));
>> norm(double(ex1-ex))
ans =
   5.5492e-016
```

19.6. Other Features

The Symbolic Math Toolbox contains many other functions, covering Fourier and Laplace transforms, special functions, conversions, and pedagogical tools. Of particular interest are functions that provide access to Maple (these are not available with the Student Edition). Function **mfun** gives access to many special functions for which MATLAB M-files are not provided; type **mfunlist** to see a list of such functions. Among these functions are the Fresnel integrals; thus commands of the form

```
x = mfun('FresnelC',t); y = mfun('FresnelS',t);
```

provide another way to evaluate the Fresnel spiral in Figure 12.2. More generally, function **maple** sends a statement to the Maple kernel and returns the result. Maple help on Maple function `mfoo` can be obtained by typing **mhelp** `mfoo`.

The **maple** command is used in the following example, in which we obtain a definite integral that evaluates to the Catalan constant; we use Maple to evaluate the constant, since it is not known to MATLAB.

```
>> int('log(x)/(1+x^2)',0,1)
ans =
-Catalan

>> maple('evalf(Catalan)')
ans =
.91596559417721901505460351493238
```

Useful functions for postprocessing are **ccode**, **fortran**, and **latex**, which produce C, Fortran, and LaTeX representations, respectively, of a symbolic expression.

[Babbage's Analytical Engine] can arrange and combine its numerical quantities
exactly as if they were letters or any other general symbols;
and in fact it might bring out its results in algebraical notation,
were provisions made accordingly.

— AUGUSTA ADA BYRON, *Countess of Lovelace* (1843)

I'm very good at integral and differential calculus,
I know the scientific names of beings animalculous;
In short, in matters vegetable, animal, and mineral,
I am the very model of a modern Major-General.

— WILLIAM SCHWENCK GILBERT, *The Pirates of Penzance. Act 1* (1879)

Maple will sometimes "go away" for quite a while to do its calculations.

— ROB CORLESS, *Essential Maple 7* (2002)

The particular form obtained by applying an analytical integration method
may prove to be unsuitable for practical purposes.
For instance, evaluating the formula may be
numerically unstable (due to cancellation, for instance) or even
impossible (due to division by zero).

— ARNOLD R. KROMMER and CHRISTOPH W. UEBERHUBER,
Computational Integration (1998)

Maple has bugs. It has always *had bugs . . .*
Every other computer algebra system also has bugs;
often different ones,
but remarkably many of these bugs are seen
throughout all computer algebra systems,
as a result of common design shortcomings.
Probably the most useful advice I can give for dealing with this is
be paranoid.
Check your results at least two ways (the more the better).

— ROB CORLESS, *Essential Maple 7* (2002)

Chapter 20
Optimizing M-Files

Most users of MATLAB find that computations are completed fast enough that execution time is not usually a cause for concern. Some computations, though, particularly when the problems are large, require a significant time and it is natural to ask whether anything can be done to speed them up. This chapter describes some techniques that produce better performance from M-files. They all exploit the fact that MATLAB is an interpreted language with dynamic memory allocation.

MATLAB has incorporated automatic optimization of code since Release 12. This is good news for users: some of the optimizations that experienced MATLAB programmers employ, and which in the past could make their code run much faster than code written by inexperienced users, are now done automatically by the MATLAB interpreter.

Several aspects make MATLAB's optimization hard to understand and difficult to write about. First, the documentation is sketchy on what exactly is optimized and to what extent. Second, these capabilities are under continuing development and improve with each release. Third, some aspects of the optimization are machine-dependent, that is, they are used only on certain platforms (PC versions of MATLAB have the most comprehensive optimization). Consequently, we will say very little about MATLAB's automatic optimization and will concentrate instead on useful programming techniques, some of which can probably never be replaced by automatic optimization of code.

External C or Fortran codes can be called from MATLAB via the MEX facility, provided they are augmented with a suitable gateway routine that interfaces the codes with MATLAB. The `mex` command is used to compile and link the C or Fortran source into a shared library that is executable from within MATLAB (and has a platform-dependent extension). See [72], [73] for details.

MATLAB's profiler is a useful tool when you are optimizing M-files, as it can help you decide which parts of the code to focus on. See Section 16.3 for details.

All timings in this chapter are for a 3.2Ghz Pentium 4.

20.1. Vectorization

Since MATLAB is a matrix language, many of the matrix-level operations and functions are carried out internally using compiled C or assembly code and are therefore executed at near-optimum efficiency. This is true of the arithmetic operators `*`, `+`, `-`, `\`, `/` and of relational and logical operators. However, `for` loops may be executed relatively slowly—depending on what is inside the loop, MATLAB may or may not be able to optimize the loop. One of the most important tips for producing efficient M-files is to avoid `for` loops in favor of vectorized constructs, that is, to convert `for` loops into equivalent vector or matrix operations. Vectorization has important benefits beyond simply increasing speed of execution. It can lead to shorter and more readable MATLAB code. Furthermore, it expresses algorithms in terms of high-level constructs that are more appropriate for high-performance computing.

Consider the following example:

```
>> n = 5e5; x = randn(n,1);
>> tic, s = 0; for i=1:n, s = s + x(i)^2; end, toc
Elapsed time is 0.266000 seconds.

>> tic, s = sum(x.^2); toc
Elapsed time is 0.015000 seconds.
```

In this example we compute the sum of squares of the elements in a random vector in two ways: with a `for` loop and with an elementwise squaring followed by a call to `sum`. The latter vectorized approach is an order of magnitude faster.

The `for` loop in Listing 10.2 on p. 148 can be vectorized, assuming that `f` returns a vector output for a vector argument. The loop and the statement before it can be replaced by

```
x = linspace(0,1,n);
p = x*f(1) + (x-1)*f(0);
max_err = max(abs(f(x)-p));
```

For a slightly more complicated example of vectorization, consider the inner loop of Gaussian elimination applied to an `n`-by-`n` matrix `A`, which can be written

```
for i = k+1:n
    for j = k+1:n;
        A(i,j) = A(i,j) - A(i,k)*A(k,j)/A(k,k);
    end
end
```

Both loops can be avoided, simply by deleting the two `for`s and `end`s:

```
i = k+1:n;
j = k+1:n;
A(i,j) = A(i,j) - A(i,k)*A(k,j)/A(k,k);
```

The approximately $(\mathtt{n}-\mathtt{k})^2$ scalar multiplications and additions have now been expressed as one matrix multiplication and one matrix addition. With `n = 1600` and `k = 1` we timed the two-loop code at 0.72 seconds and the vectorized version at 0.11 seconds—again vectorization yields a substantial improvement.

The next example concerns premultiplication of a matrix by a Givens rotation in the (j,k) plane, which replaces rows j and k by linear combinations of themselves. It might be coded as

```
temp = A(j,:);
A(j,:) = c*A(j,:) - s*A(k,:);
A(k,:) = s*temp + c*A(k,:);
```

By expressing the computation as a single matrix multiplication we can shorten the code and dispense with the temporary variable:

```
A([j k],:) = [c -s; s c] * A([j k],:);
```

The second version is approximately 50% faster for `n = 500`.

A good principle is to maximize the use of built-in MATLAB functions. Consider, for example, this code to assign to `row_norm` the ∞-norms of the rows of `A`:

```
for i=1:n
    row_norms(i) = norm(A(i,:), inf);
end
```

It can be replaced by the single statement

```
row_norms = max(abs(A),[],2);
```

(see p. 61), which is shorter and runs much more quickly. Similarly, the factorial $n!$ is more quickly computed by `prod(1:n)` than by

```
p = 1; for i = 1:n, p = p*i; end
```

(in fact, there is a MATLAB function `factorial` that uses `prod` in this way).

As a final example, we start with the following code to generate and plot an approximate Brownian (standard Wiener) path [55], which produces Figure 20.1.

```
randn('state',20)
N = 1e4; dt = 1/N;
w(1) = 0;
for j = 2:N+1
    w(j) = w(j-1) + sqrt(dt)*randn;
end
plot([0:dt:1],w)
```

This computation can be speeded up by preallocating the array `w` (see the next section) and by computing `sqrt(dt)` outside the loop. However, we obtain a more dramatic improvement by vectorizing with the help of the cumulative sum function, `cumsum`:

```
randn('state',20)
N = 1e4; dt = 1/N;
w = sqrt(dt)*cumsum([0;randn(N,1)]);
plot([0:dt:1],w)
```

This produces Figure 20.1 roughly 5 times more quickly than the original version.

Vectorization plays an important role in the numerical methods codes of Chapter 12. These codes may require many function evaluations to solve their respective problems, and it can be much more efficient to carry out a certain number of evaluations at vectors than a larger number of scalar function evaluations, not least because of the reduced overheads. The functions `quad` and `quadl` *require* the integrand to be vectorized, while the stiff ODE solvers and `bvp4c` can take advantage of vectorized function evaluations (which the user specifies via `odeset` and `bvpset`).

20.2. Preallocating Arrays

One of the attractions of MATLAB is that arrays need not be declared before first use: assignment to an array element beyond the upper bounds of the array causes MATLAB to extend the dimensions of the array as necessary. If overused, this flexibility can lead to inefficiencies, however. Consider the following implementation of a recurrence:

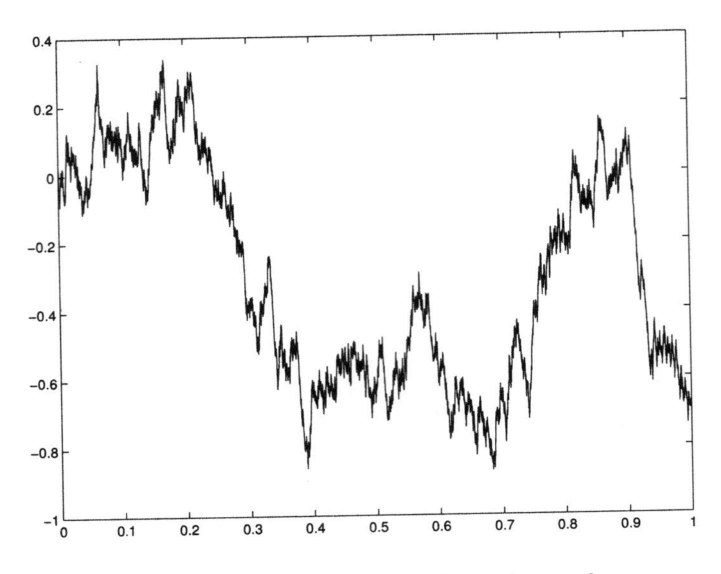

Figure 20.1. *Approximate Brownian path.*

```
% x has not so far been assigned.
x(1:2) = 1;
for i=3:n, x(i) = 0.25*x(i-1)^2 - x(i-2); end
```

On each iteration of the loop, MATLAB must increase the length of the vector `x` by 1. In the next version `x` is preallocated as a vector of precisely the length needed, so no resizing operations are required during execution of the loop:

```
% x has not so far been assigned.
x = ones(n,1);
for i=3:n, x(i) = 0.25*x(i-1)^2 - x(i-2); end
```

With `n = 1e4`, the first piece of code took 0.45 seconds and the second 0.047 seconds, showing that the first version spends most of its time doing memory allocation rather than floating point arithmetic.

Preallocation has the added advantage of reducing the fragmentation of memory resulting from dynamic memory allocation and deallocation.

You can preallocate an array structure with `repmat(struct(...))` and a cell array with the `cell` function; see Section 18.3.

20.3. Miscellaneous Optimizations

Suppose you wish to set up an `n`-by-`n` matrix of 2s. The obvious assignment is

```
A = 2*ones(n);
```

The n^2 floating point multiplications can be avoided by using

```
A = repmat(2,n);
```

The `repmat` approach is much faster for large `n`. This use of `repmat` is essentially the same as assigning

```
A = zeros(n); A(:) = 2;
```

in which scalar expansion is used to fill `A`.

There is one optimization that is automatically performed by MATLAB. Arguments that are passed to a function are not copied into the function's workspace *unless* they are altered within the function. Therefore there is no memory penalty for passing large variables to a function provided the function does not alter those variables.

20.4. Illustration: Bifurcation Diagram

For a practical example of optimizing an M-file we consider a problem from nonlinear dynamics. We wish to examine the long-term behavior of the iteration

$$y_k = F(y_{k-1}), \quad k \geq 2, \quad y_1 \text{ given},$$

where the function F is defined by

$$F(y) = y + h\left(y + \tfrac{1}{2}hy(1-y)\right)\left(1 - y - \tfrac{1}{2}hy(1-y)\right).$$

Here $h > 0$ is a parameter. (This map corresponds to the midpoint or modified Euler method [98] with stepsize h applied to the logistic ODE $dy(t)/dt = y(t)(1-y(t))$ with initial value y_1.) For a range of h values and for a few initial values, y_1, we would like to run the iteration for a "long time", say as far as $k = 500$, and then plot the next 20 iterates $\{y_i\}_{i=501}^{520}$. For each h on the x-axis we will superimpose $\{y_i\}_{i=501}^{520}$ onto the y-axis to produce a so-called bifurcation diagram.

Choosing values of h given by `1:0.005:4` and using initial values `0.2:0.5:2.7` we arrive at the M-file `bif1` in Listing 20.1. This is a straightforward implementation that uses three nested `for` loops and does not preallocate the array `y` before the first time around the inner loop. Figure 20.2 shows the result.

The M-file `bif2` in Listing 20.2 is an equivalent, but faster, implementation. Two of the loops have been removed and a single `plot` command is used. Here, we stack the iterates corresponding to all h and y_1 values into one long vector, and use elementwise multiplication to perform the iteration simultaneously on the components of this vector. The array `Ydata`, which is used to store the data for the plot, is preallocated to the correct dimensions before use. The vectorized code produces Figure 20.2 about 3 times more quickly than the original version. In MATLAB 6 the speed ratio was 200, which gives an idea of how effective MATLAB's automatic optimization is at speeding up `bif1`.

An example where a sequence of optimization steps is applied to a MATLAB code in mathematical finance may be found in [38].

Listing 20.1. *Script* `bif1`.

```
%BIF1 Bifurcation diagram for modified Euler/logistic map.
%     Computes a numerical bifurcation diagram for a map of the form
%     y_k = F(y_{k-1}) arising from the modified Euler method
%     applied to a logistic ODE.
%
%     Slower version using multiple for loops.

for h = 1:0.005:4
    for iv = 0.2:0.5:2.7
       y(1) = iv;
       for k = 2:520
          y(k) = y(k-1) + h*(y(k-1)+0.5*h*y(k-1)*(1-y(k-1)))*...
                 (1-y(k-1)-0.5*h*y(k-1)*(1-y(k-1)));
       end
       plot(h*ones(20,1),y(501:520),'.'), hold on
    end
end

title('Modified Euler/logistic map','FontSize',14)
xlabel('h'), ylabel('last 20 y')
grid on, hold off
```

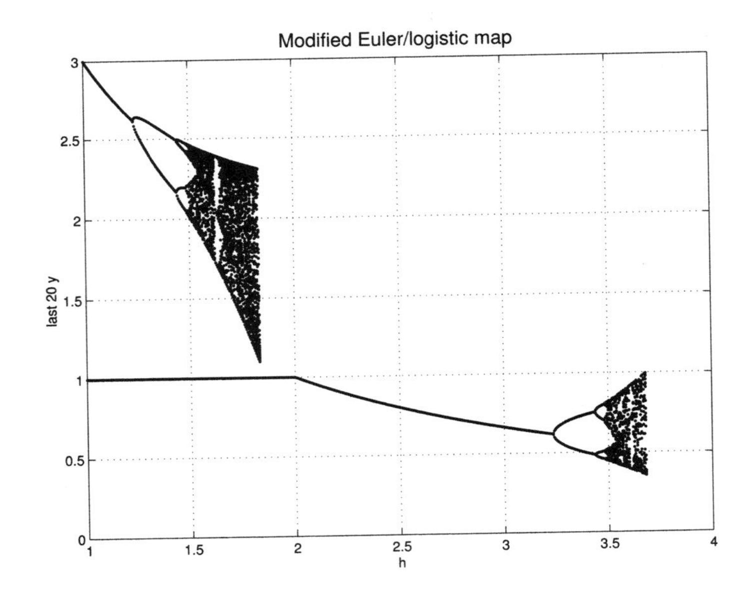

Figure 20.2. *Numerical bifurcation diagram.*

Listing 20.2. *Script* bif2.

```
%BIF2 Bifurcation diagram for modified Euler/logistic map.
%     Computes a numerical bifurcation diagram for a map of the form
%     y_k = F(y_{k-1}) arising from the modified Euler method
%     applied to a logistic ODE.
%
%     Fast, vectorized version.

h = (1:0.005:4)';
iv = 0.2:0.5:2.7;
hvals = repmat(h,length(iv),1);
Ydata = zeros((length(hvals)),20);
y = kron(iv',ones(size(h)));

for k=2:500
   y = y + hvals.*(y+0.5*hvals.*y.*(1-y)).*(1-y-0.5*hvals.*y.*(1-y));
end
for k=1:20
   y = y + hvals.*(y+0.5*hvals.*y.*(1-y)).*(1-y-0.5*hvals.*y.*(1-y));
   Ydata(:,k) = y;
end

plot(hvals,Ydata,'.')
title('Modified Euler/Logistic Map','FontSize',14)
xlabel('h'), ylabel('last 20 y'), grid on
```

Vectorization means using MATLAB language constructs to
eliminate program loops,
usually resulting in programs that
run faster and are more readable.
— STEVE EDDINS and LOREN SHURE, *MATLAB Digest* (September 2001)

Entities should not be multiplied unnecessarily.
— WILLIAM OF OCCAM (c. 1320)

Life is too short to spend writing for loops.
— *Getting Started with MATLAB* (1998)

In our six lines of MATLAB,
not a single loop has appeared explicitly,
though at least one loop is implicit in every line.
— LLOYD N. TREFETHEN and DAVID BAU, III, *Numerical Linear Algebra* (1997)

Make it right before you make it faster.
— BRIAN W. KERNIGHAN and P. J. PLAUGER,
The Elements of Programming Style (1978)

A useful rule-of-thumb is that the
execution time of a MATLAB function is
proportional to the number of statements executed,
no matter what those statements actually do.
— CLEVE B. MOLER, *MATLAB News & Notes* (Spring 1996)

Chapter 21
Tricks and Tips

Our approach in this book has been to present material of interest to the majority of MATLAB users, omitting topics of more specialized interest. In this chapter we relax this philosophy and describe some tricks and tips that, while of limited use, can be invaluable when they are needed and are of general interest as examples of more advanced MATLAB matters.

21.1. Empty Arrays

The empty matrix `[]`, mentioned in several places in this book, has dimension 0-by-0. MATLAB allows multidimensional arrays with one or more dimensions equal to zero. These are created by operations such as

```
>> 1:0
ans =
   Empty matrix: 1-by-0

>> zeros(2,0)
ans =
   Empty matrix: 2-by-0

>> ones(1,0,3)
ans =
   Empty array: 1-by-0-by-3
```

Operations on empty arrays are defined by extrapolating the rules for normal arrays to the case of a zero dimension. Consider the following example:

```
>> k = 5; A = ones(2,k); B = ones(k,3); A*B
ans =
     5     5     5
     5     5     5

>> k = 0; A = ones(2,k); B = ones(k,3); A*B
ans =
     0     0     0
     0     0     0
```

Matrix multiplication `A*B` is defined in MATLAB whenever the number of c[illegible] of `A` equals the number of rows of `B`, even if this number is zero—and in thi[illegible] elements of the product are set to zero.

Empty arrays can facilitate loop vectorization. Consider the nested [illegible]

```
for i = j-1:-1:1
    s = 0;
    for k=i+1:j-1
        s = s + R(i,k)*R(k,j);
    end
end
```

The inner loop can be vectorized to give

```
for i = j-1:-1:1
    s = R(i,i+1:j-1)*R(i+1:j-1,j);
end
```

What happens when `i` = `j-1` and the index vector `i+1:j-1` is empty? Fortunately `R(i,i+1:j-1)` evaluates to a 1-by-0 matrix and `R(i+1:j-1,j)` to a 0-by-1 matrix, and `s` is assigned the desired value 0. In versions of MATLAB prior to MATLAB 5 there was only one empty array, `[]`, and the vectorized loop in this example did not work as intended.

21.2. Exploiting Infinities

The infinities `inf` and `-inf` can be exploited to good effect.

Suppose you wish to find the maximum value of a function `f` on a grid of points `x(1:n)` and `f` does not vectorize, so that you cannot write `max(f(x))`. Then you need to write a loop, with a variable `fmax` (say) initialized to some value at least as small as any value of `f` that can be encountered. Simply assign `-inf`:

```
fmax = -inf;
for i=1:n
    fmax = max(fmax, f(x(i)));
end
```

Next, suppose that we are given p with $1 \le p \le \infty$ and wish to evaluate the dual of the vector p-norm, that is, the q-norm, where $p^{-1} + q^{-1} = 1$. If we solve for q we obtain

$$q = \frac{1}{1 - 1/p}.$$

This formula clearly evaluates correctly for all $1 < p < \infty$. For $p = \infty$ it yields the correct value 1, since $1/\infty = 0$, and for $p = 1$ it yields $q = 1/0 = \infty$. So in MATLAB ...imply write `norm(x,1/(1-1/p))` without treating the cases `p = 1` and `p =`

...ı using MATLAB for data processing and for matrix ...ın be represented as a vector or as a matrix. Consider ...roduced by (for example) the `sort` function:

```
9     8     7
```

```
>> [s,ix] = sort(x)
s =
    -1     3     7     8     9    10
ix =
     2     3     6     5     4     1
```

The output of `sort` is a sorted vector `s` and a permutation vector `ix` such that `x(ix)` equals `s`. To regenerate `x` from `s` we need the inverse of the permutation `ix`. This can be obtained as follows:

```
>> ix_inv(ix) = 1:length(ix)
ix_inv =
     6     1     2     5     4     3

>> s(ix_inv)
ans =
    10    -1     3     9     8     7
```

In matrix computations it is sometimes necessary to convert between the vector and matrix representations of a permutation. The following example illustrates how this is done, and shows how to permute the rows or columns of a matrix using either form:

```
>> p = [4 1 3 2]
p =
     4     1     3     2

>> I = eye(4);
>> P = I(p,:);
P =
     0     0     0     1
     1     0     0     0
     0     0     1     0
     0     1     0     0

>> A = magic(4)
A =
    16     2     3    13
     5    11    10     8
     9     7     6    12
     4    14    15     1

>> P*A
ans =
     4    14    15     1
    16     2     3    13
     9     7     6    12
     5    11    10     8
>> A(p,:)
ans =
```

```
     4    14    15     1
    16     2     3    13
     9     7     6    12
     5    11    10     8

>> A*P'
ans =
    13    16     3     2
     8     5    10    11
    12     9     6     7
     1     4    15    14
>> A(:,p)
ans =
    13    16     3     2
     8     5    10    11
    12     9     6     7
     1     4    15    14

>> p_from_P = (1:4)*P'
p_from_P =
     4     1     3     2
```

A random permutation vector can be generated with the function `randperm`:

```
>> randperm(8)
ans =
     2     4     1     5     8     6     3     7
```

21.4. Rank 1 Matrices

A rank 1 matrix has the form $A = xy^*$, where x and y are both column vectors. Often we need to deal with special rank 1 matrices where `x` or `y` is the vector of all 1s. For `y = ones(n,1)` we can form `A` as an outer product as follows:

```
>> n = 4; x = (1:n)';  % Example choice of n and x.
>> A = x*ones(1,n)
A =
     1     1     1     1
     2     2     2     2
     3     3     3     3
     4     4     4     4
```

Recall that `x(:,1)` extracts the first column of `x`. Then `x(:,[1 1])` extracts the first column of `x` twice, giving an `n`-by-2 matrix. Extending this idea, we can form `A` using only indexing operations:

```
A = x(:,ones(n,1))
```

(This operation is known to MATLAB afficionados as "Tony's trick", and it is used by the `meshgrid` function.) The revised code avoids the multiplication and is therefore faster.

Another way to construct the matrix is as

```
A = repmat(x,1,n);
```

See Section 20.3 for discussion of how to form an even simpler rank 1 matrix.

21.5. Set Operations

Suppose you need to find out whether any element of a vector `x` equals a scalar `a`. This can be done using `any` and an equality test, taking advantage of the way that MATLAB expands a scalar into a vector when necessary in an assignment or comparison:

```
>> x = 1:5; a = 3;

>> x == a
ans =
     0     0     1     0     0

>> any(x == a)
ans =
     1
```

More generally, `a` might itself be a vector and you need to know how many of the elements of `a` occur within `x`. The test above will not work. One possibility is to loop over the elements of `a`, carrying out the comparison `any(x == a(i))`. Shorter and faster is to use the set function `ismember`:

```
>> x = 1:5; a = [-1 3 5];
>> ismember(a,x)
ans =
     0     1     1

>> ismember(x,a)
ans =
     0     0     1     0     1
```

As this example shows, `ismember(a,x)` returns a vector with ith element `1` if `a(i)` is in `x` and `0` otherwise. The number of elements of `a` that occur in `x` can be obtained as `sum(ismember(a,x))` or `nnz(ismember(a,x))`, the latter being faster as it involves no floating point operations. MATLAB has several set functions: see `help ops`.

21.6. Subscripting Matrices as Vectors

MATLAB allows a two-dimensional array to be subscripted as though it were one-dimensional, as we saw in the example of `find` applied to a matrix on p. 68. If `A` is `m`-by-`n` and `j` is a scalar then `A(j)` means the same as `a(j)`, where `a = A(:)`; in other words, `A(j)` is the `j`th element in the vector made up of the columns of `A` stacked one on top of the other.

To see how one-dimensional subscripting can be exploited suppose we wish to assign an `n`-vector `v` to the leading diagonal of an existing `n`-by-`n` matrix `A`. This can be done by

```
A = A - diag(diag(A)) + diag(v);
```

but this code is neither elegant nor efficient. We can take advantage of the fact that the diagonal elements of `A` are equally spaced in the vector `A(:)` by writing

```
A(1:n+1:n^2) = v;
```

or

```
A(1:n+1:end) = v;
```

The main antidiagonal can be set in a similar way, by

```
A(n:n-1:n^2-n+1) = v;
```

For example,

```
>> A = spiral(5)
A =
    21    22    23    24    25
    20     7     8     9    10
    19     6     1     2    11
    18     5     4     3    12
    17    16    15    14    13

>> A(1:6:25) = -(1:5)
A =
    -1    22    23    24    25
    20    -2     8     9    10
    19     6    -3     2    11
    18     5     4    -4    12
    17    16    15    14    -5

>> A(5:4:21) = 0          % Using scalar expansion
A =
    -1    22    23    24     0
    20    -2     8     0    10
    19     6     0     2    11
    18     0     4    -4    12
     0    16    15    14    -5
```

One use of this trick is to shift a matrix by a multiple of the identity matrix: $A \leftarrow A - \alpha I$, a common operation in numerical analysis. This is accomplished with

```
A(1:n+1:end) = A(1:n+1:end)-alpha
```

It is not always easy to work out the appropriate one-dimensional subscripts with which to index a two-dimensional array. Suppose we wish to set to zero the $(1,2)$, $(1,4)$, $(2,2)$, and $(3,1)$ elements of a 4-by-4 array. Using two-dimensional subscripting this would require four separate assignment statements. Instead we can use the function `sub2ind` to convert the subscripts from two dimensions to one:

```
>> A = magic(4);
>> A(sub2ind(size(A), [1 1 2 3], [2 4 2 1])) = 0
A =
    16     0     3     0
```

```
     5     0    10     8
     0     7     6    12
     4    14    15     1
```

The input arguments of `sub2ind` are the array dimensions followed by the row subscripts, then the column subscripts.

21.7. Avoiding if Statements

Statements involving `if` can sometimes be avoided by the careful use of relational operators. Suppose we wish to code the evaluation of the function

$$f(x) = \begin{cases} \sin x, & x < 0, \\ x, & 0 \le x \le 1, \\ 1, & 1 < x, \end{cases}$$

with `x` a double precision matrix. Instead of the obvious `if-elseif-else` coding, we can write

```
y = sin(x).*(x < 0) + x.*(0 <= x & x <= 1) + (1 < x);
```

This evaluation exploits the fact that logical expressions evaluate componentwise to 1 (true) or 0 (false), and that MATLAB will happily perform arithmetic with logicals (here it automatically converts them to doubles first). A possible criticism of the evaluation is that it performs some unnecessary multiplications. The following version is more efficient, though a little less readable. It uses one-dimensional subscripting, as described in Section 21.6.

```
y = ones(size(x));
k = (x < 0);             y(k) = sin(x(k));
k = (0 < x & x <= 1); y(k) = x(k);
```

For an example of a MATLAB function that uses this approach, type `edit sind`.

21.8. Triangular and Symmetric Matrices

Some linear algebra functions perform different computations depending on the properties of their arguments. For example, when `A` is triangular MATLAB computes `A\b` by substitution, but when `A` is full it is first LU-factorized. The `eig` command performs the symmetric QR algorithm for symmetric arguments and the nonsymmetric QR algorithm for nonsymmetric arguments. If your matrix is known to be triangular or symmetric it usually makes sense to *enforce* the property in the presence of roundoff, in order to reduce computation and preserve mathematical properties. For example, the eigenvectors of a symmetric matrix can always be taken to be orthogonal and `eig` returns eigenvectors orthogonal to within rounding error when applied to a symmetric matrix. But if a matrix has a nonzero nonsymmetric part, however tiny, then `eig` applies the nonsymmetric QR algorithm and can return nonorthogonal eigenvectors:

```
>> A = ones(4); [V,D] = eig(A);  norm(V'*V-eye(4))
ans =
  4.3426e-016
```

```
>> Q = gallery('orthog',4,2); % Orthogonal matrix.
>> B = Q'*A*Q; norm(B-B')
ans =
  3.1683e-016

>> [V,D] = eig(B);  norm(V'*V-eye(4))
ans =
    0.8720
```

Here, the matrix B should be symmetric, but it is not quite symmetric because of rounding errors. There is no question of eig having performed badly here; even the exact eigenvector matrix of B is far from being orthogonal. In this situation it would normally be preferable to symmetrize B before applying eig:

```
>> C = (B + B')/2; [V,D] = eig(C);  norm(V'*V-eye(4))
ans =
  6.9010e-016
```

This discussion raises the question of how to test whether a matrix is symmetric or (upper) triangular. One possibility is to evaluate

```
norm(A-A',p), norm(A-triu(A),p)
```

taking care to use p=1 or p=inf, since the 1- and ∞-norms are much quicker to evaluate than the 2-norm. A better way is to use the logical expressions

```
isequal(A,A'), isequal(A,triu(A))
```

which do not involve any floating point arithmetic. Likewise, to check whether A is diagonal, it is better to compute

```
isequal(A,diag(diag(A)))
```

than norm(A-diag(diag(A)),1).

A technique is a trick that works.
— GIAN-CARLO ROTA

A trick used three times becomes a standard technique.
— GEORGE POLYA

For the sake of an easy extension of matrix operations,
we shall introduce one empty matrix of each size . . .
Multiplication of the empty $0 \times m$-matrix with any $m \times n$-matrix
is defined to yield the empty $0 \times n$-matrix.
The product of the empty $m \times 0$-matrix
with the empty $0 \times n$-matrix, however,
is defined to be a nonempty matrix,
namely the zero matrix of size $m \times n$.
— JOSEF STOER and CHRISTOPH WITZGALL, *Convexity and Optimization in Finite Dimensions I* (1970)

Chapter 22
Case Studies

22.1. Introduction

To supplement the short bursts of MATLAB code that appear throughout the book, we now give some larger, more realistic examples. Their purpose is to demonstrate MATLAB in use on nontrivial problems and to illustrate good programming practice. We focus on problems that are

1. easy to explain in words,
2. easy to set up mathematically,
3. suited to graphical display, and
4. solvable with around one page or less of MATLAB code.

For each case study, after summarizing the problem we list the relevant code and give a walkthrough that points out notable techniques and MATLAB functions. The walkthroughs are not intended to explain the M-files line by line. For further details on MATLAB functions used in this chapter consult the index and the online MATLAB documentation.

22.2. Brachistochrone

Suppose a particle slides down a frictionless wire. If we fix the endpoints, what shape of wire minimizes the travel time? This problem dates back to the times of Johann Bernoulli (1667–1748) and its solution involves a curve known as the *brachistochrone*. The name comes from the Greek "brachistos" (shortest) and "chronos" (time).

For this problem it is traditional to use "upside-down" coordinates so that the x-axis points in a horizontal direction but the y-axis points vertically downwards. Suppose the wire starts at the origin, $(0,0)$, and ends at (b_x, b_y). If we let $y(x)$ denote the curve followed by the wire, then the particle's sliding time takes the form

$$T = \int_0^{b_x} \sqrt{\frac{1+(dy/dx)^2}{2g\,y(x)}}\,dx,$$

where g is the constant of acceleration due to gravity. Minimizing T over an appropriate class of functions $y(x)$ produces the brachistochrone, which may be defined in terms of the parameter θ and a constant R by

$$x(\theta) = R(\theta - \sin\theta), \qquad y(\theta) = R(1-\cos\theta).$$

The curve must finish at (b_x, b_y) so we require $0 \le \theta \le \theta^\star$, where $b_x = R(\theta^\star - \sin\theta^\star)$ and $b_y = R(1 - \cos\theta^\star)$. Eliminating R, we find $\theta^\star$ by solving

$$b_y\theta^\star - b_y \sin\theta^\star + b_x \cos\theta^\star - b_x = 0, \tag{22.1}$$

and then we set $R = b_y/(1 - \cos\theta^\star)$.

Now we consider a wire formed from straight line segments. More precisely, divide the x interval $[0, b_x]$ into N equally spaced subintervals $[x_{k-1}, x_k]$ with $x_k = k\Delta x$ and $\Delta x = b_x/N$, and let y_k denote $y(x_k)$. Imagine that the wire is produced by joining the y_k heights with straight lines. Since $y_0 = 0$ and $y_N = b_y$, our join-the-dots curve is completely determined by specifying the heights $\{y_k\}_{k=1}^{N-1}$ at the internal points. For any such curve it may be shown that the particle's slide time is given by

$$T = \sum_{k=1}^{N} \frac{2\sqrt{\Delta x^2 + (y_k - y_{k-1})^2}}{\sqrt{2gy_k} + \sqrt{2gy_{k-1}}}. \tag{22.2}$$

For fixed b_x, b_y, and N, our task is to find the straight line segments that minimize the slide time. This is an optimization problem; we wish to minimize $T = T(y_1, y_2, \ldots, y_{N-1})$ in (22.2), with $y_0 = 0$ and $y_N = b_y$, over all possible $\{y_k\}_{k=1}^{N-1}$. It is then of interest to see how well the resulting curve approximates the brachistochrone. More details about the brachistochrone problem may be found, for example, at `http://mathworld.wolfram.com/BrachistochroneProblem.html`. The idea of optimizing over piecewise linear wires is taken from [67].

Code and Walkthrough

The function `brach` in Listing 22.1 compares a number of exact and join-the-dots brachistochrones, as shown in Figure 22.1. The nested function `Btime` computes the piecewise linear slide time T in (22.2) and the anonymous function `tzero` returns the left-hand side of (22.1). We set $b_x = 1$ and use ten b_y values equally spaced between 0.2 and 2 and two N values, 4 and 8. In each case the initial guess `yinit` that we pass to `fminsearch` represents the straight line from $(0, 0)$ to (b_x, b_y). After using `fminsearch` to find the optimal curve, we plot `-[0 y by]`, rather than `[0 y by]`, to account for the upside-down coordinate system. The true brachistochrone is computed via `fzero`. The `N = 4` and `N = 8` cases are plotted in the `subplot(1,2,1)` and `subplot(1,2,2)` regions, respectively.

22.3. Small-World Networks

A network, or undirected graph, is defined by a list of nodes and a list of edges connecting pairs of nodes. The information may be represented in an *adjacency matrix*. A network of N nodes may be stored in a symmetric N-by-N array A, with $a_{ij} = 1$ if nodes i and j have an edge between them, and $a_{ij} = 0$ otherwise. In the case where A represents a *social network*, for example, the nodes are people and the edges represent acquaintanceships: $a_{ij} = a_{ji} = 1$ if persons i and j know each other. The *pathlength* between nodes i and j is the minimum number of edges that must be crossed in order to get from i to j.

A sparse network may be said to have *small-world* characteristics if both

(a) it is highly clustered—if i knows j and j knows k, then, with high frequency, i knows k—and

Listing 22.1. *Function* brach.

```
function brach
%BRACH  Brachistochrone illustration.
%       Computes and plots approximate brachistochrone by optimization,
%       using FMINSEARCH, and exact brachistochrone, using FZERO.

bx = 1; g = 9.81;
byvals = linspace(0.2,2,10);
Nvals = [4 8];
for i = 1:2
   N = Nvals(i);
   subplot(1,2,i)
   for k = 1:length(byvals)

      % Approximate brachistochrone.
      by = byvals(k);
      dy = by/N; dx = bx/N;
      yinit = [dy:dy:by-dy];
      y = fminsearch(@Btime,yinit);

      plot([0:dx:bx],-[0 y by],'ro-')
      hold on

      % True brachistochrone.
      tzero = @(theta)(by*theta-by*sin(theta)+bx*cos(theta)-bx);
      tstar = fzero(tzero,pi);
      R = by/(1-cos(tstar));
      thetavals = linspace(0,tstar,100);
      xcoord = R*(thetavals-sin(thetavals));
      ycoord = R*(1-cos(thetavals));
      plot(xcoord,-ycoord,'g--','Linewidth',2)

   end
   title(sprintf('N = %1.0f',N),'FontSize',14)
   xlim([0,bx]), axis off
end
hold off

  function T = Btime(y)
  %BTIME Travel time for a particle.
  %      Piecewise linear path with equispaced y between (0,0) and (bx,by).

  yvals = [0 y by];          % End points do not vary.
  N = length(y)+1; d = bx/N;
  T = sum(2*sqrt( d^2 + (diff(yvals)).^2 )./( sqrt(2*g*yvals(2:end)) + ...
          sqrt(2*g*yvals(1:end-1) )));
  end

end
```

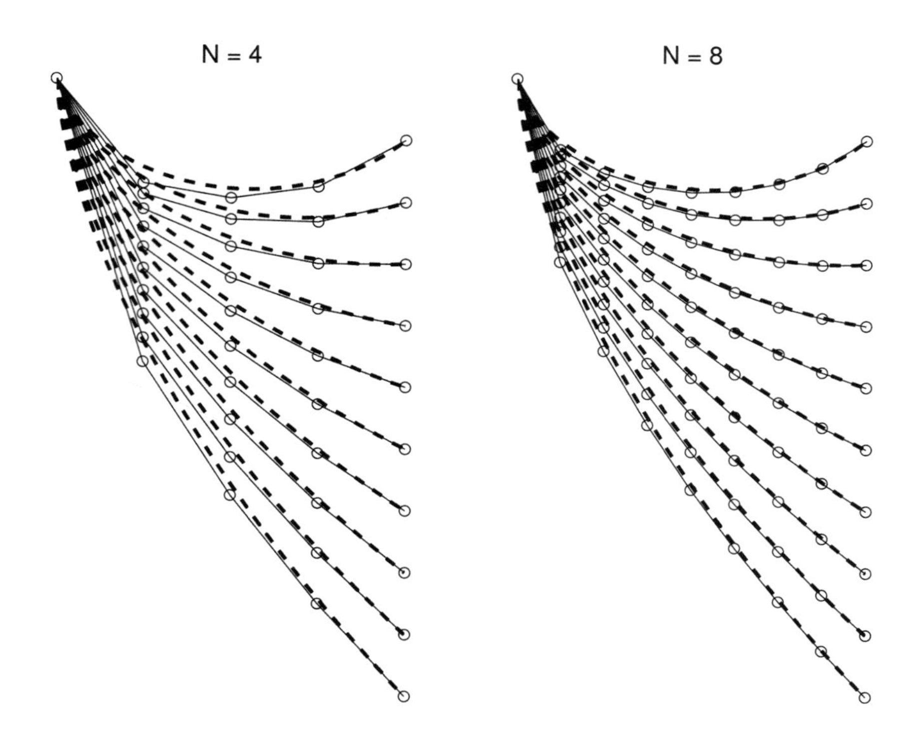

Figure 22.1. *Output from* `brach`.

(b) it has a small average pathlength.

Watts and Strogatz [124] coined the phrase small-world network and found several real-life examples. They also showed that randomly rewiring a regular lattice is a mechanism for creating a small world. Here, we will focus on property (b), the average pathlength, for a variation of the Watts–Strogatz model in the spirit of [87] that uses shortcuts rather than rewiring.

We begin with a k-nearest-neighbor ring network. Arranging the N nodes like the hours on a clock, we set $a_{ij} = 1$ if j can be reached by moving at most k steps away from i, either clockwise or counterclockwise. In the case $N = 7$, $k = 2$, the adjacency matrix is

$$\begin{bmatrix} 0 & 1 & 1 & 0 & 0 & 1 & 1 \\ 1 & 0 & 1 & 1 & 0 & 0 & 1 \\ 1 & 1 & 0 & 1 & 1 & 0 & 0 \\ 0 & 1 & 1 & 0 & 1 & 1 & 0 \\ 0 & 0 & 1 & 1 & 0 & 1 & 1 \\ 1 & 0 & 0 & 1 & 1 & 0 & 1 \\ 1 & 1 & 0 & 0 & 1 & 1 & 0 \end{bmatrix}.$$

In general, A could be defined by the MATLAB commands

```
r = zeros(1,N); r(2:k+1) = 1; r(N-k+1:N) = 1;
A = toeplitz(r);
```

Now we superimpose random shortcuts on the network; that is, we add nonzeros to the matrix at random locations, according to the following process. We look at N

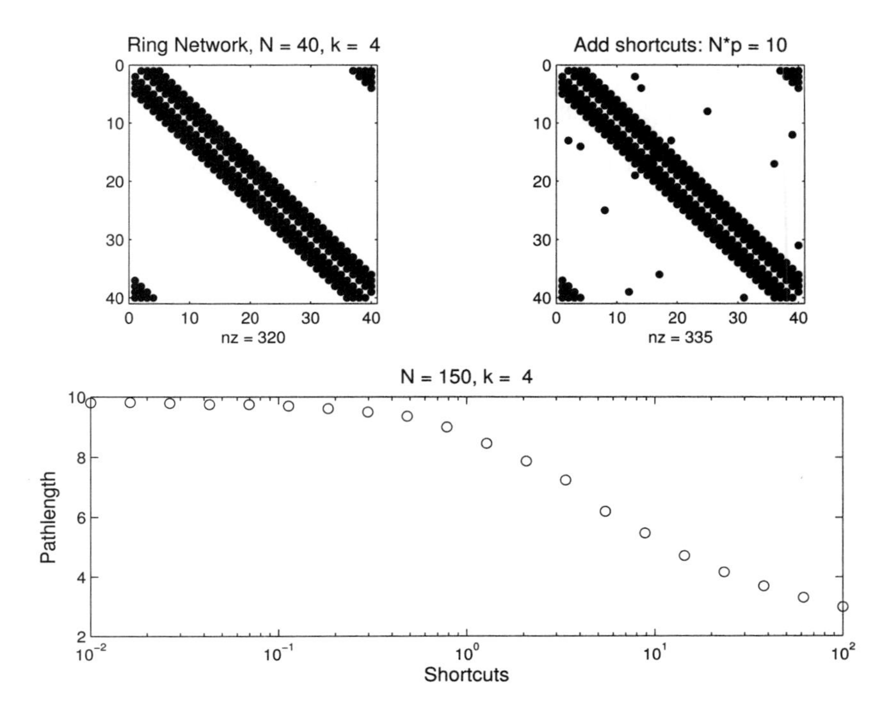

Figure 22.2. *Output from the small-world simulations of* `small_world`. *Upper: adjacency matrices. Lower: pathlength decay.*

flips of a biased coin that lands heads with probability p. If the ith flip shows heads then we choose a column $1 \leq j \leq N$ uniformly and set $a_{ij} = a_{ji} = 1$. (In other words, we add a new link—a shortcut—from the ith node to a randomly chosen jth node.) On average, the overall number of shortcuts that we create is Np.

Generating shortcuts in this way makes it easier to get around the network, and hence decreases the average pathlength. Our aim is to investigate, for fixed N and k, how sharply the average pathlength decays as the average number of shortcuts is increased.

For computational purposes, we may use the characterization that, for $r > 1$, the pathlength between nodes i and j is r if and only if $(A^r)_{ij} > 0$ and $(A^{r-1})_{ij} = 0$. Hence, we can find all pathlengths by raising the adjacency matrix to increasingly higher powers, until no zeros remain. For each pair of nodes, i and j, we must record the power at which the (i, j) element first becomes nonzero.

Code and Walkthrough

The script `small_world` in Listing 22.2 produces Figure 22.2. The upper left `spy` plot shows the adjacency matrix for a 4-nearest-neighbor ring of 40 nodes. In the upper right picture, we see an instance of the same network with shortcuts added, using $Np = 10$. The lower picture gives the results of a large-scale computation. Here, we show the average pathlength of a 150-node, 4-nearest-neighbor ring as a function of the average number of shortcuts, Np.

The first part of the code sets up an adjacency matrix for the ring using `toeplitz`.

We then generate shortcuts with the `sparse` facility. The line `v = find(rand(N,1)<p);` simulates the coin flips. Then

```
Ashort = sparse(v,ceil(N*rand(size(v))),ones(size(v)),N,N);
```

finds a column index for each successful row index and inserts the appropriate edges. Because an edge from i to j automatically implies an edge from j to i, we apply `spy` to `A+Ashort+Ashort'`. The shortcut matrix in this case is

```
>> Ashort
Ashort =
   (13,2)        1
   (14,4)        1
   (39,12)       1
   (19,13)       1
   (16,16)       1
   (36,17)       1
   (20,22)       1
    (8,25)       1
   (31,40)       1
```

We see that nine shortcuts have been added. Two of these, at `(16,16)` and `(20,22)`, will have no effect on the pathlengths: the former links node `16` to itself and the latter repeats an existing edge.

The second part of the script performs the big simulation. The outer loop, `for i = 1:M`, runs over the Np values and the inner loop, `for s = 1:Smax`, drives a Monte Carlo simulation—for each Np value, we approximate the average pathlength over all networks by the computed average over `Smax` networks. Although, for convenience, the shortcuts are created with the `sparse` function, we compute with a full matrix `Bnew` because we know that it will eventually fill in completely.

We include the identity matrix in the assignment

```
Bnetwork = B + Bshort + Bshort' + eye(N);   % Full array.
```

to make the diagonal nonzero. This allows `any(any(Bnew==0))` to be used as the termination criterion for the `while` loop. The line `L = sign(Bnetwork);` ensures that any multiply assigned edges are only counted once. The `while` loop powers up the adjacency matrix until it is full of nonzeros. If the (i, j) element first becomes nonzero at level `power`, then we enter this value in `L(i,j)`. On leaving the `while` loop, we compute the average over the off-diagonal entries of `L` using `mean(mean(L-diag(diag(L))))*N/(N-1)`. The resulting plot shows that the pathlength starts to drop significantly when an average of $O(1)$ shortcuts are added to the $O(N)$ network. Upping the network size to `N = 1000`, which of course increases the runtime, produces results that agree qualitatively with the related computations in [124].

Listing 22.2. *Script* small_world.

```
%SMALL_WORLD  Small-world network example.
%             Display ring and small-world adjacency matrices.
%             Then compute average pathlengths.

rand('state',100)

N = 40; k = 4; short_ave = 10; p = short_ave/N;
r = zeros(1,N); r(2:k+1) = 1; r(N-k+1:N) = 1;
A = toeplitz(r);
subplot(2,2,1), spy(A)
title(sprintf('Ring Network, N = %2.0f, k = %2.0f',N, k),'FontSize',12)

subplot(2,2,2)
v = find(rand(N,1)<p);
Ashort = sparse(v,ceil(N*rand(size(v))),ones(size(v)),N,N);
spy(A+Ashort+Ashort')
title(sprintf('Add shortcuts: N*p = %2.0f',N*p),'FontSize',12)
h = waitbar(0,'Computing average pathlengths');

%%%%% Average pathlength as a function of N*p %%%%

N = 150; k = 4; M = 20; Smax = 150; Np = logspace(-2,2,M);
r = zeros(1,N); r(2:k+1) = 1; r(N-k+1:N) = 1;
B = toeplitz(r);
lmean = zeros(M,1);
for i = 1:M
    waitbar(i/M)
    p = Np(i)/N;
    smean = zeros(Smax,1);
    for s = 1:Smax
        v = find(rand(N,1)<p);
        Bshort = sparse(v,ceil(N*rand(size(v))),ones(size(v)),N,N);
        Bnetwork = B + Bshort + Bshort' + eye(N);  % Full array.
        L = sign(Bnetwork);  % Convert to matrix of 0s and 1s.
        power = 1;
        Bnew = Bnetwork;
        while any(any(Bnew==0))
           power = power + 1;
           Bold = Bnew;
           Bnew = Bnew*Bnetwork;
           L = L + ( (L == 0) & (Bnew > 0) )*power;
        end
        smean(s) = mean(mean(L-diag(diag(L))))*N/(N-1);
    end
    lmean(i) = mean(smean);
end
close(h)

subplot(2,2,3:4)
semilogx(Np,lmean,'ro')
xlabel('Shortcuts','FontSize',12), ylabel('Pathlength','FontSize',12)
title(sprintf('N = %2.0f, k = %2.0f',N, k),'FontSize',12)
```

22.4. Performance Profiles

A common task in scientific computing is to compare several competing methods on a set of test problems. Assuming a scalar measure of performance has been chosen (typically speed or accuracy), how best to present the results from the tests is a nontrivial question. Some natural approaches have drawbacks. Plotting the average performance of the methods tends to make difficult problems dominate the results, and it is unclear how to handle problems that a method failed to solve. Ranking the solvers, by plotting the number of times a solver came in kth place, for k from 1 to the number of solvers, provides no information on the size of the improvement between one place and the next.

A new way of presenting results called a *performance profile* overcomes these disadvantages. This technique, introduced by Dolan and Moré [21], is not to be confused with an older technique of the same name that has been applied mainly in the context of quadrature [68].

Suppose we have a set P of m test problems and a set S of n solvers (we use the term "solver" instead of "method" to emphasize that we are considering a particular implementation in software of a method). Let $t_s(p)$ measure the performance of solver $s \in S$ on problem $p \in P$, where the smaller the value of $t_s(p)$ the better the performance. Typically $t_s(p)$ is run time, the flop count, the reciprocal of the flop rate, or a measure of accuracy or stability. Define the performance ratio

$$r_{p,s} := \frac{t_s(p)}{\min\{\, t_\sigma(p) : \sigma \in S\,\}} \geq 1,$$

which is the performance of solver s on problem p divided by the best performance of all the solvers on this problem. The performance profile of solver s is the function

$$\phi_s(\theta) = \frac{1}{m} \times \text{number of } p \in P \text{ such that } r_{p,s} \leq \theta,$$

which is monotonically increasing. In words, $\phi_s(\theta)$ is the probability that the performance of solver s is within a factor θ of the best performance over all solvers on the given set of test problems. Technically, $\phi_s(\theta)$ is the (cumulative) distribution function for the performance ratio of solver s.

The formulae above reduce to simple array arithmetic. Let the performance data be an m-by-n array A, where a_{ij} is the performance of solver j on problem i. Then

$$\phi_j(\theta) = \frac{1}{m} \times \text{number of } i \text{ among } 1\!:\!m \text{ such that } a_{ij} \leq \theta \min\{\, a_{ik} : k = 1\!:\!n\,\}. \quad (22.3)$$

To view the performance profiles we simply plot $\phi_j(\theta)$ against θ for all solvers j.

Code and Walkthrough

Function `perfprof` in Listing 22.3 computes and plots performance profiles. This function could be written in several ways. The shortest approach would be to make use of the MATLAB function `stairs`, but it is more instructive to code the necessary computations directly, as we have done here.

Note first that $\phi_j(\theta)$ is a piecewise constant function whose possible values are 0, $1/m$, $2/m$, $\dots$, 1, and whose value changes when $\theta = a_{ij}/\min\{\, a_{ik} : k = 1\!:\!n\,\}$ for some i. We will exploit the latter property but not the former.

Listing 22.3. *Function* perfprof.

```
function th_max = perfprof(A,th_max)
%PERFPROF  Performance profile.
%         TH_MAX = PERFPROF(A,TH_MAX) produces a
%         peformance profile for the data in the M-by-N matrix A,
%         where A(i,j) > 0 measures the performance of the j'th solver
%         on the i'th problem, with smaller values of A(i,j) denoting
%         "better".  For each solver theta is plotted against the
%         probability that the solver is within a factor theta of
%         the best solver over all problems, for theta on the interval
%         [1, TH_MAX].
%         Set A(i,j) = NaN if solver j failed to solve problem i.
%         TH_MAX defaults to the smallest value of theta for which
%         all probabilities are 1 (modulo any NaN entries of A).

minA = min(A,[],2);
if nargin < 2, th_max = max( max(A,[],2)./minA ); end
tol = sqrt(eps);  % Tolerance.

[m,n] = size(A);  % m problems, n solvers.

for j = 1:n                    % Loop over solvers.

    col = A(:,j)./minA;        % Performance ratios.
    col = col(~isnan(col));    % Remove NaNs.
    if isempty(col), continue; end
    theta = unique(col)';      % Unique elements, in increasing order.
    r = length(theta);
    prob = sum( col(:,ones(r,1)) <= theta(ones(length(col),1),:) ) / m;
    % Assemble data points for stairstep plot.
    k = [1:r; 1:r]; k = k(:)';
    x = theta(k(2:end)); y = prob(k(1:end-1));

    % Ensure endpoints plotted correctly.
    if x(1) >= 1 + tol, x = [1 x(1) x]; y = [0 0 y]; end
    if x(end) < th_max - tol, x = [x th_max]; y = [y y(end)]; end
    plot(x,y), hold on

end
hold off
xlim([1 th_max])
```

To understand the code, consider the jth solver and a given scalar θ_k. We need to compute $\phi_j(\theta_k)$, which is m^{-1} times the number of i for which $\texttt{col}_i \le \theta_k$, where $\texttt{col}_i = a_{ij}/\min\{\, a_{ik} : k = 1{:}\,n \,\}$. In MATLAB notation, exploiting scalar expansion, $\phi_j(\theta_k)$ is[11]

```
sum( col <= theta(k) )/m
```

We need to carry out this computation for each element in the `1`-by-`r` row vector `theta`, which can be done with the loop

```
for k = 1:r
    prob(k) = sum( col <= theta(k) )/m
end
```

Using the indexing trick from Section 21.4, this loop can be vectorized to

```
prob = sum( col(:,ones(r,1)) <= theta(ones(length(col),1),:) ) / m
```

The reason for writing `length(col)` rather than `m` is that `perfprof` first needs to remove NaNs from `col` and hence when `prob` is formed `col` may have fewer than `m` elements. The elements of `theta`, which are the distinct and sorted elements of `col`, are obtained with the MATLAB function `unique`.

Some further work is needed to produce a plot that properly displays the piecewise linear nature of the curves $\phi_j(\theta)$ (i.e., a stairstep plot). The last few lines of the loop construct the data pairs to be passed to `plot`.

To make the plots as readable as possible it is necessary to set line styles, marker types, a legend, and so on. Achieving this within `perfprof` via input arguments would be clumsy. Instead it is left for the user to set the relevant properties using Handle Graphics after calling `perfprof`.

Function `ode_pp` in Listing 22.4 illustrates the use of `perfprof`. It times the three MATLAB nonstiff ODE solvers (see Table 12.1) on six test problems. (Function `fox1` is the function in Listing 12.4.) The function illustrates various advanced MATLAB programming techniques that have been discussed in the book. In particular, it includes both nested functions and subfunctions (with a nested function inside a subfunction). Note that for the timings from `ode_pp` to be meaningful we need to run the function twice, as the first time it is run there is the unwanted overhead of compiling `ode_pp` and the solvers into MATLAB's internal format.

The performance profile plot from `ode_pp` is shown in Figure 22.3. We now explain how to interpret the figure. But, first, we emphasize that **this example is purely illustrative and the results should not be taken at face value.** Indeed some of the test problems are difficult and this experiment does not check the correctness of the solutions computed. The experiment was designed simply to give an interesting performance profile. The numbers on which the figure is based are shown in Table 22.1 (this is the transpose of the array `T` returned by `ode_pp`).

We now explain how to interpret Figure 22.3. Note first that since there are only $m = 6$ test problems we have explicitly set the y-axis tick marks to the possible values of $\phi_j(\theta)$ (0, $1/m$, $2/m$, $\ldots$, 1) and then assigned appropriately short tick labels; for larger m, MATLAB's automatic tick marks will probably give better readability.

[11] Since the argument to `sum` is a vector of `0`s and `1`s, it would be more efficient to replace `sum` by `nnz` here, but `nnz` does not produce the desired result in the vectorized expression used in `perfprof`.

Table 22.1. *Data in transpose of array* `T` *from* `ode_pp`.

Problem	1	2	3	4	5	6
`ode23`	1.26e-2	2.41e-1	3.74e-2	3.37e0	1.44e-1	5.06e-1
`ode45`	6.20e-3	1.53e-1	5.00e-2	6.45e0	1.56e-1	1.07e0
`ode113`	1.56e-2	1.97e-1	6.68e-2	7.86e0	3.76e-2	1.50e0

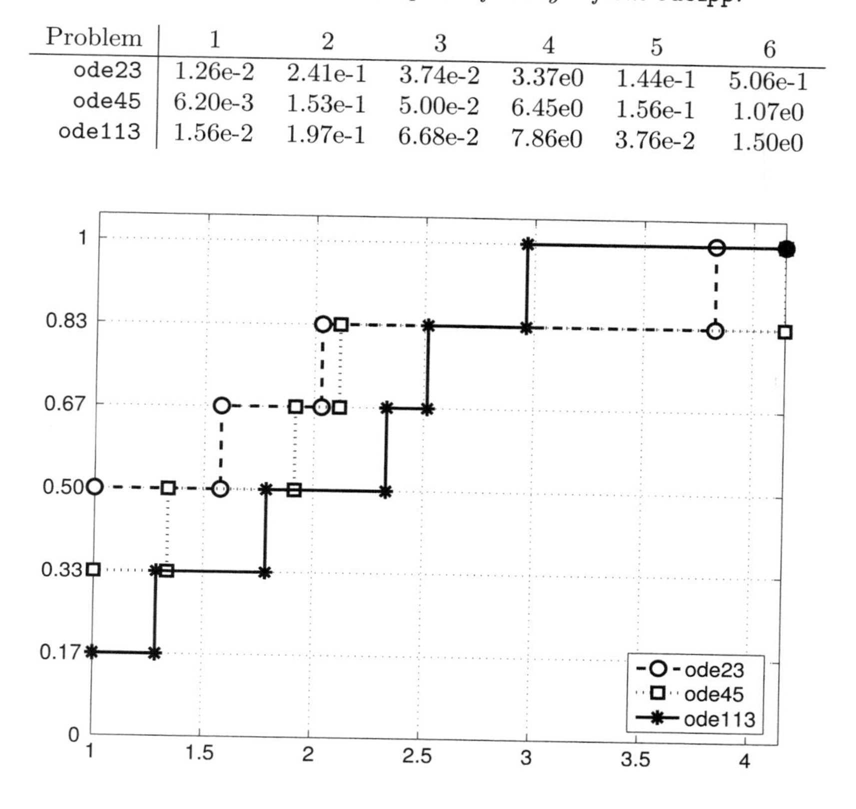

Figure 22.3. *Performance profile produced by* `ode_pp`.

- Left-hand side of plot, $\phi_s(1)$: `ode23` is the fastest solver on 50% of the problems, with `ode45` and `ode113` being fastest on 33% and 17% of the problems, respectively.

- Middle of plot, where the curves all cross: If our criterion for choosing a solver is that it has an 83% chance of being within a factor 2.5 of the fastest solver then all three solvers are equally good.

- Middle to right-hand side of plot, looking where the curves first hit probability 1: `ode113` is within a factor θ of being the fastest solver on every problem for $\theta \approx 3$. For the same to be true for `ode23` and `ode45` we need to increase θ to 3.8 and 4.1, respectively.

The performance profile therefore answers several different aspects of the question, "Which is the best solver?". It shows that `ode23` is most often the fastest, `ode113` is the most reliable in the sense of being the least likely to be much slower than the fastest, and `ode45` treads a middle ground between the two (each statement applying only to this very small and unrepresentative set of test problems).

Listing 22.4. *Function* ode_pp.

```
function T = ode_pp
%ODE_PP    Performance profile of three ODE solvers.

solvers = {@ode23, @ode45, @ode113};  nsolvers = length(solvers);
nproblems = 6;
nruns = 5;  % Number of times to run solver to get more reliable timing.

for j = 1:nsolvers
    code = solvers{j}
    for i = 1:nproblems

        options = [];
        switch i
           case 1
               fun = @fox1; tspan = [0 10]; yzero = [3;0];
           case 2
               fun = @rossler; tspan = [0 100]; yzero = [1;1;1];
               options = odeset('AbsTol',1e-7,'RelTol',1e-4);
           case 3
               fun = @fvdpol; tspan = [0 20]; yzero = [2;1]; mu = 10;
           case 4
               fun = @fvdpol; tspan = [0 20]; yzero = [2;1]; mu = 1000;
           case 5
               fun = @drug_transport; tspan = [0 6]; yzero = [0;0];
           case 6
               fun = @knee; tspan = [0 2]; yzero = 1;
        end

        t0 = clock;
        for k = 1:nruns
            [t,y] = code(fun,tspan,yzero,options);
        end
        T(i,j) = etime(clock,t0)/nruns;

    end
end

perfprof(T);
ylim([0 1.05]), grid
yvals = 0:1/nproblems:1;
set(gca,'YTick',yvals)
set(gca,'YTickLabel',['   0 ';num2str(yvals(2:end-1)','%4.2f ');'   1 '])
f = findall(gcf,'type','line');       % Handles of the three lines.
legend('ode23','ode45','ode113','Location','SE')
set(f,{'Marker'},{'*','s','o'}')       % Vectorized set.
set(f,'MarkerSize',10)
set(f,'MarkerFaceColor','auto') % Make marker interiors non-transparent.
set(f,{'LineStyle'},{'-',':','--'}')  % Vectorized set.
set(f,'LineWidth',2)
set(gca,'FontSize',14)
```

```
        function yprime = fvdpol(x,y)
        %FVDPOL   Van der Pol equation written as first order system.
        %         Parameter MU.
        yprime = [y(2); mu*y(2)*(1-y(1)^2)-y(1)];
        end

end

function yprime = rossler(t,y)
%ROSSLER    Rossler system, parameterized.
a = 0.2; b = 0.2; c = 2.5;
yprime = [-y(2)-y(3); y(1)+a*y(2); b+y(3)*(y(1)-c)];
end

function yprime = drug_transport(t,y)
%DRUG_TRANSPORT  Two-compartment pharmacokinetics example.
%                Reference: Shampine (1994, p. 105).
yprime = [-5.6*y(1) + 48*pulse(t,1/48,0.5); 5.6*y(1) - 0.7*y(2)];

      function pls = pulse(t,w,p)
      %PULSE   Pulse of height 1, width W, period P.
      pls = (rem(t,p) <= w);
      end

end

function yprime = knee(t,y)
%KNEE        Knee problem.
%            Reference: Shampine (1994, p. 115).
epsilon = 1e-4;
yprime = (1/epsilon)*((1-t)*y - y^2);
end
```

For a well-chosen set of test problems, the inequality $\phi_i(\theta) \leq \phi_j(\theta)$ holding for all θ is certainly strong evidence that solver j is superior to solver i, but this inequality does not imply that solver j performs better than solver i on every test problem. This is illustrated by `ode45` and `ode23` in Table 22.1 and Figure 22.3.

In some applications a solver may fail to solve a problem. For example, an optimization code may fail to converge or it may converge to a nonoptimal point. The failure of solver `j` to solve problem `i` can be accounted for by setting `A(i,j) = NaN`. To illustrate, consider this M-file, based on entirely fictitious data:

```
A = [1     2     3     4
     2     4     6     8
   NaN     3   NaN     2
     1     2     5     2
   NaN    10   NaN    20
     1     1     4     6
     3   NaN     4     5
     4     2     3     1
   NaN     2     2     2
   NaN     3     5     5
```

```
    NaN   NaN   NaN    5
    NaN     2     1    3];

th_max = 7;
perfprof(A,th_max*1.1);
xlim([1 th_max]), ylim([0 1.05])
f = findall(gcf,'type','line');
legend('Column 1','Column 2','Column 3','Column 4','Location','SE')
set(f,{'LineStyle'},{'-','-.','--',':'}')    % Vectorized set.
set(f,{'Marker'},{'*','o','s','+'}')          % Vectorized set.
set(f,'MarkerFaceColor','auto') % Marker interiors non-transparent.
set(f,'LineWidth',2), set(f,'MarkerSize',10)
set(gca,'FontSize',14)
% Uncomment next line to produce colored lines:
% set(f,{'Color'},{'r','g','b','k'}')
```

Figure 22.4 is produced. The intersection of the curves with the right-hand axis shows the proportion of problems that could be solved, ranging from 0.5 to 1. The solver corresponding to the dotted line and the '+' marker (and represented by the first column of `A`) was most often the best but solved the fewest problems. Such information would be hard to discern from a large array of data, but it is immediately apparent from the figure. This example illustrates the use of the second input argument `th_max` of `perfprof`. By default, the x-axis would cover the range $[0, 6]$. We have called `perfprof` with a larger value of `th_max` in order to better display the "flatlining" effect. The reason we pass a second argument to `perfprof` that is slightly larger than the intended x-axis upper limit is to avoid markers being plotted where the lines meet the right-hand edge of the plot, since these intersections are not data points. With a more complicated data set it may be necessary to call `perfprof` twice: once to compute the default `th_max` and again with an increased value (as in this example).

Finally, note that for data sets that produce a large `th_max`, setting a logarithmic scale on the x-axis can help make the performance profile more readable. This can be done after calling `perfprof` with the command `set(gca,'XScale','log')` (see Section 17.1).

22.5. Multidimensional Calculus

The calculus features of the Symbolic Math Toolbox are very useful for solving problems that are too tedious to treat by hand yet small and simple enough that symbolic manipulation is feasible. We illustrate by finding and classifying the stationary points of the function

$$F(x, y) = 4x^2 - 3x^4 + x^6/3 + xy - 4y^2 + 4y^4. \tag{22.4}$$

This is a slight variation of the function pictured in Figure 8.15.

The stationary points are the points where the gradient vector

$$\nabla F(x, y) = \begin{bmatrix} \dfrac{\partial F}{\partial x} \\ \dfrac{\partial F}{\partial y} \end{bmatrix}$$

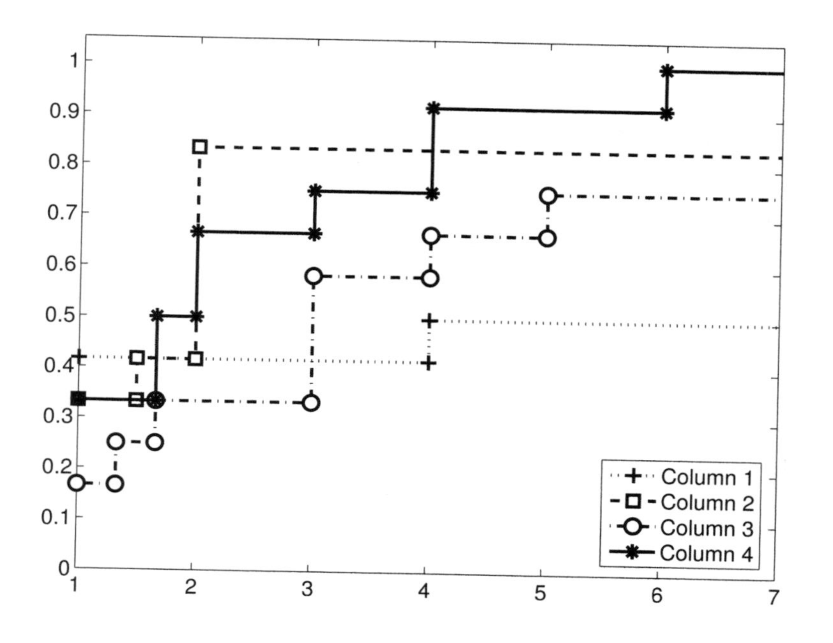

Figure 22.4. *Performance profile for fictitious data in 12-by-4 array* `A`.

is zero. The nature of a stationary point—minimum, maximum, or saddle point—can be determined from the signs of the eigenvalues of the Hessian matrix,

$$\nabla^2 F(x,y) = \begin{bmatrix} \dfrac{\partial^2 F}{\partial x^2} & \dfrac{\partial F}{\partial x \partial y} \\ \dfrac{\partial F}{\partial y \partial x} & \dfrac{\partial^2 F}{\partial y^2} \end{bmatrix},$$

provided the matrix is nonsingular.

Code and Walkthrough

Script `camel_solve` in Listing 22.5 symbolically computes the gradient and then solves to find the points where the gradient is zero. A general principle is that results from a symbolic manipulation package should always be tested and should not automatically be trusted. For each putative stationary point, the code checks numerically that the gradient is small enough to be regarded as zero, and it discards complex solutions, which are not of interest.

The Hessian is computed symbolically using the appropriate Maple function, for conciseness of code. The eigenvalues of the numerically evaluated Hessians are then used to classify the stationary points. Finally, the stationary points are printed, arranged by type, and the contour plot shown in Figure 22.5 is produced. The output from the script is as follows:

```
fx =
8*x-12*x^3+2*x^5+y
fy =
x-8*y+16*y^3
```

Listing 22.5. *Script* camel_solve.

```
%CAMEL_SOLVE  Find stationary points of the camel function.
%             This script requires the Symbolic Math Toolbox.

format short e
syms x y
f = 4*x^2 - 3*x^4 + x^6/3 + x*y - 4*y^2 + 4*y^4;

fx = diff(f,x), fy = diff(f,y)
disp('Original solutions:')
s = solve(fx,fy)

H = maple('hessian', f, '[x,y]')
n = length(s.x); j = 1; minx = []; maxx = []; saddlex = [];

for i = 1:n  % Loop over stationary points.
    fprintf('Point %2.0f:  ',i)
    xi = s.x(i); yi = s.y(i);
    pointi = double([xi yi]);
    gi = double([subs(fx,{x,y},{xi,yi}) subs(fy,{x,y},{xi,yi})]);
    % Filter out nonreal points and points where gradient not zero.
    if norm(gi) > eps
       fprintf('gradient is nonzero!\n')
    elseif ~isreal(pointi)
       fprintf('is nonreal!\n')
    else
       fprintf('(%10.2e,%10.2e)   ', pointi)
       Hi = double(subs(H,{x,y},{xi,yi}));
       eig_Hi = eig(Hi);
       if all(eig_Hi > 0)
          minx = [minx; pointi]; fprintf('minimum\n')
       elseif all(eig_Hi < 0)
          maxx = [maxx; pointi]; fprintf('maximum\n')
       elseif prod(eig_Hi) < 0
          saddlex = [saddlex; pointi]; fprintf('saddle point\n')
       else
          fprintf('nature of stationary point unclear\n')
       end
    end
end
minx, maxx, saddlex

plot(minx(:,1),minx(:,2),'*',...
     maxx(:,1),maxx(:,2),'o',...
     saddlex(:,1),saddlex(:,2),'x','MarkerSize',8)
hold on
a = axis;
[x,y] = meshgrid(linspace(a(1),a(2),200),linspace(a(3),a(4),200));
z = subs(f);  % Replaces symbolic x and y with numeric values from workspace.
contour(x,y,z,30)
xlim([-2.5 2.5]) % Fine tuning.
legend('Min', 'Max', 'Saddle')
g = findall(gca,'type','axes'); set(g,'Fontsize',14)
hold off
```

```
Original solutions:
s =
    x: [15x1 sym]
    y: [15x1 sym]
H =
[ 8-36*x^2+10*x^4,                1]
[                1,        -8+48*y^2]
Point  1:  ( 0.00e+000, 0.00e+000)  saddle point
Point  2:  ( 8.82e-001, 1.13e-001)  maximum
Point  3:  (-8.82e-001,-1.13e-001)  maximum
Point  4:  ( 9.18e-001, 6.41e-001)  saddle point
Point  5:  (-9.18e-001,-6.41e-001)  saddle point
Point  6:  (-9.02e-002, 7.13e-001)  minimum
Point  7:  ( 9.02e-002,-7.13e-001)  minimum
Point  8:  (-8.14e-001, 7.53e-001)  saddle point
Point  9:  ( 8.14e-001,-7.53e-001)  saddle point
Point 10:  (-2.30e+000, 8.21e-001)  minimum
Point 11:  ( 2.30e+000,-8.21e-001)  minimum
Point 12:  is nonreal!
Point 13:  is nonreal!
Point 14:  is nonreal!
Point 15:  is nonreal!
minx =
  -9.0183e-002  7.1268e-001
   9.0183e-002 -7.1268e-001
  -2.2969e+000  8.2144e-001
   2.2969e+000 -8.2144e-001
maxx =
   8.8223e-001  1.1318e-001
  -8.8223e-001 -1.1318e-001
saddlex =
             0            0
   9.1833e-001  6.4063e-001
  -9.1833e-001 -6.4063e-001
  -8.1410e-001  7.5335e-001
   8.1410e-001 -7.5335e-001
```

Fifteen stationary points are found symbolically, eleven of which are verified to be real and have a zero gradient.

There is no guarantee that the `solve` function yields all the solutions of the system it is asked to solve, so further analysis is needed to determine whether `camel_solve` has found all the stationary points. More sophisticated methods for solving this type of problem are explained in Chapter 4, titled "Think Globally, Act Locally", of [12], which contains some MATLAB code.

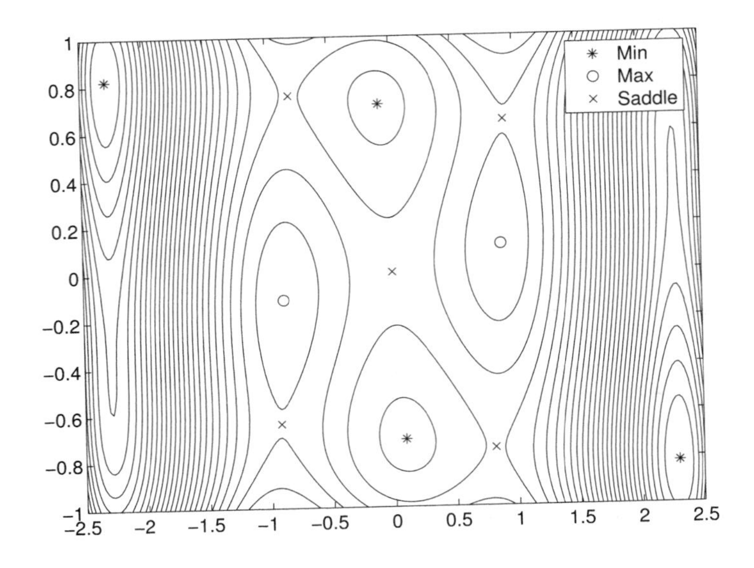

Figure 22.5. *Contours and stationary points of camel function* (22.4).

22.6. L-Systems and Turtle Graphics

The L-system formulation provides a simple means to draw plant-like objects. We will consider the case where such objects are represented by strings from an alphabet of five characters: `F`, `[`, `]`, `+`, `-`. Here, the `[` and `]` characters must appear in matching pairs. We may view a string formally using the *turtle graphics* idea. Imagine a turtle equipped with a pen. The turtle reads the characters in the string sequentially, from left to right, interpreting them as instructions, and thereby draws a picture. At any given stage, the turtle has a *current position*, (x, y), and a *current move vector*, (dx, dy). The characters have the following precise meanings.

`F` means perform the current move; that is, draw a line from (x, y) to $(x+dx, y+dy)$. Update the current position to $(x + dx, y + dy)$. Keep the current move vector as (dx, dy).

`+` means turn clockwise through a prespecified angle θ^+; that is, change the current move vector from (dx, dy) to $(\cos(\theta^+)dx + \sin(\theta^+)dy, -\sin(\theta^+)dx + \cos(\theta^+)dy)$.

`-` means turn counterclockwise through a prespecified angle θ^-; that is, change the current move vector from (dx, dy) to $(\cos(\theta^-)dx - \sin(\theta^-)dy, \sin(\theta^-)dx + \cos(\theta^-)dy)$.

`[` means record the current values of (x, y) and (dx, dy); that is, push them onto a stack. Then scale (dx, dy) by a prespecified factor. The turtle does not move. When the matching `]` marker is reached, that position (x, y) and move vector (dx, dy) are popped off the stack; the turtle returns to (x, y) (without drawing) and resets its current move vector to (dx, dy).

In order to create our strings, we must define an *initial state* and a *production rule*. We will always take the initial state to be `F`. Then, in general, to get from one generation to the next we replace every occurrence of `F` by the production rule. For example, with the production rule `F[+F]F[-F]F` we have

Initial state `F`

1st generation `F[+F]F[-F]F`

2nd generation `F[+F]F[-F]F[+F[+F]F[-F]F]F[+F]F[-F]F[-F[+F]F[-F]F]`
`F[+F]F[-F]F`

The process is akin to using the "search and replace all" facility available in a typical text editor, with `F` being searched for and replaced by the production rule. Our aim is now to draw the picture that arises when the rule, generation level, turning angles, and scale factor are specified.

The book [94] gives a very readable discussion of the ideas behind L-systems, which are named after the Swedish biologist Aristid Lindenmayer (1925–1989).

Code and Walkthrough

The recursive function `lsys` in Listing 22.6 combines the string production and string interpretation phases. The input variable `rule` is the required production rule. `lsys` uses the `switch` syntax to parse the rule, taking the appropriate action for each character. In particular, `F` results in a recursive call to `lsys` with the generation decremented by one. Note that the arrays `cstack` and `dstack` are not preallocated. Since the stack is usually just a few levels deep this is not a major inefficiency. Called with `gen` equal to `1`, `lsys` draws the first generation plant.

The script `lsys_run` in Listing 22.7 calls `lsys` with four different sets of parameters, producing the pictures in Figure 22.6.

Listing 22.6. *Function* lsys.

```
function [coord,mov] = lsys(rule,coord,mov,angle,scale,gen)
%LSYS    Recursively generated L-system.
%        LSYS(RULE,COORD,MOV,ANGLE,SCALE,GEN) generates the L-system
%        produced by GEN generations of the production rule given
%        in the string RULE.
%        COORD and MOV are the initial (x,y) and (dx,dy) values.
%        ANGLE is a 2-vector, with ANGLE(1) specifying the clockwise
%        rotations and ANGLE(2) the counterclockwise rotations.
%        SCALE is the scale factor for branch length.

%        During recursion, GEN, COORD, and MOV record the current state.

if gen == 0
   % Draw line, then update location.
   plot([coord(1),coord(1)+mov(1)],[coord(2),coord(2)+mov(2)])
   coord = coord + mov;
   hold on
else
   stack = 0;
   for k=1:length(rule)
     switch rule(k)
         case 'F'
                [coord,mov] = lsys(rule,coord,mov,angle,scale,gen-1);
         case '+'
                mov = [cos(angle(1)) sin(angle(1));
                      -sin(angle(1)) cos(angle(1))]*mov;
         case '-'
                mov = [cos(angle(2)) -sin(angle(2));
                       sin(angle(2)) cos(angle(2))]*mov;
         case '['
                stack = stack + 1;
                cstack(1:2,stack) = coord;
                dstack(1:2,stack) = mov;
                mov = scale*mov;
         case ']'
                coord = cstack(1:2,stack);
                mov = dstack(1:2,stack);
                stack = stack - 1;
     end
   end
end
```

Listing 22.7. *Script* lsys_run.

```
%LSYS_RUN    Runs lsys function to draw L-systems.

subplot(2,2,1)
rule = 'F[+F][-F][++F][--F]';
[c,d] = lsys(rule,[0;0],[0;1],[pi/8;pi/5],0.6,5);
title(rule,'FontWeight','Bold')
axis equal, axis off

subplot(2,2,2)
rule = 'F[+F]F[-F][F]';
[c,d] = lsys(rule,[0;0],[0;1],[pi/6;pi/6],1,5);
title(rule,'FontWeight','Bold')
axis equal, axis off

subplot(2,2,3)
rule = 'F[+F][-F][++F]F[+F][-F]';
[c,d] = lsys(rule,[0;0],[0;1],[pi/5;pi/6],0.8,4);
title(rule,'FontWeight','Bold')
axis equal, axis off

subplot(2,2,4)
rule = 'FF-[-F+F+F]+[+F-F-F]';
[c,d] = lsys(rule,[0;0],[0;1],[pi/6;pi/6],0.7,4);
title(rule,'FontWeight','Bold')
axis equal, axis off
```

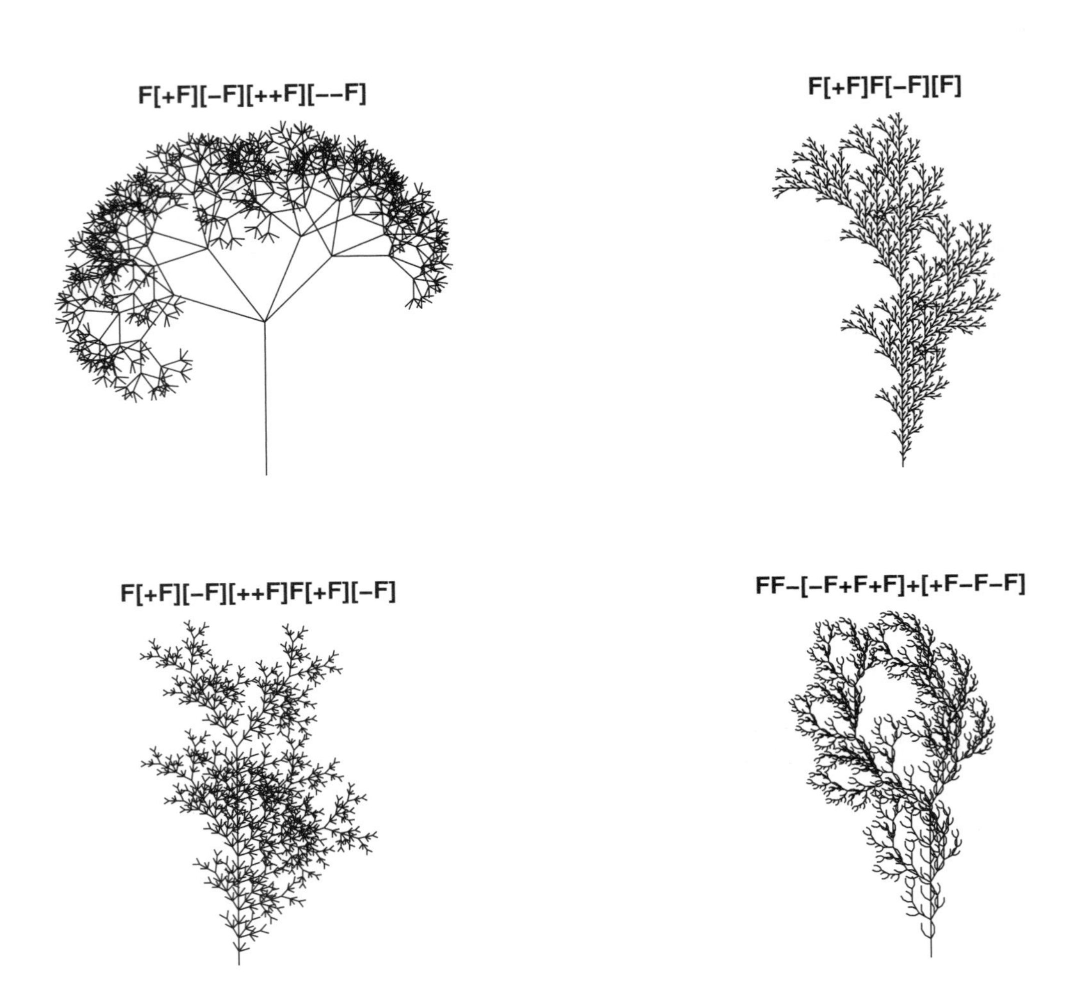

Figure 22.6. *Members of the genus* Matlabius Floribundum *produced by* `lsys`.

22.7. Black–Scholes Delta Surface

A European call option is a financial product that gives its holder the right to purchase from its writer an asset at a specified price, known as the *exercise price*, at some specified time in the future, known as the *expiry date*. A seminal paper by Black and Scholes shows how the writer of an option can eliminate risk by dynamically hedging with a portfolio of asset and cash. The amount of asset that the writer must hold is known as the *delta* of the option. Black and Scholes' formula for the delta is $N(d_1)$, where

$$d_1 = \frac{\log(S(t)/E) + (r + \frac{1}{2}\sigma^2)(T-t)}{\sigma\sqrt{T-t}}. \tag{22.5}$$

Here,

- t denotes time, with $t = 0$ and $t = T$ specifying the start and expiry dates,
- $S(t)$ is the asset price at time t,
- E is the exercise price,
- r is the interest rate,
- σ is the asset volatility,
- $N(\cdot)$ is the distribution function for a standard normal random variable, defined as

$$N(x) := \frac{1}{\sqrt{2\pi}} \int_{-\infty}^{x} e^{-\frac{s^2}{2}}\, ds.$$

MATLAB has a function `erf` that evaluates the *error function*

$$\mathrm{erf}(x) := \frac{2}{\sqrt{\pi}} \int_0^x e^{-t^2}\, dt,$$

from which the normal distribution function may be obtained as

$$N(x) = \frac{1 + \mathrm{erf}\big(x/\sqrt{2}\big)}{2}.$$

The Black–Scholes theory is derived under the assumption that the asset price $S(t_i)$ at time t_i evolves into $S(t_{i+1})$ at time $t_{i+1} > t_i$ according to

$$S(t_{i+1}) = S(t_i)\exp\big((\mu - \tfrac{1}{2}\sigma^2)(t_{i+1} - t_i) + \sigma\sqrt{t_{i+1} - t_i Z_i})\big), \tag{22.6}$$

where Z_i is a standard normal random variable. Here, μ is a constant that governs the expected increase in the asset.

Our aim is to plot the delta value $N(d_1)$ as a function of asset price S and time t, with all the other parameters, T, E, r, σ, μ, fixed. This will give a surface above the (S, t) plane. We will then generate three asset paths over $[0, T]$, using (22.6) to update the price between finely spaced time points, with the Z_i generated by calls to MATLAB's normal pseudo-random number generator, `randn`. We will sit these paths on the delta surface, that is, plot the path of $N(d_1)$ when d_1 in (22.5) takes the values given by (S_i, t_i). This illustrates the amount of asset that the option writer must maintain as the asset price evolves, in each of the three cases.

For further details on financial option valuation, see, for example, [39] and [127].

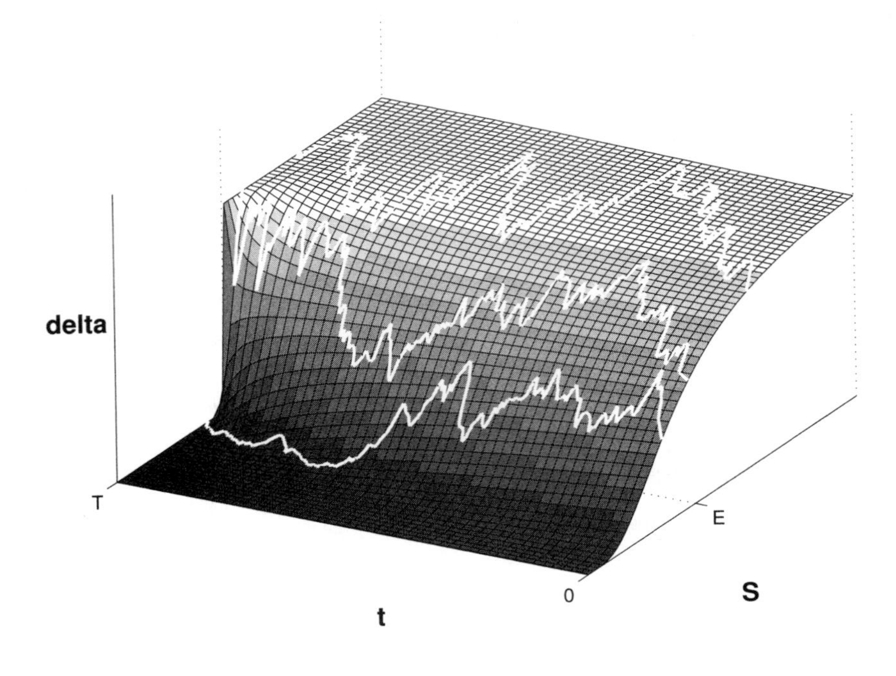

Figure 22.7. *Black–Scholes delta picture from* `bsdelta`.

Code and Walkthrough

The script M-file `bsdelta` in Listing 22.8 produces the picture in Figure 22.7. A key issue to address here is the division-by-zero arising when $t = T$ in (22.5). This difficulty is avoided by defining $N(d_1)$ at $t = T$ in terms of its limit from below:

$$\lim_{t \to T^-} N(d_1) = \begin{cases} 1 & \text{if } S(T) > E, \\ \frac{1}{2} & \text{if } S(T) = E, \\ 0 & \text{if } S(T) < E. \end{cases} \tag{22.7}$$

After initializing parameters, we use `meshgrid` to set up the arrays `Sgrid` and `tgrid` that are needed by `surf`, with the final time level omitted. We compute d_1 and $N(d_1)$ at these points, and then fill in the $t = T$ values separately, using (22.7). We call `surf` to display the surface, and apply some Handle Graphics commands to mark and label the axes.

The second part of the code generates the three asset paths. We take starting values of 1.5, 0.95, and 0.7. The cumulative product function, `cumprod`, is used to apply (22.6) over all time points. We then use the formula (22.5) for $t < T$ and (22.7) for $t = T$ to give a path of surface heights, `Npath`. These are superimposed by the 3D line plotter, `plot3`, with an increment of `deltaN = 0.1` added to the heights so that the paths are visible above the surface. Finally, we alter the default view to give a clearer picture.

Listing 22.8. *Script* bsdelta.

```
%BSDELTA  Black--Scholes delta surface with three asset paths superimposed.

randn('state',51)
E = 1; r = 0.05; sigma = 0.6; mu =0.05; T = 1;
N1 = 50; Dt = T/N1; N2 = 60;

tvals = [0:Dt:T-Dt];                          % Avoid division by zero.
Svals = linspace(.01,2.5,N2);
[Sgrid,tgrid] = meshgrid(Svals,tvals);

d1grid = (log(Sgrid/E) + ...
         (r+0.5*sigma^2)*(T-tgrid))./(sigma*sqrt(T-tgrid));
Ngrid = 0.5*(1+erf(d1grid/sqrt(2)));

tvals = [0:Dt:T];                             % Add expiry date.
[Sgrid,tgrid] = meshgrid(Svals,tvals);        % Extend the grid.
Ngrid(end+1,:) = 0.5*(sign(Svals - E) + 1);   % Append final time values.

surf(Sgrid,tgrid,Ngrid)
xlabel('S','FontWeight','Bold','FontSize',16)
ylabel('t','FontWeight','Bold','FontSize',16)
zlabel('delta','FontWeight','Bold','FontSize',16,...
       'Rotation',0,'HorizontalAlignment','right')
ylim([0 T]), xlim([0 2.5])
set(gca,'ZTick',[])
set(gca,'YTick',[0,T]), set(gca,'YTickLabel','0|T','FontSize',12)
set(gca,'XTick',E), set(gca,'XTickLabel','E','FontSize',12)

% Superimpose asset paths.
hold on
L = 200; Dt = T/L;
tpath = [0:Dt:T-Dt]';
Szero = [1.5;0.95;0.7];
for k = 1:3
   factors = exp((mu-0.5*sigma^2)*Dt+sigma*sqrt(Dt)*randn(L,1));
   Spath = [Szero(k);Szero(k)*cumprod(factors)];
   dpath = (log(Spath(1:end-1)/E) + ...
            (r+0.5*sigma^2)*(T-tpath))./(sigma*sqrt(T-tpath));
   Npath = 0.5*(1+erf(dpath/sqrt(2)));
   Npath = [Npath;0.5*(sign(Spath(end)-E)+1)];
   deltaN = 0.1;
   Npath(2:end) = Npath(2:end)+deltaN;
   plot3(Spath,[tpath;T],Npath,'w-','Linewidth',2)
end
hold off, view(-60,35)
```

22.8. Chutes and Ladders

In the game of Chutes and Ladders (also known as Snakes and Ladders) a player moves between squares on a board according to the roll of a die. On each turn, the number rolled, 1, 2, 3, 4, 5, or 6, determines how many squares to advance, with the constraints that

- if the new location is the foot of a ladder, the player automatically jumps up to the square at the top of that ladder,
- if the new location is the top of a chute (head of a snake), the player automatically jumps down to the square at the end of that chute (tail of that snake),
- if the player would progress beyond the final square, that turn is discarded, and the player's location is unchanged.

The game typically involves two or more players, with the first to reach the final square being deemed the winner. However, as there is no interaction between players, it is informative to study the single-player game, as we do here. For a given board, we wish to compute

1. the probability of finishing in exactly n rolls, and
2. the probability of finishing in at most n rolls.

Using a Markov chain formulation, this becomes a matrix computation problem.

Given the player's current location, it is possible to write down the probability of a move to any other square on the next roll. Suppose there are N squares on the board, ordered from 1 to N. We specify that the player starts "off the board"; so we introduce a fictitious zeroth square to represent the player's location before the first roll. For convenience we will refer to *states* instead of squares, with state i representing square $i-1$, so the states are ordered from 1 to $N+1$. In the case where there are no chutes or ladders on the board, the transition matrix is (with blank entries denoting zeros)

$$P = \frac{1}{6}\begin{bmatrix} 0 & 1 & 1 & 1 & 1 & 1 & 1 & & & & \\ & 0 & 1 & 1 & 1 & 1 & 1 & 1 & & & \\ & & 0 & \ddots & \ddots & \ddots & \ddots & \ddots & \ddots & & \\ & & & \ddots & \ddots & \ddots & \ddots & \ddots & \ddots & \ddots & \\ & & & & 0 & 1 & 1 & 1 & 1 & 1 & 1 \\ & & & & & 1 & 1 & 1 & 1 & 1 & 1 \\ & & & & & & 2 & 1 & 1 & 1 & 1 \\ & & & & & & & 3 & 1 & 1 & 1 \\ & & & & & & & & 4 & 1 & 1 \\ & & & & & & & & & 5 & 1 \\ & & & & & & & & & & 6 \end{bmatrix} \in \mathbb{R}^{(N+1)\times(N+1)}. \quad (22.8)$$

Here, for a player currently at state i, p_{ij} is the probability of occupying state j after the next roll. Now suppose we add a chute or ladder to the board so that a player landing at state r is automatically transported to state s. Then state r is no longer a possible location, and we should alter the transition matrix by (a) adding column r to column s and (b) removing row and column r. Having done this for all chutes and ladders, letting P now denote the resulting transition matrix, a standard

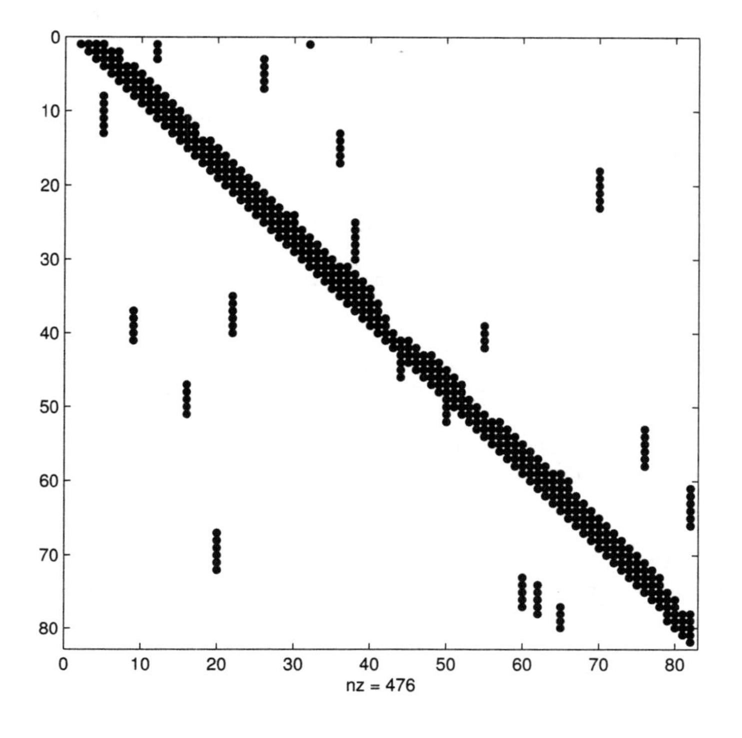

Figure 22.8. `spy` *plot of transition matrix from* `chute`.

Markov chain result tells us that the probability of reaching the final state in n rolls or less is given by the last entry in the first row of P^n. The appropriate Markov chain background theory may be found, for example, in [88].

Code and Walkthrough

The script M-file `chute` in Listing 22.9 produces the pictures in Figures 22.8 and 22.9. Here, we have a board with `N = 100` squares and 19 chutes/ladders. The `k`th chute/ladder starts at state `top(k)` and terminates at state `bot(k)`; if `top(k)` is less than `bot(k)`, we have a ladder, otherwise we have a chute. We begin by setting up P in (22.8); since P has Toeplitz structure except for the main diagonal, this is conveniently done using the `toeplitz` function. Colon notation and empty array assignments are used in accounting for the chutes and ladders. Note that the row/column deletions take place after all column updates have been done—if the operations `P(top,:) = []; P(:,top) = [];` appeared inside the `for k = 1:length(top)` loop, then we would no longer be able to use the original numbering system for `top` and `bot`.

Having constructed the final transition matrix, which has dimension $N+1-19 = 82$, we use `spy` to reveal the nonzero structure, as shown in Figure 22.8. We then compute the array `cumprob`, whose nth entry stores the probability of reaching the final state in n rolls or less, for $1 \leq n \leq 200$. Applying `diff` to `[0;cumprob]` gives an array whose nth entry stores the probability of reaching the final state in exactly n rolls. The appropriate histograms are shown in Figure 22.9.

Listing 22.9. *Script* chute.

```
%CHUTE  Chutes and ladders analysis.
%       Probability of finishing in exactly n moves and in at least n moves.

N = 100;   % Start at square zero, finish at square N.

% "+1" translates square to state.
top = [ 1  4  9 16 21 28 36 47 49 51 56 62 64 71 80  87 93 95 98] + 1;
bot = [38 14 31  6 42 84 44 26 11 67 53 19 60 91 100 24 73 75 78] + 1;

P = toeplitz(zeros(1,N+1),[0 ones(1,6) zeros(1,N-6)]);
for k = N-4:N+1, P(k,k) = k-N+5; end
P = P/6;

for k = 1:length(top)
    r = top(k); s = bot(k);       % Chute or ladder from r to s.
    P(:,s) = P(:,s) + P(:,r);     % Add column r to column s.
end
P(top,:) = []; P(:,top) = [];   % Remove starts of chutes and ladders.

figure(1)
spy(P)

M = 200;
cumprob = zeros(M,1);
cumprob(1) = P(1,end);
v = P(1,:);
for n = 2:M,
    v = v*P;
    cumprob(n) = v(end);
end

figure(2)
colormap([0.6,0.6,0.6])
subplot(2,1,1)
bar(diff([0;cumprob]))
title('Probability for Game Length','FontSize',12,'FontWeight','Bold')
grid on
xlim([0 M])
subplot(2,1,2)
bar(cumprob)
title('Cumulative Probability for Game Length',...
      'FontSize',12,'FontWeight','Bold')
grid on
xlim([0 M])
```

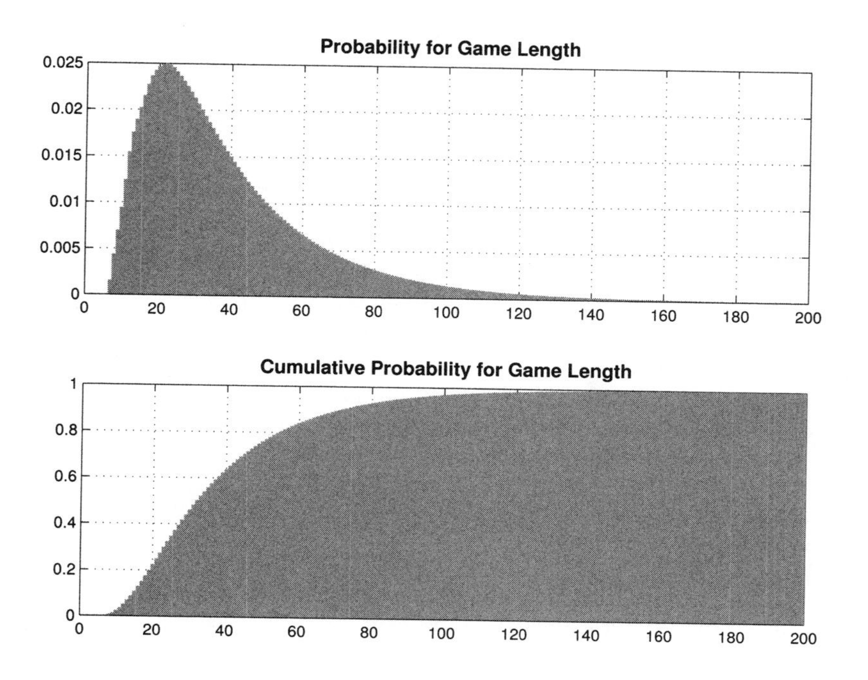

Figure 22.9. *Probability of finishing chutes and ladders game in exactly* n *rolls* (*upper*) *and at most* n *rolls* (*lower*).

22.9. Pythagorean Sum

The Pythagorean sum $\sqrt{a^2+b^2}$ of two scalars is a commonly occurring quantity and can be regarded as an "atomic operation" on a par with the four elementary operations $+$, $-$, $*$, $/$ and the square root of a single scalar. Fast, reliable ways of computing Pythagorean sums are therefore needed. It is desirable to avoid explicitly computing a square root, since the square root is a relatively expensive operation, and also to avoid squaring a or b, since the squares could overflow or underflow despite $\sqrt{a^2+b^2}$ being within the range of the arithmetic.

Given $x_0 \geq 0$ and $y_0 \geq 0$ the following iteration computes $p = \sqrt{x_0^2+y_0^2}$:

$$x_{n+1} = x_n \left(1 + 2\frac{y_n^2}{4x_n^2+y_n^2}\right), \tag{22.9a}$$

$$y_{n+1} = \frac{y_n^3}{4x_n^2+y_n^2}. \tag{22.9b}$$

To be precise, it can be shown that as $n \to \infty$, x_n converges monotonically to p from below, and y_n decreases monotonically to 0 from above, with $\sqrt{x_n^2+y_n^2} = p$. The rate of convergence of the iteration is cubic, which means that ultimately the error in x_n and y_n is bounded by a multiple of the cube of the error in x_{n-1} and y_{n-1}, respectively.

The iteration was originally suggested by Moler and Morrison, who develop an elegant floating point arithmetic implementation that avoids overflow [82]. Our interest is in using the iteration to compute the Pythagorean sum to arbitrary precision, and for simplicity we will not scale to avoid overflow. It is when working to high precision that iterations with cubic or higher orders of convergence are particularly attractive.

Code and Walkthrough

Function **pythag** in Listing 22.10 implements iteration (22.9) in the Symbolic Math Toolbox's vpa arithmetic. It has a third input argument, d, that specifies the required number of significant digits. The computations are done with d + 10 digits, since rounding errors can be expected to make the last few digits incorrect. The first two input arguments can be symbolic expressions, so they are converted to vpa form before beginning the iteration; absolute values are also taken, to allow the routine to work for negative arguments. The **error** message **error(nargchk(2,4,nargin))** invokes the **nargchk** function, which returns an empty string if the number of input arguments lies between 2 and 4 inclusive (in which case the call to **error** has no effect), or an error message string:

```
>> pythag(2)
??? Error using ==> pythag
Not enough input arguments.
```

For efficiency of the iteration it is important to order the starting values so that $y_0 \leq x_0$, since otherwise the first few iterations are spent making x_n (which tends to a positive value) greater than y_n (which tends to zero). Since the Symbolic Math Toolbox does not overload the max and min functions, we need to call Maple directly to achieve the required ordering. Note also the use of the variables **yn2** and **temp** to reduce the amount of computation.

The convergence test checks whether the dth significant digit has changed since the previous iteration. The absolute value of the relative change in two successive iterates can be smaller than the smallest positive double precision number (**realmax**), in which case it underflows to zero if converted to double precision. Hence in implementing the convergence test we compute the base 10 logarithm of the relative change, which *is* representable in double precision. The function prints to the screen information that shows the convergence of the iteration.

To illustrate, we compute $\sqrt{(1/10)^2 + e^2}$ to 2500 significant digits and check the result against the answer computed directly in vpa arithmetic. The code

```
d = 2500;
x = sym('0.1');
y = sym('exp(1)');
z = pythag(x,y,d);
z = char(z); [z(1:60) '...']

% Check:
p = vpa(sqrt(x^2+y^2), d+10);
p = char(p);
test_equal = strcmp(z(1:d),p(1:d))
```

produces the output

```
>> dopythag
log(rel_change_x_n):     -3, log(y_n):      -4
log(rel_change_x_n):    -10, log(y_n):     -15
```

Listing 22.10. *Function* pythag.

```
function [xn,k] = pythag(x,y,d,noprnt)
%PYTHAG     Pythagorean sum in variable precision arithmetic.
%           P = PYTHAG(X,Y,D) computes the Pythagorean sum
%           SQRT(X^2+Y^2) of the real numbers X and Y
%           correct to about D significant digits,
%           using an iteration that avoids computing square roots.
%           D defaults to 50.
%           By default, the progress of the iteration is printed;
%           the call PYTHAG(X,Y,D,1) suppresses this.
%           [X,K] = PYTHAG(X,Y,D) returns also the number of
%           iterations, K.

%           This function requires the Symbolic Math Toolbox.

error(nargchk(2,4,nargin))      % Check number of input arguments.
if nargin < 4, noprnt = 0; end
if nargin < 3, d = 50; end

d_old = digits;
% Work with slightly more accuracy than requested for final result.
digits(d+10)
x = abs(vpa(x)); y = abs(vpa(y));

xn = maple('max',x,y); % Take max since xn increases to Pyth. sum.
yn = maple('min',x,y);

k = 0;
x_change = 0;

while abs(x_change) < d
      k = k +1;
      yn2 = yn^2;
      temp = yn2/(4*xn^2+yn2);
      xnp1 = xn*(1 + 2*temp);
      ynp1 = yn*temp;
      x_change = double( log10(abs((xnp1-xn)/xnp1)) );
      y_exp = double( log10(ynp1) );
      if ~noprnt
        fprintf('log(rel_change_x_n): %6.0f, log(y_n): %6.0f\n', ...
                  x_change, y_exp)
      end
      xn = xnp1; yn = ynp1;
end
xn = vpa(xn,d); % Return requested number of digits.
digits(d_old)   % Restore original value.
```

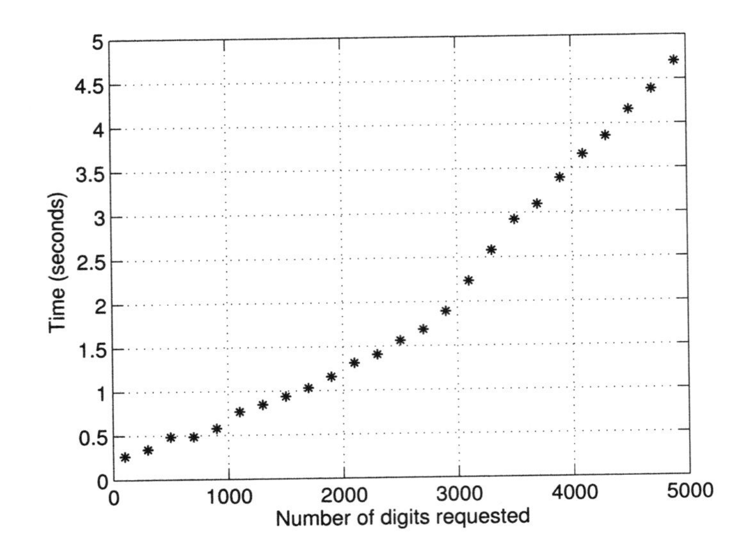

Figure 22.10. *Execution time of* `pythag` *versus requested accuracy.*

```
log(rel_change_x_n):     -31, log(y_n):     -46
log(rel_change_x_n):     -93, log(y_n):    -140
log(rel_change_x_n):    -281, log(y_n):    -421
log(rel_change_x_n):    -843, log(y_n):   -1264
log(rel_change_x_n):    -Inf, log(y_n):   -3795
ans =
2.7201206037473136693255563235433096109025216407472208575162...
test_equal =
      1
```

The `-Inf` is a result of `xn` and `xnp1` being exactly equal. The cubic convergence is evident in the increase in size of the logarithms by a factor approximately 3 from one line to the next.

Figure 22.10 plots the execution time of `pythag` versus the number of requested digits, for `x = vpa('1/3'), y = vpa('1/7')`. We see approximately linear growth of time with the number of digits. The number of iterations increases very slowly because of the cubic convergence, varying from 5 to 9.

22.10. Fisher's Equation

Fisher's equation, a PDE of the form

$$\frac{\partial u}{\partial t} = \frac{\partial^2 u}{\partial x^2} + u(1-u),$$

is used as a model for various biological phenomena. It is common to pose the equation over the whole x-axis, $-\infty < x < \infty$, and to specify boundary conditions $u(x,t) \to 1$ as $x \to -\infty$ and $u(x,t) \to 0$ as $x \to \infty$. In this context, traveling wave solutions of the form $u(x,t) = f(x - ct)$ have been widely studied. For such a solution, the function

f defines a fixed profile that is transported along the x-axis as time evolves. If we let $z = x - ct$, then the solution $u(z,t)$ becomes stationary (independent of time) in the moving coordinate system (z,t). In the hope of catching a traveling wave, we will take a large space interval, $-50 \le x \le 50$, and specify Neumann boundary conditions $\partial u/\partial x = 0$ at $x = \pm 50$. We will solve the PDE for $0 \le t \le 20$ with two different initial conditions: the step function

$$u(x,0) = \begin{cases} 0.99, & x \le -20, \\ 0, & x > -20, \end{cases} \tag{22.10}$$

and the small hump

$$u(x,0) = \begin{cases} \frac{1}{4}\cos^2(\frac{\pi x}{10}), & |x| \le 5, \\ 0, & |x| > 5. \end{cases} \tag{22.11}$$

Code and Walkthrough

The function **fisher** in Listing 22.11 uses MATLAB's **pdepe** to solve Fisher's equation, producing the pictures in Figure 22.11 and 22.12. The subfunction **fica** implements the first initial condition, (22.10), and a mesh plot of the resulting solution is displayed in the upper left region of the first figure window. We set **view(30,30)** in order to get a more revealing perspective. It appears that the solution is indeed evolving into a fixed profile that progresses linearly in time. To investigate further, the upper right picture shows **contour** applied to the solution, specifying contour levels of 0.2, 0.4, 0.6, and 0.8. The contours appear to settle into equally spaced straight lines. Although Fisher's equation admits traveling waves of any speed $c \ge 2$, it may be argued (see, for example, [84]) that a wave of speed $c = 2$ is the most likely to be observed. A reference triangle of this slope has been added with a basic **plot** command to give a visual check.

The second initial condition, (22.11), implemented in **ficb**, gives rise to the pictures in the lower half of Figure 22.11. In this case, two wavefronts are generated, emanating from each side of the initial hump, and the **contour** plot is again consistent with wave speed $c = 2$.

In the second figure window, as shown in Figure 22.12, we **waterfall** the solution for (22.10) in the moving coordinate system $(x - 2t, t)$ to give further visual confirmation that a traveling wave of speed $c = 2$ has emerged.

Listing 22.11. *Script* fisher.

```
function fisher
%FISHER    Displays solutions to Fisher PDE.

m = 0; a = -50; b = 50; t0 = 0; tf = 20;
xvals = linspace(a,b,101); tvals = linspace(t0,tf,51);
[xmesh, tmesh] = meshgrid(xvals,tvals);

figure(1), subplot(2,2,1)
sol = pdepe(m,@fpde,@fica,@fbc,xvals,tvals);
ua = sol(:,:,1); mesh(xmesh,tmesh,ua)
xlabel('x'), ylabel('t'), zlabel('u','Rotation',0), title('u(x,t)')
text_set, view(30,30)

subplot(2,2,2), contour(xmesh,tmesh,ua,[0.2:0.2:0.8])
xlabel('x'), ylabel('t','Rotation',0), title('Contour Plot')
text_set, hold on
plot([10,20,20,10],[8,13,8,8],'r--'), text(0,6,'Ref. slope = 2')
hold off

subplot(2,2,3), sol = pdepe(m,@fpde,@ficb,@fbc,xvals,tvals);
ub = sol(:,:,1); mesh(xmesh,tmesh,ub)
xlabel('x'), ylabel('t'), zlabel('u','Rotation',0), title('u(x,t)')
text_set, view(30,30)

subplot(2,2,4), contour(xmesh,tmesh,ub,[0.2:0.2:0.8])
xlabel('x'), ylabel('t','Rotation',0), title('Contour Plot')
text_set, hold on
plot([25,35,35,25],[5,10,5,5],'r--'), text(15,3,'Ref. slope = 2')
hold off

figure(2), zmesh = xmesh - 2*diag(tvals)*ones(size(xmesh));
waterfall(zmesh,tmesh,ua)
xlabel('x-2t'), ylabel('t'), zlabel('u','Rotation',0), title('u(x-2t,t)')
zlim([0 1]), text_set, view(15,30)

%-------------------------- Subfunctions -----------------------------%
function [c,f,s] = fpde(x,t,u,DuDx)
%FDE  Fisher PDE.
c = 1; f = DuDx; s = u*(1-u);

function u0 = fica(x)
%FIC  Fisher initial condition: 1st case.
u0 = 0.99*(x<=-20);

function [pa,qa,pb,qb] = fbc(xa,ua,xb,ub,t)
%FBC  Fisher  boundary conditions.
pa = 0; qa = 1; pb = 0; qb = 1;

function u0 = ficb(x)
%FIC2  Fisher initial condition: 2nd case.
u0 = 0.25*(cos(0.1*pi*x).^2).*(abs(x)<=5);

function text_set
h = findall(gca,'type','text'); set(h,'FontSize',12,'FontWeight','Bold')
```

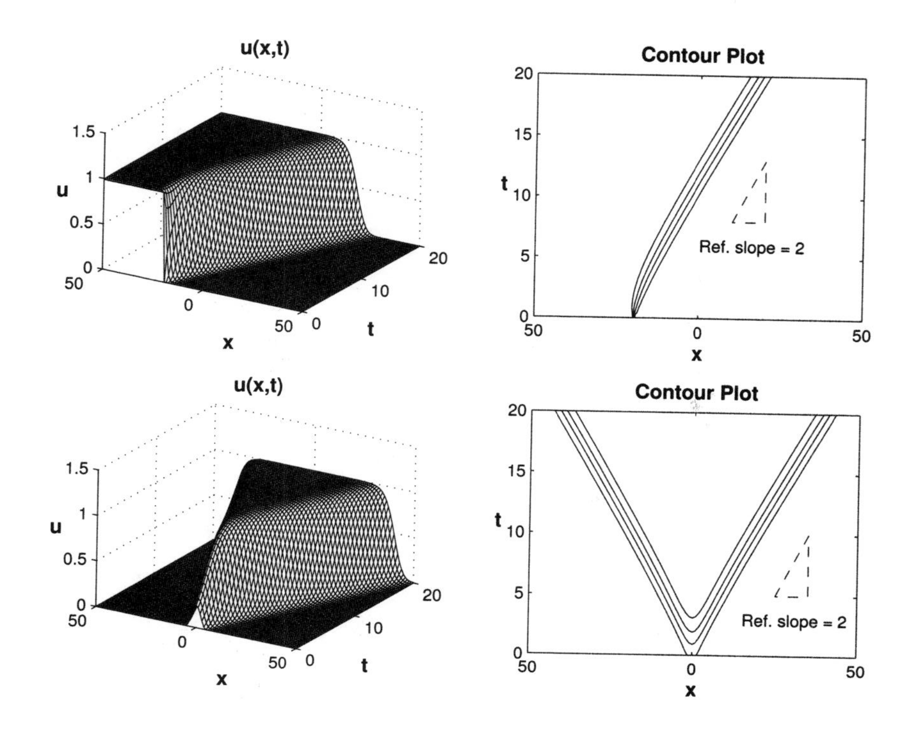

Figure 22.11. *Traveling wave solutions for Fisher's equation, from* `fisher`.

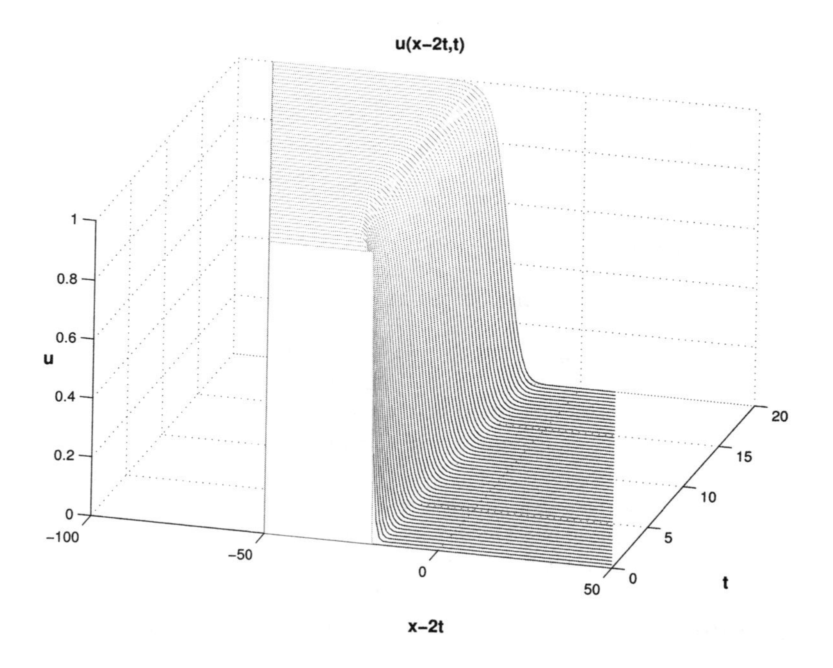

Figure 22.12. *Solution of Fisher's equation for initial conditions* (22.10) *in moving coordinate system, from* `fisher`.

Example is always more efficacious than precept.
— SAMUEL JOHNSON (1759)

The computation of $\mathrm{sqrt}(a^2 + b^2)$ *is required in many matrix algorithms,*
particularly those involving complex arithmetic.
A new approach to carrying out this operation is described by
Moler and Morrison. . . .
In MATLAB, the algorithm is used for complex modulus,
Euclidean vector norm, plane rotations,
and the shift calculation in the eigenvalue and singular value iterations.
— CLEVE B. MOLER, *MATLAB Users' Guide* (1982)

Performance profiles can be used to compare the performance of two solvers,
but performance profiles are most useful in comparing several solvers.
Because large amounts of data are generated in these situations,
trends in performance are often difficult to see.
— ELIZABETH D. DOLAN and JORGE J. MORÉ, *Benchmarking Optimization Software with Performance Profiles* (2002)

And none of this would have been any fun without MATLAB.
— NOËL M. NACHTIGAL, SATISH C. REDDY, and LLOYD N. TREFETHEN, *How Fast Are Nonsymmetric Matrix Iterations?* (1992)

Appendix A
The Top 111 MATLAB Functions

This appendix lists the 111 MATLAB functions that we believe are the most useful for the typical MATLAB user. Information about these functions can be found by looking in the index of this book, or by using MATLAB's online documentation.

Table A.1. *Elementary and specialized vectors and matrices.*

`zeros`	Zeros array
`ones`	Ones array
`eye`	Identity matrix
`rand`	Uniformly distributed random numbers
`randn`	Normally distributed random numbers
`gallery`	Test matrices
`linspace`	Linearly spaced vector

Table A.2. *Special variables and constants.*

`ans`	Most recent answer
`eps`	Floating point relative accuracy
`i, j`	Imaginary unit ($\sqrt{-1}$)
`inf`	∞
`NaN`	Not a Number
`pi`	π

Table A.3. *Array information and manipulation.*

`size`	Array dimensions
`length`	Length of array (size of longest dimension)
`reshape`	Change size of array
`:`	Regularly spaced vector and index into matrix
`end`	Last index in an indexing expression
`diag`	Diagonal matrices and diagonals of matrix
`tril`	Extract lower triangular part
`triu`	Extract upper triangular part
`repmat`	Replicate and tile array

Table A.4. *Logical operators.*

`all`	Test for all nonzeros
`any`	Test for any nonzeros
`find`	Find indices of nonzero elements
`isempty`	Test for empty array
`isequal`	Test if arrays are equal

Table A.5. *Flow control.*

`error`	Display error message and abort function
`for`	Repeat statements a specific number of times
`if`	Conditionally execute statements
`switch, case`	Choose among several cases
`while`	Repeat statements indefinitely

Table A.6. *Basic data analysis.*

`max`	Largest component
`min`	Smallest component
`mean`	Average or mean value
`std`	Standard deviation
`sum`	Sum of elements
`prod`	Product of elements
`sort`	Sort elements

Table A.7. *Graphics.*

`plot`	x-y plot
`fplot`	Function plotter
`ezplot`	Easy-to-use function plotter
`semilogy`	Plot with logarithmically scaled y-axis
`bar`	Bar graph
`hist`	Histogram
`axis`	Axis control
`xlim`	Set x-axis limits
`ylim`	Set y-axis limits
`grid`	Grid lines
`xlabel`	Label x-axis
`ylabel`	Label y-axis
`title`	Title graph
`legend`	Display legend
`text`	Text annotation
`subplot`	Create axes in grid pattern
`hold`	Hold current graph
`contour`	Contour plot
`mesh`	Wireframe surface
`surf`	Solid surface
`spy`	Visualize sparsity pattern
`print`	Print figure to disk or printer
`clf`	Clear current figure
`close`	Close figure

Table A.8. *Linear algebra.*

`norm`	Norm of vector or matrix
`cond`	Condition number of matrix (with respect to inversion)
`\`	Solve linear system of equations
`eig`	Eigenvalues and eigenvectors
`lu`	LU factorization
`qr`	QR factorization
`svd`	Singular value decomposition

Table A.9. *Functions connected with M-files.*

`edit`	Invoke MATLAB editor
`lookfor`	Search H1 line (first comment line) of all M-files for keyword
`nargin`	Number of function input arguments
`nargout`	Number of function output arguments
`type`	List file in Command Window
`which`	Display full pathname of M-file

Table A.10. *Miscellaneous.*

`clc`	Clear Command Window
`demo`	Demonstrations
`diary`	Save Command Window text to file
`dir`	Display directory listing
`doc`	Display HTML documentation in Help browser
`help`	Display help in Command Window
`tic, toc`	Start/stop stopwatch timer
`what`	List MATLAB files in current directory grouped by type

Table A.11. *Data types and conversions.*

`double`	Convert to double precision
`char`	Create or convert to character array (string)
`cell`	Create cell array
`num2str`	Convert number to string
`sparse`	Create sparse matrix
`struct`	Create or convert to structure array

Table A.12. *Managing the workspace.*

`clear`	Clear items from workspace
`who, whos`	List variables in workspace
`load`	Load workspace variables from disk
`save`	Save workspace variables to disk
`exit, quit`	Terminate MATLAB session

Table A.13. *Input and output.*

`disp`	Display text or array
`format`	Set output format
`fprintf`	Write formatted data to screen or file
`sprintf`	Write formatted data to string
`input`	Prompt for user input

Table A.14. *Numerical methods.*

`bvp4c`	Solve two-point boundary value problem
`fft`	Discrete Fourier transform
`fminbnd`	Minimize function of one variable on fixed interval
`fzero`	Find zero of function of one variable
`interp1`	One-dimensional interpolation (several methods)
`ode45`	Explicit Runge–Kutta pair for nonstiff differential equations
`polyfit`	Least squares polynomial fit
`quadl`	Numerical integration
`roots`	Roots of polynomial
`spline`	Cubic spline interpolation

Appendix B
Changes in MATLAB

Recent releases of MATLAB have introduced many changes, which are documented in the Release Notes available from the Help browser. The changes include new language features, new functions, and alterations to function names or syntax that require M-files to be rewritten for future compatibility. In this appendix we give a highly selective summary of changes introduced in versions 5.0 onwards of MATLAB. Our aim is to point out important changes that may be overlooked by users upgrading from earlier versions and that may cause M-files written for earlier versions to behave differently in MATLAB 7.

B.1. MATLAB 5.0

- New, improved random number generators introduced. Previously, the state of the random number generators was set with

  ```
  rand('seed',j), randn('seed',j)
  ```

  If this syntax is used now it causes the old generators to be used. The state should now be set with

  ```
  rand('state',j), randn('state',j)
  ```

B.2. MATLAB 5.3

Functions renamed as follows: `fmin` → `fminbnd`, `fmins` → `fminsearch`, `nnls` → `lsqnonneg`.

B.3. MATLAB 6

- Matrix computations based on LAPACK, rather than LINPACK as previously. MATLAB now takes advantage of Hermitian structure when computing the eigensystem of a complex Hermitian matrix, and of Hermitian definite structure when computing the eigensystem of a Hermitian definite generalized eigenvalue problem. Eigenvalues may be returned in a different order than with earlier versions of MATLAB, eigenvectors may be normalized differently, and the columns of unitary matrices may differ by scale factors of modulus unity.
- As a result of the switch to LAPACK, the `flops` function is no longer operative. (In earlier versions of MATLAB, `flops` provided a count of the total number of floating point operations performed.)

- Some of the arguments of the `eigs` function have changed (this function is now an interface to ARPACK rather than an M-file).
- The precedence of the logical `and` and `or` operators, which used to be the same, has been changed so that `and` has higher precedence.
- Function handles ("`@fun`") have been introduced for passing functions as arguments; they are preferred to the passing of function names in strings.
- The quadrature function `quad8` has been superseded by `quadl`. The default error tolerance for `quad` is now of order `eps` rather than 10^{-3}.
- The way in which the ODE solvers are called has been changed to exploit function handles (the "ODE file" format, documented in `help odefile`, is no longer used).

B.4. MATLAB 6.5

- The logical attribute of a numeric data type has been upgraded to be a class, on the same level as character and cell arrays; previously, logical was an attribute of any numeric data type. Hence `double` now needs to be inserted around logical expressions where the result is required to be of type `double`.
- `sparse`, previously a class, is now an attribute of an array.

B.5. MATLAB 7.0

- All M-file, function, and directory names are now case sensitive on all platforms. In earlier versions of MATLAB, M-file names and directory names were not case sensitive under Windows. To allow backward compatibility, MATLAB under Windows allows an M-file to be called with incorrect case, but prints a warning message (search for "case sensitivity" in the online MATLAB documentation for full details of the possible behaviors).
- In searching for functions MATLAB now treats function M-files just like built-in functions. If you have a function M-file on the path that has the same name as a built-in function it will now override the latter, unlike in earlier releases.
- In earlier versions of MATLAB, the "function-function" routines expected parameters to a function passed to them to be specified as trailing arguments `p1,p2,...`. This usage is now superseded by the appropriate use of anonymous functions and nested functions.
- The format of MAT-files has changed (data is now stored in Unicode character encoding and compressed). To write MAT-files that can be read by MATLAB 6 it is necessary to append `-v6` to the `save` command.

Appendix C
Resources

The first port of call for information about MATLAB resources should be the Web page of The MathWorks, at

`http://www.mathworks.com`

This Web page can be accessed from the Help-Web Resources menu item of the MATLAB desktop. It includes documentation for MATLAB and all the toolboxes, in both html and PDF form, technical notes, a search facility, and MATLAB Central, "an open exchange for the MATLAB and Simulink user community" that includes a large collection of user-contributed M-files,

The newsgroup `comp.soft-sys.matlab` is devoted to MATLAB. It contains problems and solutions from MATLAB users, with contributions from MathWorks employees.

Full contact details for The MathWorks can be obtained by typing `info` at the MATLAB prompt. For reference we give the details here:

```
The MathWorks, Inc.
3 Apple Hill Drive
Natick, MA 01760-2098 USA

                    Phone:  +508-647-7000
                      Fax:  +508-647-7001
                      Web:  www.mathworks.com
                Newsgroup:  comp.soft-sys.matlab
                      FTP:  ftp.mathworks.com

E-mail:
         info@mathworks.com  Sales, pricing, and general information
      support@mathworks.com  Technical support for all products
          doc@mathworks.com  Documentation error reports
         bugs@mathworks.com  Bug reports
      service@mathworks.com  Order status, invoice, and license issues
     renewals@mathworks.com  Renewal/subscription pricing
      pricing@mathworks.com  Product and pricing information
       access@mathworks.com  MATLAB Access Program
      suggest@mathworks.com  Product enhancement suggestions
   news-notes@mathworks.com  MATLAB News & Notes Editor
  connections@mathworks.com  MATLAB Connections Program
```

Glossary

Array Editor. A tool allowing array contents to be viewed and edited in tabular format.

Command History. A tool that lists MATLAB commands previously typed in the current and past sessions and allows them to be copied or executed.

Command Window. The window in which the MATLAB prompt `>>` appears and in which commands are typed. It is part of the MATLAB desktop.

Current Directory browser. A browser for viewing M-files and other files and performing operations on them.

Editor/Debugger. A tool for creating, editing, and debugging M-files.

FIG-file. A file with a `.fig` extension that contains a representation of a figure that can be reloaded into MATLAB.

figure. A MATLAB window for displaying graphics.

function M-file. A type of M-file that can accept input arguments and return output arguments and whose variables are local to the function.

Handle Graphics. The object-oriented graphics system that underlies MATLAB's graphics. It employs a hierarchical organization of objects that are manipulated via their handles.

Help browser. A browser that allows you to view and search the documentation for MATLAB and other MathWorks products.

IEEE arithmetic. A standard for floating point arithmetic [44], to which MATLAB's arithmetic conforms.

LAPACK. A Fortran 77 library of programs for linear equation, least squares, eigenvalue, and singular value computations [3]. Many of MATLAB's linear algebra functions are based on LAPACK.

M-file. A file with a `.m` extension that contains a sequence of MATLAB commands. It is of one of two types: a function or a script.

MAT-file. A file with a `.mat` extension that contains MATLAB variables. Created and accessed with the `save` and `load` commands.

MATLAB desktop. A user interface for managing files, tools, and applications associated with MATLAB.

MATLAB Web browser. A Web browser that is part of the MATLAB system. Used for displaying `profile` reports, for example.

MEX-file. A subroutine produced from C or Fortran code whose name has a platform-specific extension. It behaves like an M-file or built-in function.

script M-file. A type of M-file that takes no input or output arguments and operates on data in the workspace.

toolbox. A collection of M-files that extends MATLAB's capabilities, usually in a particular application area.

Workspace browser. A browser that lists variables in the workspace and allows operations to be performed on them.

Bibliography

MATLAB documents marked "online version" can be accessed via the MATLAB Help browser.

[1] Forman S. Acton. *Numerical Methods That Work.* Harper and Row, New York, 1970. xviii+541 pp. Reprinted by Mathematical Association of America, Washington, D.C., with new preface and additional problems, 1990. ISBN 0-88385-450-3. (Cited on pp. 141, 170.)

[2] D. E. Amos. Algorithm 644: A portable package for Bessel functions of a complex argument and nonnegative order. *ACM Trans. Math. Software*, 12(3):265–273, 1986. See also remark in same journal, 16 (1990), p. 404. (Cited on p. 40.)

[3] E. Anderson, Z. Bai, C. H. Bischof, S. Blackford, J. W. Demmel, J. J. Dongarra, J. J. Du Croz, A. Greenbaum, S. J. Hammarling, A. McKenney, and D. C. Sorensen. *LAPACK Users' Guide.* Third edition, Society for Industrial and Applied Mathematics, Philadelphia, PA, USA, 1999. xxvi+407 pp. ISBN 0-89871-447-8. (Cited on pp. 121, 359.)

[4] Uri M. Ascher and Linda R. Petzold. *Computer Methods for Ordinary Differential Equations and Differential-Algebraic Equations.* Society for Industrial and Applied Mathematics, Philadelphia, PA, USA, 1998. xvii+314 pp. ISBN 0-89871-412-5. (Cited on p. 191.)

[5] Russell Ash. *The Top 10 of Everything.* Dorland Kindersley, London, 1994. 288 pp. ISBN 0-7513-0137-X. (Cited on p. 264.)

[6] Kendall E. Atkinson. *An Introduction to Numerical Analysis.* Second edition, Wiley, New York, 1989. xvi+693 pp. ISBN 0-471-50023-2. (Cited on pp. 159, 171, 174.)

[7] Kendall E. Atkinson. *Elementary Numerical Analysis.* Second edition, Wiley, New York, 1993. xiii+425 pp. ISBN 0-471-60010-5. (Cited on p. 159.)

[8] Zhaojun Bai, James W. Demmel, Jack J. Dongarra, Axel Ruhe, and Henk A. van der Vorst, editors. *Templates for the Solution of Algebraic Eigenvalue Problems: A Practical Guide.* Society for Industrial and Applied Mathematics, Philadelphia, PA, USA, 2000. xxix+410 pp. ISBN 0-89871-471-0. (Cited on p. 121.)

[9] Richard Barrett, Michael Berry, Tony F. Chan, James Demmel, June Donato, Jack Dongarra, Victor Eijkhout, Roldan Pozo, Charles Romine, and Henk van der Vorst. *Templates for the Solution of Linear Systems: Building Blocks for Iterative Methods.* Society for Industrial and Applied Mathematics, Philadelphia, PA, USA, 1994. xiii+112 pp. ISBN 0-89871-328-5. (Cited on p. 137.)

[10] Zachary Battles and Lloyd N. Trefethen. An extension of MATLAB to continuous functions and operators. *SIAM J. Sci. Comput.*, 25(5):1743–1770, 2004. (Cited on p. xx.)

[11] Jon L. Bentley. *More Programming Pearls: Confessions of a Coder.* Addison-Wesley, Reading, MA, USA, 1988. viii+207 pp. ISBN 0-201-11889-0. (Cited on p. 118.)

[12] Folkmar Bornemann, Dirk Laurie, Stan Wagon, and Jörg Waldvogel. *The SIAM 100-Digit Challenge: A Study in High-Accuracy Numerical Computing.* Society for Industrial and Applied Mathematics, Philadelphia, PA, USA, 2004. xi+306 pp. ISBN 0-89871-561-X. (Cited on p. 329.)

[13] K. E. Brenan, S. L. Campbell, and L. R. Petzold. *Numerical Solution of Initial-Value Problems in Differential-Algebraic Equations.* Society for Industrial and Applied Mathematics, Philadelphia, PA, USA, 1996. x+256 pp. Corrected republication of work first published in 1989 by North-Holland, New York. ISBN 0-89871-353-6. (Cited on p. 191.)

[14] James L. Buchanan and Peter R. Turner. *Numerical Methods and Analysis.* McGraw-Hill, New York, 1992. xv+751 pp. ISBN 0-07-008717-2, 0-07-112922-7 (international paperback edition). (Cited on p. 93.)

[15] Zizhong Chen, Jack Dongarra, Piotr Luszczek, and Kenneth Roche. Self-adapting software for numerical linear algebra and LAPACK for clusters. *Parallel Comput.*, 29: 1723–1743, 2003. (Cited on p. 170.)

[16] Robert M. Corless. *Essential Maple 7: An Introduction for Scientific Programmers.* Springer-Verlag, New York, 2002. ISBN 0-387-95352-3. (Cited on p. 296.)

[17] Germund Dahlquist and Åke Björck. *Numerical Methods.* Prentice-Hall, Englewood Cliffs, NJ, USA, 1974. xviii+573 pp. Translated by Ned Anderson. ISBN 0-13-627315-7. (Cited on pp. 159, 171.)

[18] Harold T. Davis. *Introduction to Nonlinear Differential and Integral Equations.* Dover, New York, 1962. xv+566 pp. ISBN 0-486-60971-5. (Cited on p. 183.)

[19] Timothy A. Davis. UMFPACK. `http://www.cise.ufl.edu/research/sparse/umfpack`. (Cited on p. 232.)

[20] James W. Demmel. *Applied Numerical Linear Algebra.* Society for Industrial and Applied Mathematics, Philadelphia, PA, USA, 1997. xi+419 pp. ISBN 0-89871-389-7. (Cited on p. 121.)

[21] Elizabeth D. Dolan and Jorge J. Moré. Benchmarking optimization software with performance profiles. *Math. Programming*, 91:201–213, 2002. (Cited on pp. 320, 348.)

[22] I. S. Duff, A. M. Erisman, and J. K. Reid. *Direct Methods for Sparse Matrices.* Oxford University Press, 1986. xiii+341 pp. ISBN 0-19-853408-6. (Cited on p. 235.)

[23] Alan Edelman. Eigenvalue roulette and random test matrices. In *Linear Algebra for Large Scale and Real-Time Applications*, Marc S. Moonen, Gene H. Golub, and Bart L. De Moor, editors, volume 232 of *NATO ASI Series E*, Kluwer Academic Publishers, Dordrecht, The Netherlands, 1993, pages 365–368. (Cited on pp. 75, 76.)

[24] Alan Edelman, Eric Kostlan, and Michael Shub. How many eigenvalues of a random matrix are real? *J. Amer. Math. Soc.*, 7(1):247–267, 1994. (Cited on p. 76.)

[25] Mark Embree and Lloyd N. Trefethen. Growth and decay of random Fibonacci sequences. *Proc. Roy. Soc. London Ser. A*, 455:2471–2485, 1999. (Cited on p. 8.)

[26] Bengt Fornberg. *A Practical Guide to Pseudospectral Methods.* Cambridge University Press, 1995. x+231 pp. ISBN 0-521-49582-2. (Cited on p. 256.)

[27] George E. Forsythe, Michael A. Malcolm, and Cleve B. Moler. *Computer Methods for Mathematical Computations.* Prentice-Hall, Englewood Cliffs, NJ, USA, 1977. xi+259 pp. ISBN 0-13-165332-6. (Cited on pp. 159, 168.)

[28] Matteo Frigo and Steven G. Johnson. FFTW. `http://www.fftw.org/`. (Cited on p. 170.)

[29] Walter Gander and Walter Gautschi. Adaptive quadrature—revisited. *BIT*, 40(1): 84–101, 2000. (Cited on p. 172.)

ohn R. Gilbert, Cleve B. Moler, and Robert S. Schreiber. Sparse matrices in ATLAB: Design and implementation. *SIAM J. Matrix Anal. Appl.*, 13(1):333–356, 92. (Cited on p. 235.)

[31] Gene H. Golub and Charles F. Van Loan. *Matrix Computations*. Third edition, Johns Hopkins University Press, Baltimore, MD, USA, 1996. xxvii+694 pp. ISBN 0-8018-5413-X (hardback), 0-8018-5414-8 (paperback). (Cited on pp. 121, 130.)

[32] Anne Greenbaum. *Iterative Methods for Solving Linear Systems*. Society for Industrial and Applied Mathematics, Philadelphia, PA, USA, 1997. xiii+220 pp. ISBN 0-89871-396-X. (Cited on p. 137.)

[33] David F. Griffiths and Desmond J. Higham. *Learning LaTeX*. Society for Industrial and Applied Mathematics, Philadelphia, PA, USA, 1997. x+84 pp. ISBN 0-89871-383-8. (Cited on pp. 94, 95, 116.)

[34] E. Hairer and G. Wanner. *Analysis by Its History*. Springer-Verlag, New York, 1996. x+374 pp. ISBN 0-387-94551-2. (Cited on p. 172.)

[35] E. Hairer and G. Wanner. *Solving Ordinary Differential Equations II: Stiff and Differential-Algebraic Problems*. Second edition, Springer-Verlag, Berlin, 1996. xv+614 pp. ISBN 3-540-60452-9. (Cited on pp. 187, 189.)

[36] Leonard Montague Harrod, editor. *Indexers on Indexing: A Selection of Articles Published in* The Indexer. R. K. Bowker, London, 1978. x+430 pp. ISBN 0-8352-1099-5. (Cited on p. 364.)

[37] Piet Hein. *Grooks*. Number 85 in Borgens Pocketbooks. Second edition, Borgens Forlag, Copenhagen, Denmark, 1992. 53 pp. First published in 1966. ISBN 87-418-1079-1. (Cited on p. 227.)

[38] Desmond J. Higham. Nine ways to implement the binomial method for option valuation in MATLAB. *SIAM Rev.*, 44(4):661–677, 2002. (Cited on p. 301.)

[39] Desmond J. Higham. *An Introduction to Financial Option Valuation: Mathematics, Stochastics and Computation*. Cambridge University Press, 2004. xxi+273 pp. ISBN 0-521-83884-3. (Cited on pp. 118, 335.)

[40] Nicholas J. Higham. Algorithm 694: A collection of test matrices in MATLAB. *ACM Trans. Math. Software*, 17(3):289–305, September 1991. (Cited on p. 48.)

[41] Nicholas J. Higham. The Test Matrix Toolbox for MATLAB (version 3.0). Numerical Analysis Report No. 276, Manchester Centre for Computational Mathematics, Manchester, England, September 1995. 70 pp. (Cited on p. 48.)

[42] Nicholas J. Higham. *Accuracy and Stability of Numerical Algorithms*. Second edition, Society for Industrial and Applied Mathematics, Philadelphia, PA, USA, 2002. xxx+680 pp. ISBN 0-89871-521-0. (Cited on pp. 37, 40, 51, 254.)

[43] Francis B. Hildebrand. *Advanced Calculus for Applications*. Second edition, Prentice-Hall, Englewood Cliffs, NJ, USA, 1976. xiii+733 pp. ISBN 0-13-011189-9. (Cited on p. 200.)

[44] *IEEE Standard for Binary Floating-Point Arithmetic, ANSI/IEEE Standard 754-1985*. Institute of Electrical and Electronics Engineers, New York, 1985. Reprinted in SIGPLAN Notices, 22(2):9–25, 1987. (Cited on pp. 37, 359.)

[45] D. S. Jones and B. D. Sleeman. *Differential Equations and Mathematical Biology*. CRC Press, Boca Raton, FL, USA, 2003. 408 pp. ISBN 1-58488-296-4. (Cited on p. 210.)

[46] William M. Kahan. Handheld calculator evaluates integrals. *Hewlett-Packard Journal*, August:23–32, 1980. (Cited on p. 213.)

[47] David K. Kahaner, Cleve B. Moler, and Stephen G. Nash. *Numerical Methods and Software*. Prentice-Hall, Englewood Cliffs, NJ, USA, 1989. xii+495 pp. ISBN 0-13-627258-4. (Cited on p. 159.)

[48] Irving Kaplansky. Reminiscences. In *Paul Halmos: Celebrating 50 Years of Mathematics*, John H. Ewing and F. W. Gehring, editors, Springer-Verlag, Berlin, 1991, pages 87–89. (Cited on p. 141.)

[49] Roger Emanuel Kaufman. *A FORTRAN Coloring Book*. MIT Press, Cambridge, MA, USA, 1978. ISBN 0-262-61026-4. (Cited on pp. 73, 157, 220.)

[50] C. T. Kelley. *Iterative Methods for Linear and Nonlinear Equations*. Society for Industrial and Applied Mathematics, Philadelphia, PA, USA, 1995. xiii+165 pp. ISBN 0-89871-352-8. (Cited on p. 137.)

[51] C. T. Kelley. *Iterative Methods for Optimization*. Society for Industrial and Applied Mathematics, Philadelphia, PA, USA, 1999. xv+180 pp. ISBN 0-89871-433-8. (Cited on p. 169.)

[52] C. T. Kelley. *Solving Nonlinear Equations with Newton's Method*. Society for Industrial and Applied Mathematics, Philadelphia, PA, USA, 2003. xiii+104 pp. ISBN 0-89871-546-6. (Cited on p. 165.)

[53] Brian W. Kernighan and P. J. Plauger. *The Elements of Programming Style*. Second edition, McGraw-Hill, New York, 1978. xii+168 pp. ISBN 0-07-034207-5. (Cited on pp. 85, 157, 220, 227, 243, 304.)

[54] Jacek A. Kierzenka and Lawrence F. Shampine. A BVP solver based on residual control and the MATLAB PSE. *ACM Trans. Math. Software*, 27(3):229–316, 2001. (Cited on p. 202.)

[55] Peter E. Kloeden and Eckhard Platen. *Numerical Solution of Stochastic Differential Equations*. Springer-Verlag, Berlin, 1992. xxxv+632 pp. ISBN 3-540-54062-8. (Cited on p. 299.)

[56] G. Norman Knight. Book indexing in Great Britain: A brief history. *The Indexer*, 6 (1):14–18, 1968. Reprinted in [36, pp. 9–13]. (Cited on p. 369.)

[57] Donald E. Knuth. Structured programming with `go to` statements. *Computing Surveys*, 6(4):261–301, 1974. Reprinted in [59]. (Cited on p. 243.)

[58] Donald E. Knuth. *The TEXbook*. Addison-Wesley, Reading, MA, USA, 1986. ix+483 pp. ISBN 0-201-13448-9. (Cited on p. 94.)

[59] Donald E. Knuth. *Literate Programming*. CSLI Lecture Notes Number 27. Center for the Study of Language and Information, Stanford University, Stanford, CA, USA, 1992. xv+368 pp. ISBN 0-9370-7380-6. (Cited on p. 364.)

[60] Donald E. Knuth. *Digital Typography*. CSLI Lecture Notes Number 78. Center for the Study of Language and Information, Stanford University, Stanford, CA, USA, 1999. xv+685 pp. ISBN 0-57586-010-4. (Cited on p. 119.)

[61] Helmut Kopka and Patrick W. Daly. *Guide to LATEX*. Fourth edition, Addison-Wesley, Boston, MA, USA, 2004. xii+597 pp. ISBN 0-321-17385-6. (Cited on pp. 94, 95, 116.)

[62] Arnold R. Krommer and Christoph W. Ueberhuber. *Computational Integration*. Society for Industrial and Applied Mathematics, Philadelphia, PA, USA, 1998. xix+445 pp. ISBN 0-89871-374-9. (Cited on p. 296.)

[63] Jeffrey C. Lagarias. The $3x+1$ problem and its generalizations. *Amer. Math. Monthly*, 92(1):3–23, 1985. (Cited on p. 10.)

[64] Leslie Lamport. *LATEX: A Document Preparation System. User's Guide and Reference Manual*. Second edition, Addison-Wesley, Reading, MA, USA, 1994. xvi+272 pp. ISBN 0-201-52983-1. (Cited on pp. 94, 95, 116.)

[65] R. B. Lehoucq, D. C. Sorensen, and C. Yang. *ARPACK Users' Guide: Solution of Large-Scale Eigenvalue Problems with Implicitly Restarted Arnoldi Methods*. Society for Industrial and Applied Mathematics, Philadelphia, PA, USA, 1998. xv+142 pp. ISBN 0-89871-407-9. (Cited on p. 138.)

[66] F. M. Leslie. Liquid crystal devices. Technical report, Institute Wiskundige Dienstverlening, Technische Universiteit Eindhoven, Eindhoven, The Netherlands, 1992. (Cited on p. 198.)

[67] Shangzhi Li, Falai Chen, Yaohua Wu, and Yunhua Zhang. *Mathematics Experiments.* World Scientific, New Jersey, 2003. ix+217 pp. ISBN 9812380493. (Cited on p. 314.)

[68] J. N. Lyness and J. J. Kaganove. Comments on the nature of automatic quadrature routines. *ACM Trans. Math. Software*, 2(1):65–81, 1976. (Cited on p. 320.)

[69] Tom Marchioro. Putting math to work: An interview with Cleve Moler. *Computing in Science and Engineering*, 1(4):10–13, Jul/Aug 1999. (Cited on p. 35.)

[70] Annik Martin and Shigui Ruan. Predator-prey models with delay and prey harvesting. *J. Math. Biol.*, 43:247–267, 2001. (Cited on p. 203.)

[71] *Creating Graphical User Interfaces.* The MathWorks, Inc., Natick, MA, USA. Online version. (Cited on p. 251.)

[72] *External Interfaces.* The MathWorks, Inc., Natick, MA, USA. Online version. (Cited on p. 297.)

[73] *External Interfaces Reference.* The MathWorks, Inc., Natick, MA, USA. Online version. (Cited on p. 297.)

[74] *Getting Started with MATLAB.* The MathWorks, Inc., Natick, MA, USA. Online version. (Cited on p. 304.)

[75] *Programming.* The MathWorks, Inc., Natick, MA, USA. Online version. Formerly part of *Using MATLAB.* (Cited on pp. 83, 270.)

[76] *Using MATLAB Graphics.* The MathWorks, Inc., Natick, MA, USA. Online version. (Cited on pp. 97, 118, 245, 251, 252.)

[77] Cleve B. Moler. Demonstration of a matrix laboratory. In *Numerical Analysis, Mexico 1981*, J. P. Hennart, editor, volume 909 of *Lecture Notes in Mathematics*, Springer-Verlag, Berlin, 1982, pages 84–98. (Cited on p. 35.)

[78] Cleve B. Moler. MATLAB users' guide. Technical Report CS81-1 (revised), Department of Computer Science, University of New Mexico, Albuquerque, New Mexico, August 1982. 60 pp. (Cited on p. 348.)

[79] Cleve B. Moler. Yet another look at the FFT. *The MathWorks Newsletter*, Spring 1992. (Cited on p. 91.)

[80] Cleve B. Moler. MATLAB's magical mystery tour. *The MathWorks Newsletter*, 7(1): 8–9, 1993. (Cited on p. 48.)

[81] Cleve B. Moler. *Numerical Computing with MATLAB.* Society for Industrial and Applied Mathematics, Philadelphia, PA, USA, 2004. xi+336 pp. Also available electronically from `www.mathworks.com`. ISBN 0-89871-560-1. (Cited on p. 22.)

[82] Cleve B. Moler and Donald Morrison. Replacing square roots by Pythagorean sums. *IBM J. Res. Develop.*, 27(6):577–581, 1983. (Cited on p. 341.)

[83] K. W. Morton and D. F. Mayers. *Numerical Solution of Partial Differential Equations.* Cambridge University Press, 1994. 227 pp. ISBN 0-521-42922-6. (Cited on p. 159.)

[84] J. D. Murray. *Mathematical Biology I. An Introduction.* Springer-Verlag, Berlin, 2002. ISBN 0-387-95223-3. (Cited on p. 345.)

[85] Noël M. Nachtigal, Satish C. Reddy, and Lloyd N. Trefethen. How fast are nonsymmetric matrix iterations? *SIAM J. Matrix Anal. Appl.*, 13(3):778–795, 1992. (Cited on p. 348.)

[86] Salih N. Neftci. *An Introduction to the Mathematics of Financial Derivatives.* Second edition, Academic Press, San Diego, CA, USA, 2000. xxvii+527 pp. ISBN 0-12-515392-9. (Cited on p. 208.)

[87] M. E. J. Newman, C. Moore, and D. J. Watts. Mean-field solution of the small-world network model. *Physical Review Letters*, 84:3201–3204, 2000. (Cited on p. 316.)

[88] J. R. Norris. *Markov Chains.* Cambridge University Press, Cambridge, UK, 1997. ISBN 0-521-48181-3. (Cited on p. 339.)

[89] Michael L. Overton. *Numerical Computing with IEEE Floating Point Arithmetic: Including One Theorem, One Rule of Thumb, and One Hundred and One Exercises.* Society for Industrial and Applied Mathematics, Philadelphia, PA, USA, 2001. xiv+104 pp. ISBN 0-89871-482-6. (Cited on p. 37.)

[90] Heinz-Otto Peitgen, Hartmut Jürgens, and Dietmar Saupe. *Fractals for the Classroom. Part One: Introduction to Fractals and Chaos.* Springer-Verlag, New York, 1992. xiv+450 pp. ISBN 0-387-97041-X. (Cited on pp. 16, 19, 105, 153.)

[91] Heinz-Otto Peitgen, Hartmut Jürgens, and Dietmar Saupe. *Fractals for the Classroom. Part Two: Complex Systems and Mandelbrot Set.* Springer-Verlag, New York, 1992. xii+500 pp. ISBN 0-387-97722-8. (Cited on p. 11.)

[92] E. Pitts. The stability of pendent liquid drops. Part 1. Drops formed in a narrow gap. *J. Fluid Mech.*, 59(4):753–767, 1973. (Cited on p. 196.)

[93] William H. Press, Saul A. Teukolsky, William T. Vetterling, and Brian P. Flannery. *Numerical Recipes in FORTRAN: The Art of Scientific Computing.* Second edition, Cambridge University Press, 1992. xxvi+963 pp. ISBN 0-521-43064-X. (Cited on p. 169.)

[94] Przemyslaw Prusinkiewicz and Aristid Lindenmayer. *The Algorithmic Beauty of Plants.* Springer-Verlag, Berlin, 1990. ISBN 0387-97297-8. (Cited on p. 331.)

[95] Siegfried M. Rump. INTLAB—INTerval LABoratory. `http://www.ti3.tu-harburg.de/~rump/intlab/`. (Cited on p. xx.)

[96] Yousef Saad. *Iterative Methods for Sparse Linear Systems.* Second edition, Society for Industrial and Applied Mathematics, Philadelphia, PA, USA, 2003. xviii+528 pp. ISBN 0-89871-534-2. (Cited on p. 137.)

[97] Robert Sedgewick. *Algorithms.* Second edition, Addison-Wesley, Reading, MA, USA, 1988. xii+657 pp. ISBN 0-201-06673-4. (Cited on p. 275.)

[98] Lawrence F. Shampine. *Numerical Solution of Ordinary Differential Equations.* Chapman and Hall, New York, 1994. x+484 pp. ISBN 0-412-05151-6. (Cited on pp. 177, 179, 187, 189, 301.)

[99] Lawrence F. Shampine. Solving $0 = F(t, y(t), y'(t))$ in MATLAB. *Journal of Numerical Mathematics*, 10(4):291–310, 2002. (Cited on p. 193.)

[100] Lawrence F. Shampine. Singular boundary value problems for ODEs. *Appl. Math. Comput.*, 138(1):99–112, 2003. (Cited on p. 202.)

[101] Lawrence F. Shampine, Richard C. Allen, Jr., and Steven Pruess. *Fundamentals of Numerical Computing.* Wiley, New York, 1997. x+268 pp. ISBN 0-471-16363-5. (Cited on pp. 159, 171, 174.)

[102] Lawrence F. Shampine, Ian Gladwell, and S. Thompson. *Solving ODEs with MATLAB.* Cambridge University Press, Cambridge, UK, 2003. viii+263 pp. ISBN 0-521-53094-6. (Cited on p. 205.)

[103] Lawrence F. Shampine, Jacek A. Kierzenka, and Mark W. Reichelt. Solving boundary value problems for ordinary differential equations in MATLAB with `bvp4c`. Manuscript, available at `http://www.mathworks.com/bvp_tutorial`, 2000. 27 pp. (Cited on pp. 201, 213.)

wrence F. Shampine and Mark W. Reichelt. The MATLAB ODE suite. *SIAM J. . Comput.*, 18(1):1–22, 1997. (Cited on pp. 189, 213.)

[105] Lawrence F. Shampine and S. Thompson. Solving DDEs in MATLAB. *Appl. Numer. Math.*, 37:441–458, 2001. (Cited on p. 205.)

[106] Lawrence F. Shampine, S. Thompson, and J. Kierzenka. Solving delay differential equations with `dde23`. Manuscript, available at `http://www.mathworks.com/dde_tutorial`, 2002. 27 pp. (Cited on p. 205.)

[107] G. W. Stewart. *Matrix Algorithms. Volume I: Basic Decompositions.* Society for Industrial and Applied Mathematics, Philadelphia, PA, USA, 1998. xx+458 pp. ISBN 0-89871-414-1. (Cited on p. 121.)

[108] G. W. Stewart. *Matrix Algorithms. Volume II: Eigensystems.* Society for Industrial and Applied Mathematics, Philadelphia, PA, USA, 2001. xix+469 pp. ISBN 0-89871-503-2. (Cited on p. 121.)

[109] Josef Stoer and Christoph Witzgall. *Convexity and Optimization in Finite Dimensions I.* Springer-Verlag, Berlin, 1970. ix+293 pp. (Cited on p. 312.)

[110] Gilbert Strang. *Introduction to Linear Algebra.* Third edition, Wellesley-Cambridge Press, Wellesley, MA, USA, 2003. viii+568 pp. ISBN 0-9614088-9-8. (Cited on p. 121.)

[111] Steven H. Strogatz. *Nonlinear Dynamics and Chaos: With Applications to Physics, Biology, Chemistry, and Engineering.* Addison-Wesley, Reading, MA, USA, 1994. xi+498 pp. ISBN 0-201-54344-3. (Cited on pp. 11, 178, 179, 182.)

[112] Test set for IVP solvers, release 2.2. `http://pitagora.dm.uniba.it/~testset`. (Cited on pp. 191, 193.)

[113] Lloyd N. Trefethen. *Spectral Methods in MATLAB.* Society for Industrial and Applied Mathematics, Philadelphia, PA, USA, 2000. xvi+165 pp. ISBN 0-89871-465-6. (Cited on pp. xxiii, 118, 264.)

[114] Lloyd N. Trefethen and David Bau III. *Numerical Linear Algebra.* Society for Industrial and Applied Mathematics, Philadelphia, PA, USA, 1997. xii+361 pp. ISBN 0-89871-361-7. (Cited on p. 121.)

[115] Edward R. Tufte. *The Visual Display of Quantitative Information.* Graphics Press, Cheshire, CT, USA, 1983. 197 pp. (Cited on pp. 118, 119, 264.)

[116] Edward R. Tufte. *Envisioning Information.* Graphics Press, Cheshire, CT, USA, 1990. 126 pp. (Cited on p. 118.)

[117] Edward R. Tufte. *Visual Explanations: Images and Quantities, Evidence and Narrative.* Graphics Press, Cheshire, CT, USA, 1997. 158 pp. ISBN 0-9613921-2-6. (Cited on p. 118.)

[118] Charles F. Van Loan. *Computational Frameworks for the Fast Fourier Transform.* Society for Industrial and Applied Mathematics, Philadelphia, PA, USA, 1992. xiii+273 pp. ISBN 0-89871-285-8. (Cited on p. 170.)

[119] Charles F. Van Loan. Using examples to build computational intuition. *SIAM News*, 28:1, 7, October 1995. (Cited on p. xxiii.)

[120] Charles F. Van Loan. *Introduction to Scientific Computing: A Matrix-Vector Approach Using MATLAB.* Prentice-Hall, Englewood Cliffs, NJ, USA, 2000. xi+367 pp. ISBN 0-13-949157-0. (Cited on p. 159.)

[121] D. Viswanath. Random Fibonacci sequences and the number 1.3198824.... *Math. Comp.*, 69(231):1131–1155, 2000. (Cited on p. 8.)

[122] Stan Wagon. *Mathematica in Action.* Second edition, TELOS division of Springer-Verlag, New York, NY, USA, 2000. xvi+592 pp. ISBN 0-387-98684-7. (Cited on p. 284.)

[123] David S. Watkins. *Fundamentals of Matrix Computations.* Second edition, Wiley, New York, 2002. xiii+618 pp. ISBN 0-471-21394-2. (Cited on p. 121.)

[124] Duncan J. Watts and Steven H. Strogatz. Collective dynamics of 'small-world' networks. *Nature*, 393:440–442, 4 June 1998. (Cited on pp. 316, 318.)

[125] Junjie Wei and Shigui Ruan. Stability and bifurcation in a neural network model with two delays. *Physica D*, 130:255–272, 1999. (Cited on p. 205.)

[126] Maurice V. Wilkes. *Memoirs of a Computer Pioneer.* MIT Press, Cambridge, MA, USA, 1985. viii+240 pp. ISBN 0-262-23122-0. (Cited on p. 227.)

[127] Paul Wilmott, Sam Howison, and Jeff Dewynne. *The Mathematics of Financial Derivatives: A Student Introduction.* Cambridge University Press, 1995. xiii+317 pp. ISBN 0-521-49699-3. (Cited on pp. 208, 335.)

[128] Thomas G. Wright. Eigtool. `http://www.comlab.ox.ac.uk/pseudospectra/eigtool/`. (Cited on p. xx.)

Index

The 18th century saw the advent of the professional indexer.
He was usually of inferior status—a Grub Street hack—
although well-read and occasionally a university graduate.
— G. NORMAN KNIGHT, *Book Indexing in Great Britain: A Brief History* (1968)

I find that a great part of the information I have was acquired by
looking up something and finding something else on the way.
— FRANKLIN P. ADAMS

A suffix "t" after a page number denotes a table, "f" a figure, and "n" a footnote. Entries in typewriter font beginning with lower case letters are MATLAB functions; those beginning with capital letters are Handle Graphics property names.